GCSE geography

AQA

Series editor
Simon Ross

Nicholas Rowles

David Holmes

Bob Digby

OXFORD
UNIVERSITY PRESS

OXFORD
UNIVERSITY PRESS

Great Clarendon Street, Oxford, OX2 6DP, United Kingdom

Oxford University Press is a department of the University of Oxford. It furthers the University's objective of excellence in research, scholarship, and education by publishing worldwide. Oxford is a registered trade mark of Oxford University Press in the UK and in certain other countries

Series editor: Simon Ross

Authors: Nicholas Rowles, David Holmes, Bob Digby

The moral rights of the authors have been asserted

Database right of Oxford University Press (maker) 2016

First published in 2016

British Library Cataloguing in Publication Data
Data available

ISBN 978-0-19-836661-4

10 9 8 7 6 5 4

Paper used in the production of this book is a natural, recyclable product made from wood grown in sustainable forests. The manufacturing process conforms to the environmental regulations of the country of origin.

Printed in Italy by L.E.G.O. SpA

Acknowledgements

The publisher and authors would like to thank the following for their permission to use photographs and other copyright material:

Cover: watchara/Shutterstock; fzd.it/Shutterstock; **p6:** Omar Havana/Getty Images; **p8:** Esdras Ndikumana/AFP/Getty Images; **p12:** Richard Roscoe/Barcroft Media/Getty Images; **p15:** © David Gysel/Alamy; think4photop/Shutterstock; **p16:** MARTIN BERNETTI/AFP/Getty Images; Jonathan Saruk/Getty Images; **p17:** Manish Swarup/AP/Press Association Images; ZUMA/REX; **p18:** Matyas Rehak/Shutterstock; shahreen/Shutterstock; **p19:** Michael Zysman/Shutterstock; **p23:** Jeff Gilbert/Alamy Stock Photo; robertharding/Alamy Stock Photo; Paul Springett C/Alamy Stock Photo; **p24:** Harvepino/Shutterstock; © Dan Callister/Alamy; **p28:** © imagegallery2/Alamy; **p29:** © imagegallery2/Alamy; © ShelterBox 2015; **p30:** © Neil Cooper/Alamy; **p31:** National Hurricane Center/National Oceanic and Atmospheric Administration; Helen Sharp; **p32:** © RooM the Agency/Alamy; Apex News and Pictures; **p33:** CHRIS BALCOMBE/REX; **p34:** © Adrian Sherratt/Alamy; **p35:** © SWNS/Alamy; **p36:** Ordinance Survey; **p37:** © SWNS/Alamy; **p38:** © Robert Stainforth/Alamy; © DAVID MOIR/Reuters/Corbis; **p39:** OLI SCARFF/AFP/Getty Images; © epa european pressphoto agency b.v./Alamy; **p41:** Nick Cobbing/REX; NASA; **p42:** NASA; **p43:** © FineArt/Alamy; Contemporary illustration of the eruption of Tambora volcano in the Moluccas islands, 1815 (engraving), German School, (19th century)/© SZ Photo/Bridgeman Images; **p46:** EDF Energy; **p48:** © Jack Sullivan/Alamy; © Keith Shuttlewood/Alamy; **p49:** © Emma Stoner/Alamy; © Pix/Alamy; **p50:** Photoshot License Ltd/Alamy Stock Photo; **p52:** © Angela Hampton Picture Library/Alamy; **p54:** © martin meehan/Alamy; **p55:** Land and Water Services; **p56:** Marques/Shutterstock; AlxYago/Shutterstock; Volodymyr Burdiak/Shutterstock; Francesco Dazzi/Shutterstock; **p58:** Wolfgang Kaehler/LightRocket via Getty Images; **p60:** © Prisma Bildagentur AG/Alamy; **p61:** Nigel Dickinson; © Bazuki Muhammad/Reuters/Reuters/Corbis; © FLPA/Alamy; **p62:** © Eye Ubiquitous/Alamy; Ahmad Yahaya/EyeEm/Getty Images; **p63:** Mattias Klum/Getty Images; **p65:** pyzata/Shutterstock; **p65:** Andy Isaacson; **p66:** Fred Benenson; **p67:** Brian Bailey/Getty Images; Alex Hipkiss/RSPB; **p68:** Simon Ross; **p69:** Geo-Zlat/Shutterstock; **p70:** © Ewen Bell/Alamy; **p71:** © Tibor Bognar/Corbis; © Franck METOIS/Alamy; **p72:** © Bartek Wrzesniowski/Alamy; **p73:** © Frederic Soltan/Sygma/Corbis; © David Cumming/Eye Ubiquitous/Corbis; **p74:** © PhotoAlto sas/Alamy; **p75:** © Photoshot Holdings Ltd/Alamy; **p76:** Simon Ross; fabernova/Shutterstock; **p77:** Biosphoto/Xavier Eichaker; Environmental Images/Universal Images Group/REX; **p78:** Shchipkova Elena/Shutterstock; Nicram Sabod/Shutterstock; **p79:** © Unai Peña/Alamy; **p80:** Trond Viken, Norwegian Ministry of Trade, Industry and Fisheries; © Frans Lanting/Corbis; **p81:** Ttphoto/Shutterstock; Tyler Olson/Shutterstock; **p82:** © Renato Granieri/Alamy; **p83:** © Robert Harding Picture Library Ltd/Alamy; Dmitry Chulov/Shutterstock; © Rolf Adlercreutz/Alamy; **p84:** © Design Pics Inc/Alamy; © Paul Andrew Lawrence/Alamy; © ASK Images/Alamy; **p85:** Yongyut Kumsri/Shutterstock; **p86:** © Robert Harding Picture Library Ltd/Alamy; **p87:** MarcAndreLeTourneux/Shutterstock; **p88:** David Angel/Alamy Stock Photo; **p90:** David Woods/Shutterstock; Matthew Dixon/Shutterstock; **p91:** ARTWORK; **p92:** Francesco Carucci/Shutterstock; AFLO / MAINICHI NEWSPAPER/epa/Corbis; **p94:** Jason Hawkes/CORBIS; Peter Smith Photography; **p96:** RooM the Agency / Alamy Stock Photo; **p98:** Simon Ross; **p99:** Alan Novelli/Alamy Stock Photo; **p100:** © Pat Downing/Alamy; Historic England Archive; **p101:** Skyscan; **p102:** Mike Charles/Shutterstock; **p103:** © dbphots/Alamy; **p104:** © Angie Sharp/Alamy; ian woolcock/Shutterstock; **p105:** Ordinance Survey; **p106:** robertharding/Alamy Stock Photo; © robertharding/Alamy; **p107:** UIG via Getty Images; Alamy; **p108:** © Mick House/Alamy; © David Chapman/Alamy; Mick House/Alamy Stock Photo; **p109:** Westend61/Getty Images; Dave McAleavy Images/Alamy Stock Photo; **p110:** Simon Ross; **p110:** The Environment Agency; **p111:** incamerastock/Alamy Stock Photo; **p113:** Balfour Beatty; **p113:** Dorset Councils' Partnership; **p115:** Kevin Eaves/Shutterstock; **p116:** Simon Ross; **p117:** i4lcocl2 / Shutterstock; **p118:** © Paul Heinrich/Alamy; **p120:** Eag1eEyes/Shutterstock; **p122:** Don Johnston/Alamy; **p123:** Ordinance Survey; **p124:** North news & Pictures Ltd; **p126:** © David Angel/Alamy; **p128:** © Nick Hanna/Alamy; Ian Shepherd/Wild Trout Trust; **p129:** © Jeff Morgan 01/Alamy; **p130:** David HartleyREX/Shutterstock; **p131:** Rebecca Mileham; **p132:** © All Canada Photos/Alamy; **p133:** © Danita Delimont/Alamy; © Alan Majchrowicz/Alamy; **p134:** Andrew Rawcliffe/Alamy Stock Photo; **p135:** The Photolibrary Wales/Alamy Stock Photo; **p136:** Simon Ross; **p137:** © Ashley Cooper/Alamy; © David Robertson/Alamy; **p138:** Travel Ink/Getty images; Septemberlegs/Alamy Stock Photo; **p139:** Ordinance Survey; **p140:** Simon Ross; © Jeff Morgan 08/Alamy; **p141:** Yorkman/Shutterstock; ©Natural Retreats; **p142:** © incamerastock/Alamy; © Ashley Cooper pics/Alamy; **p143:** Terry Abraham; **p144:** Philip Birtwistle/Shutterstock; **p145:** © Ian Dagnall/Alamy; © JAMIE SMITH/Alamy; National Trust/Fix the Fells; **p146:** Trong Nguyen/Shutterstock; **p150:** © Fredrik Renander/Alamy; Daniel Berehulak/Getty Images; **p152:** © imageBROKER/Alamy; © Prisma Bildagentur AG/Alamy; **p153:** Mario Tama/Getty Images; **p154:** © nobleIMAGES/Alamy; **p155:** Rio 2016/Alex Ferro; © BrazilPhotos.com/Alamy; **p156:** Cesar Duarte/Argosfoto; **p157:** Steve Outram/Alamy Stock Photo; **p159:** © Peter Treanor/Alamy; **p160:** Elena Mirage/Shutterstock; **p161:** © Peter Tsai Photography/Alamy; Image broker/Alamy; **p162:** Donatas Dabravolskas/Shutterstock; lazyllama/Shutterstock; **p163:** Mario Tama/Getty Images; Ruy Barbosa Pinto/Getty Images; **p166:** Courtesy of The Bristol Port Company; p167: robertharding/Alamy Stock Photo; **p168:** UWE Bristol; **p169:** © Doug Houghton/Alamy; pjhpix/Shutterstock; **p170:** © david a eastley/Alamy; **p171:** © Pictorial Press Ltd/Alamy; © Russell Binns/Alamy; **p172:** http://www.bristol.gov.uk/; **p173:** © Scott Hortop Travel/Alamy; PHOTOGRAPHER/Avon Wildlife Trust, www.avonwildlifetrust.org.uk; **p174:** pjhpix/Shutterstock; pjhpix/Shutterstock; **p175:** Andrew Beattie/Alamy Stock Photo; Jane Tregelles/Alamy Stock Photo; **p178:** © Jeff Morgan 04/Alamy; **p179:** Courtesy of Leese and Nagle; **p180:** antb/Shutterstock; Ordinance Survey; **p181:** © Bennett Dean/Eye Ubiquitous/Eye Ubiquitous/Corbis; **p182:** PLEIADES © CNES 2016, Distribution Airbus DS; Ordinance Survey; **p183:** Bristol City Council; Phil Wills/Alamy Stock Photo; **p184:** PLEIADES © CNES 2016, Distribution Airbus DS; **p185:** Bristol City Council; **p186:** Westend61 GmbH/Alamy Stock Photo; **p187:** © Hemis/Alamy; **p188:** LOOK Die Bildagentur der Fotografen GmbH/Alamy Stock Photo; Rolf Disch Solar Architecture, Germany; © Agencja Fotograficzna Caro/Alamy; **p189:** Daniel Schoenen/LOOK-foto/Getty Images; **p190:** © allOver images/Alamy; **p191:** © AsiaStock/Alamy; PETER PARKS/AFP/Getty Images; **p192:** jordi clave garsot/Alamy Stock Photo; **p194:** © Jake Lyell/Alamy; **p195:** © Copyright Sasi Group (University of Sheffield) and Mark Newman (University of Michigan); **p197:** Zurijeta/Shutterstock; Grigory Kubatyan/Shutterstock; **p199:** Anton_Ivanov/Shutterstock; Jasmin Awad/Shutterstock; **p202:** © Mike Goldwater/Alamy; © Jenny Matthews/Alamy; **p204:** Kjell Nilsson-Maki, cartoonstock.com; **p205:** Kopirin/Shutterstock; **p206:** © Michael Honegger/Alamy; **p207:** © epa european pressphoto agency b.v./Alamy; © Nick Turner/Alamy; **p209:** © Asia Images Group Pte Ltd/Alamy; © Danita Delimont/Alamy; **p210:** Richard Hanson/TEARFUND; **p211:** Design Pics Inc/REX Shutterstock; Sean Sprague; **p212:** © flowerphotos/Alamy; **p213:** Fair Trade; Simon Rawles; **p215:** Majority World/REX Shutterstock; **p216:** © robertharding/Alamy; **p217:** Andrew Park/Shutterstock; **p218:** epa european pressphoto agency b.v./Alamy Stock Photo; **p219:** Chris Hondros/Getty Images; © Friedrich Stark/Alamy; **p220:** IBM; **p221:** c.Everett Collection/REX Shutterstock; Stringer/Getty Images; **p222:** jeremy sutton-hibbert/Alamy Stock Photo; **p223:** epa; **p224:** © powderkeg stock/Alamy; **p225:** Eye UbiquitousAlamy; **p226:** Khalil Senosi/Associated Press; Unilever Nigeria; **p227:** Eye Ubiquitous/Alamy Stock Photo; **p228:** Malcolm Linton/Getty Images; **p229:** Giacomo Pirozzi/PANOS; **p230:** Merwelene van der Merwe/Mimages; Carlos Cazalis/Corbis; **p231:** powderkeg stock/Alamy Stock Photo; Amnesty International; **p232:** © jordi clave garsot/Alamy; **p233:** © Stephen Chung/Alamy; **p234:** James Brooks/Alamy Stock Photo; **p235:** © World Pictures/Alamy; **p236:** Monty Rakusen/Corbis; auremar/Shutterstock; **p237:** A and N photography/Shutterstock; BAS; **p238:** Chris Brunnen; **p239:** Ordinance Survey; www.morecobalt.co.uk; **p240:** © Martin Cooke/Alamy; © Washington Imaging/Alamy; **p241:** © Construction Photography/Alamy; AGGREGATE INDUSTRIES UK LIMITED; **p242:** Philip Bird LRPS CPAGB/Shutterstock; **p243:** © David Gowans/Alamy; **p244:** Art Directors & TRIP/Alamy Stock Photo; **p245:** Lionel Derimais/Corbis; **p246:** Peel Ports Group Ltd; **p250:** Paul Lovelace/REX Shutterstock; **p251:** © PASCAL ROSSIGNOL/Reuters/Corbis; **p253:** Art Widak/Demotix/Corbis; **p254:** r.classen n / Shutterstock; **p258:** Art Directors & TRIP/Alamy Stock Photo; **p263:** © Conrad Elias/Alamy; **p265:** © Jack Sullivan/Alamy; **p266:** © dpa picture alliance archive/Alamy; Images of Africa Photobank/Alamy Stock Photo; **p267:** Abbie Trayler-Smith/Panos; **p268:** Kristian Buus/Alamy Stock Photo; **p269:** Maya Pedal; **p271:** © Christine Osborne/CORBIS; Jim Holmes/Panos; **p272:** ussr/Shutterstock; Septemberlegs Creative/Alamy Stock Photo; **p273:** © Jim West/Alamy; Jeff Morgan 10/Alamy; **p274:** © Nick Turner/Alamy; **p275:** Seyi; **p277:** © Dinodia Photos/Alamy; **p278:** Dinodia Photos/Alamy Stock Photo; **p280:** © Greenshoots Communications/Alamy; **p283:** AdeleD/Shutterstock; **p284:** AP India; **p285:** © Martin Bond/Alamy; Q2A Media; **p286:** Dieter Telemans/panos; **p287:** Pacific press/Alamy; **p290:** © imageBROKER/Alamy; **p291:** ActionAid; **p295:** Denys Prykhodov/Shutterstock; **p296:** Highshot/Malmö stad; **p297:** © Archimage/Alamy; Darren Brode/Shutterstock; Victor R. Caivano/PA IMAGES; **p298:** Christopher McLeod; **p299:** Practical Action/Peru; **p301:** David Holmes; **p302:** Parkes Photographic Archive/Alamy Stock Photo; **p303:** Ordinance Survey; Heritage Action; **p306:** Bailey-Cooper Photography/Alamy Stock Photo; David Holmes; **p308:** David Holmes; **p309:** David Holmes; **p313:** David Holmes; **p315:** David Holmes; **p316:** David Holmes; **p318:** David Holmes; **p323:** David Holmes; **p325:** Bob Digby; **p334:** robertharding/Alamy Stock Photo; Xinhua/Alamy Stock Photo; Viewstock/Getty Images; NASA; **p335:** OUP.

Page layout/design: Kamae Design, Oxford.

Artworks are by: Lovell Johns, Barking Dog Art, Kamae, Q2A, Giorgio Bacchin.

Every effort has been made to contact copyright holders of material reproduced in this book. Any omissions will be rectified in subsequent printings if notice is given to the publisher.

Approval message from AQA

This textbook has been approved by AQA for use with our qualification. This means that we have checked that it broadly covers the specification and we are satisfied with the overall quality. Full details of our approval process can be found on our website.

We approve textbooks because we know how important it is for teachers and students to have the right resources to support their teaching and learning. However, the publisher is ultimately responsible for the editorial control and quality of this book.

Please note that when teaching the AQA GCSE Geography course, you must refer to AQA's specification as your definitive source of information. While this book has been written to match the specification, it cannot provide complete coverage of every aspect of the course.

A wide range of other useful resources can be found on the relevant subject pages of our website: www.aqa.org.uk.

Contents

Contents

Section A The challenge of natural hazards

Collapsed buildings in Bhaktapur, Nepal, following the 2015 earthquake

Unit 1 Living with the physical environment is about physical processes and systems, how they change, and how people interact with them at a range of scales and in a range of places. It is split into three sections.

Section A The challenge of natural hazards includes:

- an introduction to natural hazards
- tectonic hazards
- weather hazards
- climate change.

You need to study all the topics in Section A – in your final exam you will have to answer questions on all of them.

What if...

1 you were caught up in a volcanic eruption?

2 your home was flattened in an earthquake?

3 you lived in the path of a tropical storm?

4 the UK was hit by a tsunami?

5 the world ran out of fossil fuels?

Your key skills

To be a good geographer, you need to develop important geographical skills – in this section you will learn the following skills:

- Using different graphical techniques to present information.
- Carrying out personal research.
- Drawing and annotating diagrams and sketches.
- Describing and interpreting information from maps and graphs.
- Finding evidence from photographs.
- Using OS maps.

Your key words

As you go through the chapters in this section, make sure you know and understand the key words shown in bold. Definitions are provided in the Glossary on pages 346–9. To be a good geographer, you need to use good subject terminology.

Your exam

Section A makes up part of Paper 1 – a one and a half hour written exam worth 35 per cent of your GCSE.

1 Natural hazards

1.1 What are natural hazards?

On this spread you will find out about the risks from natural hazards

A Landslides affecting Bujumbura, Burundi, 2015

What is a natural hazard?

In March 2015 landslides struck Bujumbura in Western Burundi, Central Africa, killing several people and leaving thousands homeless. Following a period of heavy rain, mud and rocks plunged down hillsides destroying houses and damaging roads (photo **A**).

This event is an example of a natural hazard. It is a natural event that has had a huge **social impact**. If the landslide had occurred in a remote area where it did not pose any threat to people it would not be considered a hazard.

Landslides are not major killers. The most deadly natural hazards are floods, storms, earthquakes and droughts. Between 2002 and 2012, an average of 100 000 people worldwide were killed each year by natural hazards. In most years, flooding caused the greatest number of deaths.

Diagram **B** is called a Venn diagram. Notice that a natural hazard occurs when a natural event overlaps with human activities.

What are the different types of natural hazard?

There is a huge range of natural hazards. These include:

◆ volcanic eruptions

◆ earthquakes

◆ storms

◆ tsunami (huge waves caused by earthquakes)

◆ landslides

◆ floods.

Diagram **C** shows how natural hazards can be sorted into three main groups.

No hazard

Hazard or disaster

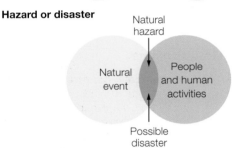

B When is a natural event a hazard?

> **Did you know?**
> Hurricane Patricia (2015) was the most powerful tropical storm ever recorded, with winds reaching 320 km/h (200 mph).

C Different types of natural hazard

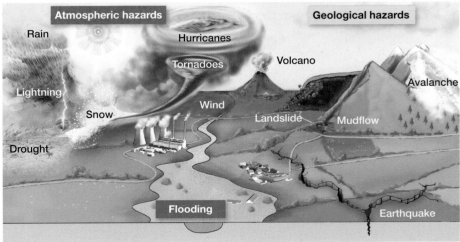

What is 'hazard risk'?

Hazard risk is the chance or probability of being affected by a natural event. People who choose to live close to a river may be at risk from flooding. Those who live close to the sea may be at risk from tropical cyclones or tsunami.

So why do people put themselves at risk by living in such places? They weigh up the advantages and disadvantages and, because such events don't happen very often, they may decide to accept the risk. Some people may have little choice of where to live or knowledge that where they are living is dangerous.

Think about it

Are natural hazards occurring more frequently today than 100 years ago? Use diagram B to help you.

What factors affect risk?

There are several factors that have led to an increase in the number of people at risk from natural events.

Poverty
In poorer parts of the world poverty may force people to live in areas at risk. This is especially true in cities such as Lima in Peru or Caracas in Venezuela. Here, a shortage of housing has led to people building on unstable slopes prone to floods and landslides.

Urbanisation
Over 50 per cent of the world's population now live in cities. Some of the world's largest cities (for example, Tokyo, Istanbul and Los Angeles) are at risk from earthquakes.
Densely populated urban areas are at great risk from natural events such as earthquakes and tropical cyclones. The 2010 Haiti earthquake destroyed much of the capital Port-au-Prince killing some 230 000 people.

Factors increasing the risk from natural hazards

Farming
When a river floods it deposits fertile silt on its floodplain, which is excellent for farming. But when people choose to live there they are putting themselves at risk. In low-lying countries many people may live on floodplains, like that of the River Ganges in Bangladesh.

Climate change
In a warmer world the atmosphere will have more energy leading to more intense storms and hurricanes. Climate change may cause some parts of the world to become wetter with an increased risk of flooding. Other areas may become drier and prone to droughts and famines.

ACTIVITIES

1 Describe what has happened in photo **A**.
2 **a** Make a copy of diagram **B**.
   b Explain in your own words how a 'natural event' becomes a 'natural hazard'.
3 **a** What are the three main groups of hazard shown in diagram **C**?
   b Why do you think more people are likely to be affected by river flooding than by landslides and mudflows?
4 In the future, why is it likely that increasing numbers of people will be at risk from natural hazards?

Stretch yourself

Find out about natural hazards in Bangladesh. What are the natural events that threaten the country? Why are so many people at risk from these events?

Maths skills

Use a divided bar chart or a pie chart to present the following information.

Percentage of fatalities (2014)

Hydrological events (e.g. floods)	66%
Meteorological events (e.g. storms)	17%
Geophysical events (e.g. earthquakes)	11%
Climatological events (e.g. drought)	6%

Practice question

Explain two human developments that would increase the risk of people being affected by natural hazards. *(4 marks)*

2 Tectonic hazards

2.1 Distribution of earthquakes and volcanoes

On this spread you will find out where earthquakes and volcanoes happen and link their location to plate tectonics

Why is there a pattern of earthquakes?

An **earthquake** is a sudden and violent period of ground shaking. It is most commonly caused by a sudden movement of rocks within the Earth's crust. This occurs mainly at the margins of *tectonic plates* (map **B**) where plates are moving and enormous pressures build up and are released.

Compare map **B** to map **A**. Notice the pattern of earthquakes along **plate margins**, for example along the western coast of North and South America. The occurrence of earthquakes around the edge of the Pacific Ocean follows the plate margins.

Some earthquakes do not occur at plate margins. These may be caused by human activity such as underground mining or oil extraction.

A *Earthquakes recorded during March 2015*

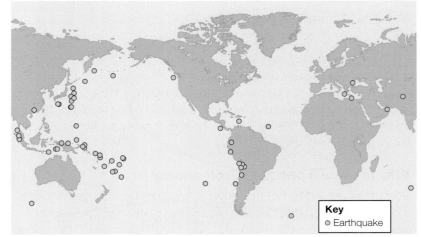

Key
○ Earthquake

Tectonic plates

◆ The Earth's crust is split into a number of plates about 100 km thick.

◆ There are two types of crust – dense, thin oceanic crust and less dense, thick continental crust.

◆ Plates move in relation to each other due to convection (heat) currents from deep within the Earth.

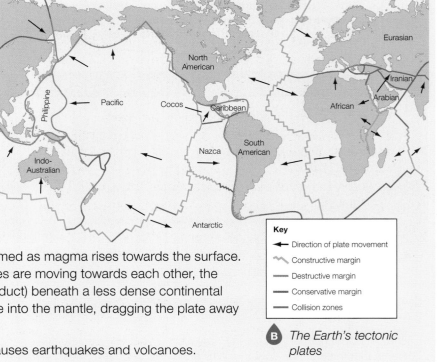

Key
← Direction of plate movement
⌇ Constructive margin
— Destructive margin
— Conservative margin
— Collision zones

B *The Earth's tectonic plates*

◆ At a *constructive* plate margin plates move apart. New crust is formed as magma rises towards the surface. At a *destructive* margin, where plates are moving towards each other, the denser oceanic plate may sink (subduct) beneath a less dense continental plate. Gravity pulls the oceanic plate into the mantle, dragging the plate away from the constructive margin.

◆ Tectonic activity at plate margins causes earthquakes and volcanoes.

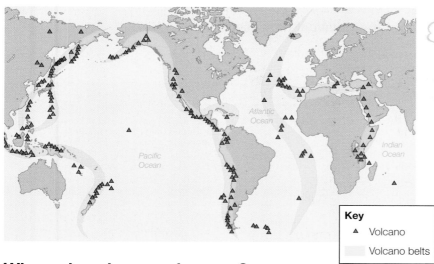

Key

▲ Volcano

    Volcano belts

C *The distribution of volcanoes*

Where do volcanoes happen?

Look at map **C** which shows the distribution of volcanoes. A **volcano** is a large and often conical-shaped landform usually formed over a long period of time by a series of eruptions. Like earthquakes, the majority of volcanoes occur in long belts that follow the plate margins, for example around the edge of the Pacific Ocean. This is known as the 'Pacific Ring of Fire'. There is also a belt of volcanoes through the middle of the Atlantic Ocean. This is the Mid-Atlantic Ridge which includes the Azores and Iceland which are volcanic islands.

Why is there a pattern of volcanoes?

Volcanoes are fed by hot molten rock (magma) from deep within the Earth. This rises to the surface at *constructive* and *destructive* plate margins. Volcanoes also form at **hot spots**, where the crust is thin and magma is able to break through to the surface. The Hawaiian Islands in the Pacific Ocean are a good example of a hot spot.

ACTIVITIES

1 Use map **A** to describe the pattern of earthquakes.
2 Use map **B** to answer the following questions.
   a Which plate is the UK on?
   b Name a country which is being split by two plates.
   c Describe the movement of the plates at the margin of the Nazca and South American plates.
3 Describe the pattern of volcanoes (map **C**). Refer to names of oceans, continents and countries in your answer.
4 Why do the majority of earthquakes and volcanoes occur at plate margins?

Maths skills

A total of 1482 earthquakes occurred in a 7-day period at the end of April 2016. Work out the average number of earthquakes per day and per hour. Can you calculate the frequency of the earthquakes?

Practice question

Explain why the majority of earthquakes and volcanoes occur at plate margins. *(4 marks)*

Stretch yourself

Use the United States Geological Survey (USGS) website to find a map of recent earthquakes. You could look at a single day or a whole week. Copy and paste the map and write a few sentences (or use text boxes) to describe the pattern of earthquakes. Use map **B** to relate this to named plate margins.

On this spread you will find out about the physical processes at plate margins

What happens at plate margins?

Iceland is a country in the North Atlantic Ocean. It is situated on the Mid-Atlantic Ridge, a plate margin where two plates are moving away from each other. There are several active volcanoes in Iceland including Eyjafjallajökull which erupted in 2010 (photo **A**). It is possible to identify three main types of plate margin:

◆ **Constructive (transform)** – where two plates are moving apart.

◆ **Destructive** – where two plates are moving towards one another.

◆ **Conservative** – where two plates are sliding alongside each other.

Maths skills

At a constructive plate margin, each plate moves at an average of 2 cm a year. Calculate the increase in the width of Iceland over a period of one million years.

A *Eruption of Eyjafjallajökull, 2010*

Did you know?

Volcanic eruptions over millions of years mean that Iceland is growing outwards from the middle!

Constructive margin

At a constructive margin two plates are moving apart. Diagram **B** shows what is happening at the constructive margin in the mid-Atlantic. Magma is forcing its way to the surface along the Mid-Atlantic Ridge. As it breaks through the overlying crust it causes earthquakes. On reaching the surface it forms volcanoes such as Eyjafjallajökull in Iceland.

The magma at constructive margins is very hot and fluid. Lava erupting from a volcano will flow a long way before cooling. This results in typically broad and flat *shield volcanoes*.

Volcanic island, for example Iceland or the Azores

Mid-Atlantic Ridge

North American Plate (continental crust)

Eurasian Plate (continental crust)

Atlantic Ocean

Volcano

Crust

Mantle

Magma from mantle

B *Constructive plate margin*

Destructive margin

At a destructive plate margin two plates are moving towards one another. Diagram **C** shows what is happening on the west coast of South America.

Where the two plates meet a deep ocean trench has formed. The oceanic Nazca Plate, which is relatively dense, is *subducted* beneath the less dense South American Plate. Friction between the two plates causes strong earthquakes. As the oceanic plate moves downwards it melts. This creates magma which is less fluid than at a constructive margin. It breaks through to the surface to form steep-sided *composite volcanoes*. Eruptions are often very violent and explosive.

Where two continental plates meet, there is no subduction. Instead, the two plates collide and the crust becomes crumpled and uplifted. This collision forms fold mountains such as the Himalayas. These mountain-building processes cause earthquakes. There are no volcanoes at these *collision* margins because there is no magma.

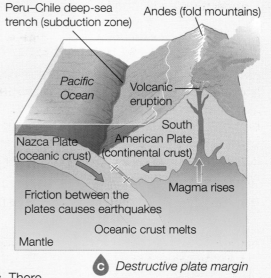

C *Destructive plate margin*

Conservative margin

At conservative plate margins, two plates are moving past each other. Friction between the plates then causes earthquakes. Map **D** shows the San Andreas Fault in California, a well-known example of a conservative margin. The faster-moving Pacific Plate is sliding in the same direction next to the slower-moving North American Plate.

Earthquakes happen along conservative margins as stresses gradually build up over many years. They can be destructive as they are close to the Earth's surface. These are released suddenly when the plates slip and shift.

There are no volcanoes because there is no magma.

 D *Conservative plate margin*

ACTIVITIES

1 a What type of plate margin runs through the middle of Iceland?

b Why do earthquakes occur in Iceland?

c Explain why there are volcanoes like Eyjafjallajökull in Iceland.

2 Explain the formation of earthquakes and volcanoes at a destructive margin (diagram **C**).

3 Make a copy of map **D**.

a Use crosses to show where you would expect earthquakes to happen.

b Why are there no volcanoes at a conservative plate margin?

Stretch yourself

Find out about the North Anatolian Fault, one of the world's most active plate margins.

• Where is it?

• What type of plate margin is it?

• What are the hazards associated with the North Anatolian Fault?

• Which major city near this fault is at greatest risk from a natural disaster?

Practice question

Explain the physical processes that happen at constructive plate margins. *(4 marks)*

The effects of earthquakes

On this spread you will find out about the effects of two earthquakes in contrasting countries – Chile and Nepal

Example

The earthquakes in Chile and Nepal

Earthquakes can have devastating effects on peoples' lives and activities. **Primary effects** are caused by ground shaking and can include deaths and injuries, and damage to roads and buildings. **Secondary effects** are the result of primary effects (ground shaking) and include tsunami, fires and landslides. Responses to earthquakes include emergency care and support and help with longer-term reconstruction.

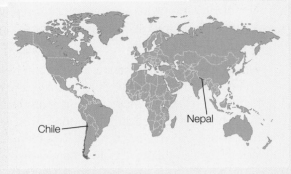

A *Contrasting Chile and Nepal*

Indicator	Chile	Nepal	UK
Gross Domestic Product (GDP) – a measure of wealth	38th out of 193 countries	109th out of 193 countries	6th out of 193 countries
Human Development Index (HDI) – a measure of the level of development	41st out of 187 countries	145th out of 187 countries	14th out of 187 countries

Chile

Imagine what it would be like if the ground shook underneath you for three minutes! This is what happened on 27 February 2010 when a very powerful earthquake measuring 8.8 on the Richter scale struck just off the coast of central Chile (map **B**). The earthquake occurred at a destructive plate margin where the Nazca Plate is moving beneath the South American Plate.

It was followed by a series of smaller aftershocks.

Because the earthquake occurred out to sea, tsunami warnings were issued as waves raced across the Pacific Ocean at speeds of up to 800 km per hour.

B *The Chile earthquake*

Nepal

On 25 April 2015 Nepal was struck by an earthquake measuring 7.9 on the Richter scale. The epicentre was about 80 km (50 miles) to the north-west of Nepal's capital Kathmandu in the foothills of the Himalayas (map **C**). This is a destructive plate margin where the Indo-Australian Plate is colliding with the Eurasian Plate at a rate of 45 mm per year. The collision and pressure at this margin are responsible for the formation of the Himalayas.

The earthquake was very shallow, just 15 km below the surface. This resulted in very severe ground shaking and widespread landslides and avalanches. The earthquake caused damage hundreds of kilometres away in India, Tibet and Pakistan.

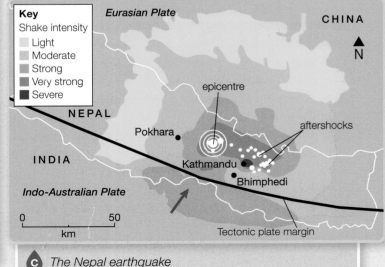

C *The Nepal earthquake*

Primary effects (caused by ground shaking)

- Around 500 people killed and 12 000 injured – 800 000 people affected.
- 220 000 homes, 4500 schools, 53 ports, 56 hospitals and other public buildings destroyed.
- Port of Talcahuanao and Santiago airport badly damaged.
- Much of Chile lost power, water supplies and communications.
- Cost of the earthquake estimated at US$30 billion.

D *The effects of the Chile earthquake*

Secondary effects (tsunamis, fires and landslides)

- 1500 km of roads damaged, mainly by landslides – remote communities cut off for many days.
- Several coastal towns devastated by tsunami waves.
- Several Pacific countries struck by tsunami – warnings prevented loss of life.
- A fire at a chemical plant near Santiago – the area had to be evacuated.

Primary effects

- 9000 people died and 20 000 injured – over 8 million people (a third of Nepal's population) affected.
- 3 million people left homeless when homes were destroyed.
- Electricity and water supplies, sanitation and communications affected.
- 1.4 million people needed food, water and shelter in the days and weeks after the earthquake.
- 7000 schools destroyed and hospitals overwhelmed.
- International airport became congested as aid arrived.
- 50% of shops destroyed, affecting food supplies and people's livelihoods.
- Cost of damage estimated at over US$5 billion.

E *The effects of the Nepal earthquake*

Secondary effects

- Ground shaking triggered landslides and avalanches, blocking roads and hampering relief efforts.
- Avalanches on Mount Everest killed at least 19 people – the greatest loss of life on the mountain in a single incident.
- An avalanche in the Langtang region left 250 people missing.
- A landslide blocked the Kali Gandaki River, 140 km (90 miles) north west of the capital, Kathmandu – many people evacuated in case of flooding.
- The earthquake occurred on land so did not cause a tsunami.

ACTIVITIES

1   a   What is the evidence that Nepal is poorer and less developed than Chile?

    b   Why did the Nepal earthquake affect such a vast area?

    c   Why did the Chile earthquake trigger a tsunami?

2   Describe the primary effects of the Nepal earthquake shown in figure **E**.

3   To what extent did the levels of wealth and development of the two countries affect the impacts of the earthquakes?

4   What were the effects of the tsunami waves caused by the Chilean earthquake?

Stretch yourself

A second powerful earthquake struck Nepal on 12 May 2015. How might this have affected the country's recovery?

Practice question

Explain how different levels of wealth and development affected the impact of the earthquakes in Chile and Nepal. *(6 marks)*

Responses to earthquakes

On this spread you will find out about responses to earthquakes in Chile and Nepal

Responding to earthquakes

There are two different types of response to natural disasters such as earthquakes:

◆ **Immediate responses** – search and rescue and keeping survivors alive by providing medical care, food, water and shelter.

◆ **Long-term responses** – re-building and reconstruction, with the aim of returning people's lives back to normal and reducing future risk.

Did you know?
The last earthquake to hit Kathmandu was in 1934 when over 10 000 people were killed.

Comparing responses in Chile and Nepal

Earthquakes in Chile are quite common. Local communities and the government were prepared and knew how to respond quickly and effectively to the earthquake. Chile had the money to support people and to rebuild.

Earthquakes in Nepal are not uncommon. Scientists have identified a pattern of large earthquakes in this region every 80 years or so. Despite these warnings and new building regulations, little had been done to prepare the city and its people for when the earthquake struck.

Chile: immediate responses

◆ Emergency services acted swiftly. International help needed to supply field hospitals, satellite phones and floating bridges.

◆ Temporary repairs made to the important Route 5 north-south highway within 24 hours, enabling aid to be transported from Santiago to affected areas.

◆ Power and water restored to 90% of homes within 10 days.

◆ A national appeal raised US$60 million – enough to build 30 000 small emergency shelters (photo **A**).

A *Temporary wooden shelters for those made homeless by the earthquake*

Chile: long-term responses

◆ A month after the earthquake Chile's government launched a housing reconstruction plan to help nearly 200 000 households affected by the earthquake.

◆ Chile's strong economy, based on copper exports, could be rebuilt without the need for much foreign aid.

◆ The President announced it could take four years for Chile to recover fully from the damage to buildings and ports (photo **B**).

B *Buildings destroyed by the Chile earthquake*

Nepal: immediate responses

◆ Search and rescue teams (photo **C**), water and medical support arrived quickly from countries such as UK, India and China.

◆ Helicopters rescued many people caught in avalanches on Mount Everest and delivered supplies to villages cut off by landslides.

◆ Half a million tents needed to provide shelter for the homeless.

◆ Financial aid pledged from many countries.

◆ Field hospitals set up to support overcrowded main hospitals.

◆ 300 000 people migrated from Kathmandu to seek shelter and support with family and friends.

◆ Social media widely used in search and rescue operations and satellites mapped damaged areas.

Rubble to be shifted   Rescue dogs   Listening for survivors   Local knowledge

Lifting equipment   Weak buildings – danger of collapse   Video cameras to see inside collapsed buildings

C   *Searching a building for survivors in Kathmandu*

Nepal: long-term responses

◆ Roads repaired and landslides cleared. Lakes, formed by landslides damming river valleys, need to be emptied to avoid flooding.

◆ Thousands of homeless people to be re-housed, and damaged homes repaired. Over 7000 schools to be re-built or repaired.

◆ Stricter controls on building codes.

◆ In June 2015 Nepal hosted an international conference to discuss reconstruction and seek technical and financial support from other countries.

◆ Tourism, a major source of income, to be boosted – by July 2015 some heritage sites re-opened and tourists were starting to return.

◆ Repairs to Everest base camp (photo **D**) and trekking routes – by August 2015 new routes had been established and the mountain re-opened for climbers.

◆ In late 2015 a blockade at the Indian border badly affected supplies of fuels, medicines and construction materials.

D   *Everest base camp*

ACTIVITIES

1 **a** Why did the Chilean government focus on repairing the main north–south highway?

    b Why was the Chilean government able to respond quickly and effectively to the earthquake?

2 Describe how search and rescue teams locate and rescue people from collapsed buildings (photo **C**).

3 What were the immediate needs of the survivors of the Nepal earthquake?

4 What needs to be done to support Nepal's recovery following the earthquake?

Stretch yourself

Investigate the latest information about recovery in Chile and Nepal.
What has been done to reduce the impacts of future earthquakes in the two countries?

Practice question

Choose either the earthquake in Chile or Nepal. Describe the immediate and long-term responses to the disaster. *(6 marks)*

On this spread you will find out why people continue to live in areas at risk from earthquakes and volcanoes

Living in the shadow of a volcano

In AD79 Mount Vesuvius in southern Italy erupted, burying the nearby cities of Pompeii and Herculaneum in volcanic ash and killing thousands of people. Today over one million people live in the shadow of the volcano, most of them in the city of Naples (photo **A**). Vesuvius last erupted in 1944. In the 300 years before then it erupted nearly every 20 years. Many people think that the next eruption is long overdue!

A *Naples in the shadow of Mount Vesuvius*

Living at risk from tectonic hazards

You have seen from the examples of Chile and Nepal how destructive earthquakes can be. So why do people choose to live in such dangerous places?

The majority of tectonic hazards occur at plate margins which criss-cross the Earth's surface. Some margins run through densely populated regions such as Japan, parts of China, and southern Europe (map **C**).

There are several reasons why people live in areas at risk from tectonic hazards, as shown below.

B *Fertile farmland on the slopes of Mount Merapi, Indonesia*

Earthquakes and volcanic eruptions don't happen very often. They are not seen as a great threat in most people's lives.

People living in poverty have other things to think about on a daily basis – money, food, security and family.

Plate margins often coincide with very favourable areas for settlement, such as coastal areas where ports have developed.

Better building design can withstand earthquakes so people feel less at risk.

Why choose to live in hazardous areas?

Some people may not be aware of the risks of living close to a plate margin.

More effective monitoring of volcanoes and tsunami waves enable people to receive warnings and evacuate before events happen.

Fault lines associated with earthquakes can allow water supplies to reach the surface. This is particularly important in dry desert regions.

Volcanoes can bring benefits such as fertile soils, rocks for building, rich mineral deposits and hot water (photo **B**).

Life on a plate margin in Iceland

Iceland lies on the Mid-Atlantic Ridge, a constructive plate margin that stretches through the middle of the Atlantic Ocean. There are several active volcanoes – an eruption occurs on average every five years. Earthquakes are common. Over 320 000 people live in Iceland and close to one million people visit the country each year.

Whilst the tectonic activity does pose a threat, the people in Iceland consider it to be a low risk. This is mainly due to effective scientific monitoring and awareness of the potential dangers. In fact, tectonic activity brings huge benefits to the country (figure **D**).

C *Tectonic plates and population density*

Key
- Densely populated
- Moderately populated
- Sparsely populated
- Plate margins

D *Geothermal power plant near Krafla volcano, Iceland*

Hot water from within the Earth's crust provides heat and hot water for nearly 90% of all buildings in Iceland

The naturally occurring hot water – some of which reaches the surface through cracks created by earthquakes – is used to heat greenhouses and swimming pools

Volcanic rocks are used in construction for roads and buildings

Geothermal energy is used to generate 25% of the country's electricity (most of the rest is generated by hydroelectric power)

Iceland's dramatic landscape with waterfalls, volcanoes and mountain glaciers has become a huge draw for tourists. Tourism provides jobs for many people.

Thousands of tourists visited Iceland after the recent eruption of Eyjafjallajökull in 2010.

ACTIVITIES

1 Why do you think one million people choose to live so close to one of Europe's most dangerous volcanoes (photo **A**)?

2 **a** Which areas of the world are most densely populated (map **C**)?

 b Which of these areas lie on active plate margins? Name some of these margins.

 c Why do you think so many people live in areas at risk from earthquakes and volcanic eruptions?

3 What evidence is there in photo **B** that people are making use of the land close to Mount Merapi?

4 How do the people of Iceland benefit from living on a plate margin (figure **D**)?

Stretch yourself

Carry out some research to find out how people in Iceland benefit from living in an area at risk from tectonic activity.

- What is geothermal energy and how is it used to generate electricity?
- How is Iceland's naturally occurring hot water used for heating?
- How has tectonic activity created attractions for tourists?

Practice question

Use figure **D** to evaluate the benefits of Iceland's location on a plate margin. *(6 marks)*

Reducing the risk from tectonic hazards

On this spread you will find out how the risks from tectonic hazards can be reduced

How can the risks from tectonic hazards be reduced?

There are four main **management strategies** for reducing the risk from tectonic hazards:

◆ **Monitoring** – using scientific equipment to detect warning signs of events such as a volcanic eruption.

◆ **Prediction** – using historical evidence and monitoring, scientists can make predictions about when and where a tectonic hazard may happen.

◆ **Protection** – designing buildings that will withstand tectonic hazards.

◆ **Planning** – identifying and avoiding places most at risk.

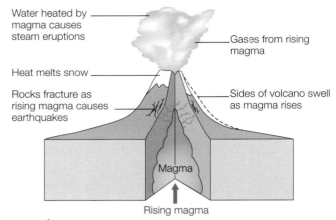

A *Warning signs of a volcanic eruption*

Monitoring

Volcanoes

As magma rises through a volcano it gives a number of warning signs that an eruption is likely to occur (diagram **A**).

All of the world's active volcanoes are closely monitored by scientists. If an eruption seems likely, warnings can be issued and action taken to evacuate surrounding areas. Modern hi-tech equipment is used, some of which is located on the volcano itself. Scientists monitor volcanoes in the following ways:

◆ *Remote sensing* – satellites detect heat and changes to the volcano's shape.

◆ *Seismicity* – seismographs record earthquakes.

◆ *Ground deformation* – changes to the shape of the volcano are measured using laser beams.

◆ *Geophysical measurements* – detect changes in gravity as magma rises to the surface.

◆ *Gas* – instruments detect gases released as magma rises.

◆ *Hydrology* – measurements of gases dissolved in water.

Earthquakes

Earthquakes generally occur without warning. Whilst there is some evidence of changes in water pressure, ground deformation and minor tremors prior to an earthquake, scientists have yet to discover reliable ways to monitor and predict earthquakes.

Prediction

Volcanoes

The prediction of a volcanic eruption is based on scientific monitoring. In 2010 an increase in earthquake activity beneath the Eyjafjallajökull ice cap in Iceland enabled scientists to make an accurate prediction about the eruptions that took place in March and April that year.

Earthquakes

It is impossible to make accurate predictions about earthquakes due to the lack of clear warning signs. However, scientists studying historical records of earthquakes at plate margins have identified locations that they believe are at greatest risk. Map **B** shows why scientists believe the city of Istanbul in Turkey is at risk from an earthquake … soon!

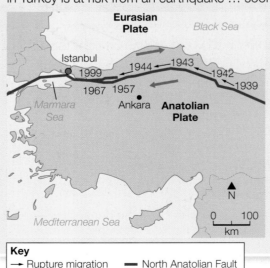

B *Migration of earthquakes along the North Anatolian Fault, Turkey*

Protection

Volcanoes

The sheer power of a volcanic eruption means that there is often little that can be done to protect people and property. However, it is possible to use earth embankments or explosives to divert lava flows away from property. This has been done on the slopes of Mount Etna in Italy.

Earthquakes

Earthquake protection is the main way to reduce risk. It is possible to construct buildings and bridges to resist the ground shaking associated with an earthquake (diagram **C**). In Chile, new buildings have reinforced concrete columns strengthened by a steel frame. Regular earthquake drills help people keep alert and be prepared.

It is possible to construct tsunami walls at the coast to protect people and important buildings like nuclear power stations.

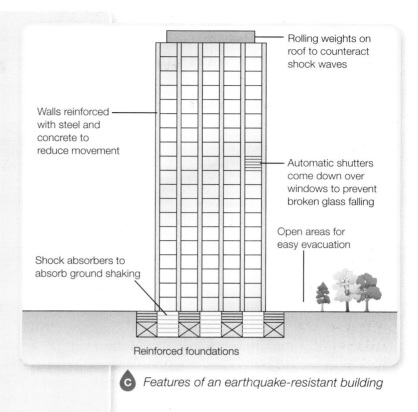

Rolling weights on roof to counteract shock waves

Walls reinforced with steel and concrete to reduce movement

Automatic shutters come down over windows to prevent broken glass falling

Open areas for easy evacuation

Shock absorbers to absorb ground shaking

Reinforced foundations

C *Features of an earthquake-resistant building*

Planning

Volcanoes

Hazard maps have been produced for many of the world's most dangerous volcanoes, showing the likely areas to be affected. They can be used in planning to restrict certain land uses or to identify which areas need to be evacuated when an eruption is about to happen.

Earthquakes

Maps can be produced to show the effects of an earthquake or identify those areas most at risk from damage. High-value land uses such as hospitals, reservoirs and office blocks can then be protected in these vulnerable areas.

3.1 Global atmospheric circulation

On this spread you will find out how global atmospheric circulation affects global weather and climate

What is global atmospheric circulation?

The cruising altitude (height) of an aeroplane is about 10 km above the ground surface. At this altitude the vast majority of the atmosphere's mass is below you (diagram **A**). The atmosphere – the air above our heads – is a highly complex swirling mass of gases, liquids and solids. These include water droplets, water vapour, ash, carbon dioxide and oxygen – just to mention a few!

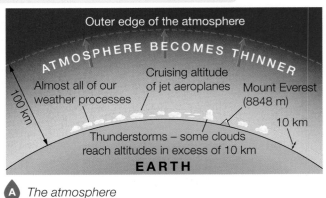

A *The atmosphere*

How does global atmospheric circulation work?

Diagram **B** shows global atmospheric circulation. This involves a number of circular air movements called cells. These cells all join together to form the overall circulation of the Earth's atmosphere.

◆ Air that is *sinking* towards the ground surface forms areas of *high pressure* (for example, at the North Pole). Winds on the ground move outwards from these areas.

◆ Air that is *rising* from the ground surface forms areas of *low pressure* on the ground, for example at the Equator. Winds on the ground move towards these areas of low pressure.

◆ Winds on the ground are distorted by the Earth's rotation. They curve as they move from areas of high pressure to areas of low pressure.

◆ Surface winds are very important in transferring heat and moisture from one place to another.

◆ The patterns of pressure belts and winds are affected by seasonal changes. The tilt and rotation of the Earth causes relative changes in the position of the overhead Sun. These seasonal changes cause pressure belts and winds to move north during our summer and then south during our winter.

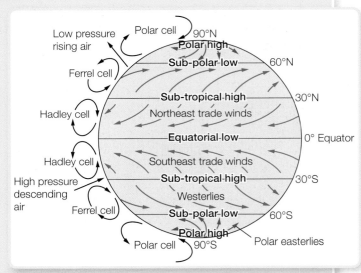

B *Global atmospheric circulation*

Did you know?
The prevailing wind for the UK comes from the south-west over the Atlantic Ocean. This is why we have a moist and mild climate.

How does global circulation affect the world's weather?

Global atmospheric circulation is what drives the world's weather. The circulation cells, pressure belts and surface winds (diagram **B**) affect the weather around the world. For example, the trade winds in the tropics are responsible for driving tropical storms (hurricanes) across these regions bringing chaos and destruction to coastal regions in their path.

Cloudy and wet in the UK

The UK is located at about 55° North just below the 60° N line of latitude. This puts the UK close to the boundary of cold polar air moving down from the north and warm sub-tropical air moving up from the south.

The boundary between these two air masses is unstable. Here there is rising air and low-pressure belts (the *sub-polar low*) on the ground. Rising air cools, condenses and forms cloud and rain. This is why it is often cloudy and wet in the UK.

Surface winds in these mid-latitudes come from the south-west. These winds bring warm and wet conditions to the UK. But sometimes the cold polar air from the north moves down over the UK bringing snow and very cold winter weather.

C *Wet weather in the UK*

D *Hot, dry weather in the desert*

Hot and dry in the desert

Most of the world's hot deserts are found at about 30° north and south of the Equator. Here the air is sinking (diagram **B**), making a belt of high pressure (the sub-tropical high). Air isn't rising here, so there are few clouds forming and little rainfall. The lack of cloud makes it very hot during the day very cold at night, as heat is quickly lost from the ground.

E *Hot, humid weather at the Equator*

Hot and sweaty at the Equator

At the Equator the air is rising (diagram **B**) and there is another low pressure belt (the *equatorial low*). This part of the world is very much hotter than the UK, with the sun directly overhead. Equatorial regions, such as central Africa and south-east Asia, experience hot, humid conditions. It is often cloudy with high rainfall. This is the region where tropical rainforests are found.

ACTIVITIES

1 Copy diagram **B**. Draw the lines of latitude and label the Equator. Add the winds and circulation cells to your diagram. Use different colours to show the high and low pressure belts.

2 What do you notice about patterns of surface winds in relation to high and low pressure belts?

3 Explain why the patterns of pressure belts and surface winds move north and south during the year.

4 How does the atmospheric circulation system explain the UK's mild, cloudy and wet weather?

5 Draw a sketch to show how atmospheric circulation accounts for the high rainfall at the Equator.

Stretch yourself

Find a map to show the tracks followed by tropical storms. Use diagram **B** to add the Equator and the tropics. Draw on the trade winds to show how they are responsible for the east-west movement of the storms.

Practice question

Explain how the global atmospheric system affects the weather and climate of the tropics. *(6 marks)*

On this spread you will find out about the distribution and formation of tropical storms

What is a tropical storm?

A tropical storm is a huge storm that develops in the Tropics (image **A**). In the USA and the Caribbean these are called **hurricanes**. In south-east Asia and Australia they are called **cyclones**, but in Japan and the Philippines they are called **typhoons**.

Tropical storms are incredibly powerful and can cause devastation to small islands and coastal regions. Photo **B** shows some of the damage caused by Hurricane Sandy on the east coast of the USA in 2012. It was the costliest and most deadly Atlantic storm of the year, killing 285 people.

Where do tropical storms form?

Map **C** shows the distribution of tropical storms. It also provides some useful clues about the formation of tropical storms.

- Tropical storms form over warm oceans (above 27 °C), which explains why they are found in the Tropics.

- They form in the summer and autumn when sea temperatures are at their highest.

- Most tropical storms form 5–15° north and south of the Equator. This is because at the Equator there is not enough 'spin' from the rotation of the Earth. The effect of the Earth's rotation is called the *Coriolis effect*. A tropical storm is a spinning mass of clouds (photo **A**).

- In tropical regions the intense heat makes the air unstable causing it to rise rapidly. These unstable conditions are important in the formation of hurricanes.

A *Satellite image of Hurricane Sandy off the coast of Florida, USA, 2012*

B *The impact of Hurricane Sandy in Queens, New York*

C *The distribution of tropical storms*

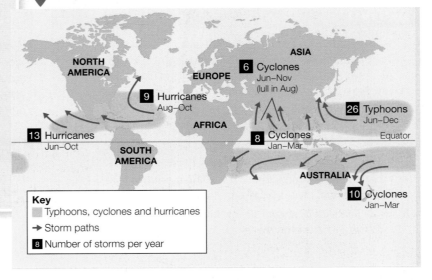

NORTH AMERICA

EUROPE

ASIA

6 Cyclones Jun–Nov (lull in Aug)

9 Hurricanes Aug–Oct

AFRICA

26 Typhoons Jun–Dec

13 Hurricanes Jun–Oct

SOUTH AMERICA

8 Cyclones Jan–Mar

Equator

AUSTRALIA

10 Cyclones Jan–Mar

Key
Typhoons, cyclones and hurricanes
➡ Storm paths
8 Number of storms per year

How do tropical storms form?

Scientists are not certain what causes the formation of a hurricane, but it involves the sequence of events shown below.

On reaching land the storm's energy supply (evaporated water) is cut off. Friction with the land slows it down and it begins to weaken. If the storm reaches warm seas after crossing the land, it may pick up strength again.

As the storm is carried across the ocean by the prevailing winds, it continues to gather strength.

The storm now develops an eye at its centre where air descends rapidly. The outer edge of the eye is the eyewall where the most intense weather conditions (strong winds and heavy rain) are felt.

Several smaller thunderstorms join together to form a giant spinning storm. When surface winds reach an average of 120 km per hour (75 miles per hour) the storm officially becomes a tropical storm.

As the air condenses it releases heat which powers the storm and draws up more and more water from the ocean.

This evaporated air cools as it rises and condenses to form towering thunderstorm clouds.

A strong upward movement of air draws water vapour up from the warm ocean surface.

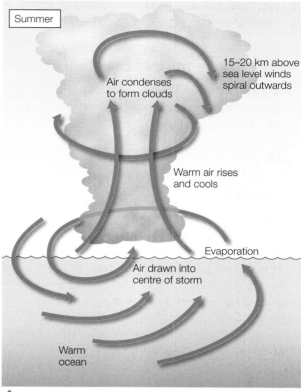

D *Formation of a tropical storm*

Summer
Air condenses to form clouds
15–20 km above sea level winds spiral outwards
Warm air rises and cools
Evaporation
Air drawn into centre of storm
Warm ocean

ACTIVITIES

1 a In which part of the world are tropical storms called cyclones?

   b During which months are hurricanes most likely to affect the east coast of the USA?

   c On average how many cyclones affect Australia each year?

   d Which countries are most likely to experience tropical storms during the year?

2 Why do tropical storms not form at the Equator?

Stretch yourself

Make a copy of diagram **D** showing how a tropical storm forms. Add detailed labels in the form of a sequence (1, 2, 3, etc.). Describe the formation of a tropical storm.

Practice question

Using map **C** and your own knowledge, describe the global distribution of tropical storms. *(4 marks)*

The structure and features of tropical storms

On this spread you will find out about the structure and features of tropical storms, and how climate change might affect tropical storms in the future.

What is the structure of a tropical storm?

Tropical storms can be huge, up to 480 km (300 miles) across. A tropical storm has a roughly symmetrical shape. Diagram **A** shows an imaginary cross-section (X–Y) through a tropical cyclone.

Did you know?
A tropical storm can release the energy of 10 atom bombs every second!

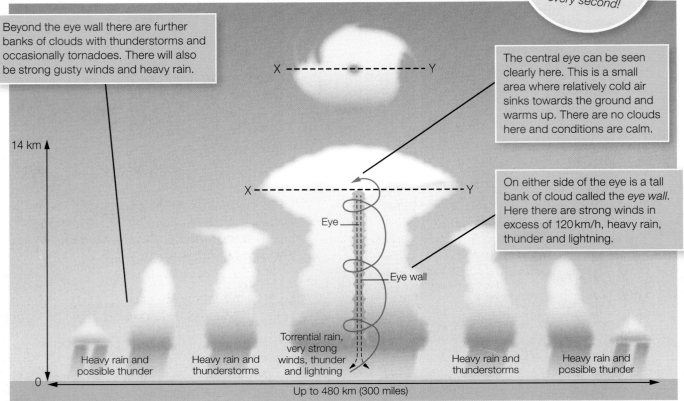

Beyond the eye wall there are further banks of clouds with thunderstorms and occasionally tornadoes. There will also be strong gusty winds and heavy rain.

The central *eye* can be seen clearly here. This is a small area where relatively cold air sinks towards the ground and warms up. There are no clouds here and conditions are calm.

On either side of the eye is a tall bank of cloud called the *eye wall*. Here there are strong winds in excess of 120 km/h, heavy rain, thunder and lightning.

14 km

Eye

Eye wall

Torrential rain, very strong winds, thunder and lightning

Heavy rain and possible thunder

Heavy rain and thunderstorms

Heavy rain and thunderstorms

Heavy rain and possible thunder

0

Up to 480 km (300 miles)

 Structure of a tropical storm

Will climate change affect tropical storms?

There is strong scientific evidence that global temperatures have risen over the last few decades. These rises may be impacting on the world's natural systems. But what impact will they have on tropical storms?

Tropical storm facts

◆ Tropical storms are the most destructive storms on Earth.

◆ They are given names for identification. Hurricanes, for example, are given alternating male and female names each 'season'. The first hurricane starts with 'A', the second 'B', and so on. In 2016, the first hurricane will be Alex, then Bonnie, then Colin...

◆ Hurricane Camille in 1969 had the highest recorded wind speed, estimated at 304 km/h (190 mph).

How strong is a hurricane?

Hurricanes are measured using the Saffir-Simpson scale.

Category	Wind speeds
5	> 252 km/h
4	209–251 km/h
3	178–208 km/h
2	154–177 km/h
1	119–153 km/h

Distribution

Over the last few decades sea surface temperatures in the Tropics have increased by 0.25–0.5°C. As patterns of sea surface temperatures change, they may affect the distribution of tropical storms.

In the future, tropical storms may affect areas outside the current hazard zone, such as the South Atlantic and parts of the sub-tropics. Hurricanes may also become more powerful.

Hurricane Catarina (2004)

In March 2004, the south-east coast of Brazil was struck by a Category 2 hurricane, the first ever recorded here. Coastal communities were taken by surprise and extensive damage was done. Some people died, 40 000 homes were damaged and 85 per cent of the region's banana plants were destroyed.

Hurricanes do not usually form in the South Atlantic (see map **C** on page 24). Cold ocean currents keep waters below the minimum temperature required for hurricane formation. Strong winds 'shear' rising air preventing storms from forming.

In March 2004, sea surface temperatures were unusually high. Conditions were right for a hurricane to form. Such events might become more common as sea surface temperatures change.

Frequency

Graph **B** shows the number of hurricanes recorded in the North Atlantic since 1878. Six of the ten most active years since 1950 have happened since the mid-1990s. Some computer models indicate that the frequency of tropical storms may decrease in the future – but, their *intensity* might increase.

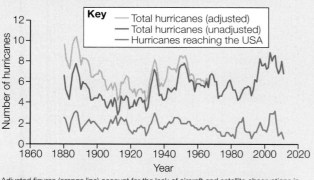

Adjusted figures (orange line) account for the lack of aircraft and satellite observations in the early years

B *Hurricanes in the North Atlantic, 1878–2013*

Intensity

Graph **C** shows hurricane intensity in the North Atlantic has risen in the last 20 years. This appears to be linked to increases in sea surface temperatures. But comparisons with the past may not be completely reliable. More data will be needed over a longer period of time.

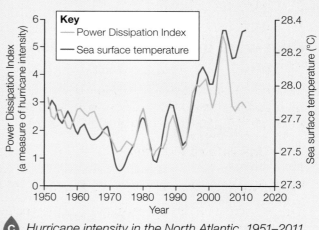

C *Hurricane intensity in the North Atlantic, 1951–2011*

ACTIVITIES

1 Make a copy of diagram **A**. Add labels to describe the main features of a tropical storm.

2 **a** What is the orange line on graph **B** and why is it important?

   b Describe the pattern of hurricanes reaching the USA since 1980.

   c Is there evidence of an overall trend since 1878?

3 Describe and explain the pattern of the Power Dissipation Index between 1950 and 2011 (graph **C**).

Stretch yourself

Carry out some research on Hurricane Catarina. Why did the formation of the storm make it so unusual?

Practice question

Study graph **C**. Has there been an increase in hurricane intensity in recent decades? Support your answer with evidence. *(4 marks)*

Typhoon Haiyan – a tropical storm

Example

On this spread you will find out about the effects of and responses to Typhoon Haiyan

Tropical storms can have devastating effects on people and property. The strong winds can tear off roofs, overturn cars and make large objects fly. Torrential rain can lead to flooding. Strong winds and low atmospheric pressure may cause the sea level to rise by several metres to form a destructive storm surge. These storm surges cause the most loss of life.

Tropical storms can be tracked and warnings given for people to evacuate coastal areas. In the aftermath, people need emergency support. Reconstruction may take many months.

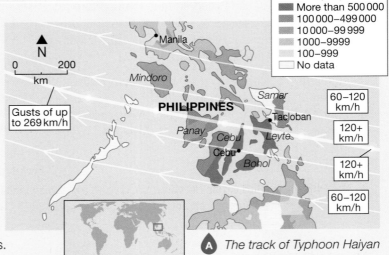

Key
Affected people
- ■ More than 500 000
- ■ 100 000–499 000
- ■ 10 000–99 999
- ■ 1000–9999
- □ 100–999
- □ No data

N   0   200   km

Gusts of up to 269 km/h

Manila

Mindoro

PHILIPPINES

Samar

Tacloban

Panay   *Cebu*   *Leyte*

Cebu

Bohol

60–120 km/h
120+ km/h
120+ km/h
60–120 km/h

A *The track of Typhoon Haiyan*

What happened?

In November 2013 'Super' Typhoon Haiyan – a category 5 storm on the Saffir-Simpson scale – hit the Philippines (map **A**). Huge areas of coastline and several towns were devastated by winds of up to 275 km/h (170 mph) and waves as high as 15 m (45 ft). It was one of the strongest storms ever recorded.

What were the effects of Typhoon Haiyan?

The province of Leyte took the full force of the storm. The city of Tacloban was one of the worst affected places, with most of the 220 000 inhabitants left homeless.

Most of the destruction in Tacloban was caused by a 5-metre high *storm surge*. This is a wall of water similar to a tsunami. The very low atmospheric pressure associated with the typhoon caused the level of the sea to rise. As the strong winds swept this water onshore, it formed a wall of water several metres high.

B *The destruction at Tacloban*

Primary effects (impacts of strong winds, heavy rain and storm surge)

- ◆ About 6300 people killed – most drowned by the storm surge.
- ◆ Over 600 000 people displaced and 40 000 homes damaged or flattened – 90% of Tacloban city destroyed.
- ◆ Tacloban airport terminal badly damaged.
- ◆ The typhoon destroyed 30 000 fishing boats.
- ◆ Strong winds damaged buildings and power lines and destroyed crops.
- ◆ Over 400 mm of rain caused widespread **flooding**.

Secondary effects (longer-term impacts resulting from primary effects)

- ◆ 14 million people affected, many left homeless and 6 million people lost their source of income.
- ◆ Flooding caused landslides and blocked roads, cutting off aid to remote communities.
- ◆ Power supplies in some areas cut off for a month.
- ◆ Ferry services and airline flights disrupted for weeks, slowing down aid efforts.
- ◆ Shortages of water, food and shelter affected many people, leading to outbreaks of disease.
- ◆ Many jobs lost, hospitals were damaged, shops and schools were destroyed, affecting people's livelihoods and education.
- ◆ Looting and violence broke out in Tacloban.

What were the responses to Typhoon Haiyan?

Immediate responses

◆ International government and aid agencies responded quickly with food aid, water and temporary shelters.

◆ US aircraft carrier *George Washington* and helicopters assisted with search and rescue and delivery of aid.

◆ Over 1200 evacuation centres were set up to help the homeless.

◆ UK government sent shelter kits (photo **D**), each one able to provide emergency shelter for a family.

◆ French, Belgian and Israeli field hospitals set up to help the injured.

◆ The Philippines Red Cross delivered basic food aid, which included rice, canned food, sugar, salt and cooking oil.

C *A survivor in Tacloban*

Long-term responses

◆ The UN and countries including the UK, Australia, Japan and the US donated financial aid, supplies and medical support.

◆ Rebuilding of roads, bridges and airport facilities.

◆ 'Cash for work' programmes – people paid to help clear debris and rebuild the city.

◆ Foreign donors, including the US, Australia and the EU, supported new livelihood opportunities.

◆ Rice farming and fishing quickly re-established. Coconut production – where trees may take five years to bear fruit – will take longer.

◆ Aid agencies such as Oxfam supported the replacement of fishing boats – a vital source of income.

◆ Thousands of homes have been built away from areas at risk from flooding.

◆ More cyclone shelters built to accommodate people evacuated from coastal areas.

D *The contents of a Shelter Box*

ACTIVITIES

1 Describe the track of the typhoon (map **A**).

2 **a** Why do you think so many buildings were destroyed (photo **B**)?

  b What are the challenges facing the authorities in rebuilding this area?

3 **a** Why do you think the man in photo **C** appears happy despite all the destruction around him?

  b What are his immediate needs and what are the challenges facing him in the future?

4 Describe the purpose of each of the items in the Shelter Box (photo **D**).

Stretch yourself

How has the city of Tacloban been rebuilt since the disaster struck? What is the situation like now? Is the city in a better position to cope with a future typhoon?

Practice question

Describe the primary and secondary effects of a tropical storm. Use a named example and your own knowledge.
(9 marks)

On this spread you will find out how the effects of tropical storms can be reduced

Monitoring and prediction

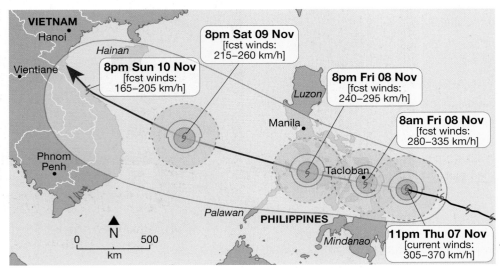

VIETNAM
Hanoi
Hainan
Vientiane

8pm Sat 09 Nov
[fcst winds: 215–260 km/h]

8pm Sun 10 Nov
[fcst winds: 165–205 km/h]

Luzon

8pm Fri 08 Nov
[fcst winds: 240–295 km/h]

Manila

8am Fri 08 Nov
[fcst winds: 280–335 km/h]

Phnom Penh
Tacloban

Palawan **PHILIPPINES**

N
0   500
km

Mindanao

11pm Thu 07 Nov
[current winds: 305–370 km/h]

Key
← Forecast track
— Previous track
⌐ Potential track area
⚡ Typhoon or super typhoon
◉ Area of 120 km/h winds
○ Area of 95 km/h winds
○ Area of 65 km/h winds
⟨⟩ Rain area

A *The predicted track of Typhoon Haiyan*

Map **A** shows the location of Typhoon Haiyan at 11 pm local time on Thursday 7 November 2013. This was just 9 hours before it struck Leyte and flattened most of the city of Tacloban. The map shows the predicted course (track) of the tropical storm across the Philippines. Notice that the area showing the predicted track becomes wider with time. This is because the future track of the tropical storm is uncertain.

Developments in technology have made it possible to predict and monitor tropical storms more accurately and effectively.

In the North Atlantic, there are two levels of warning issued by the National Hurricane Center in Miami:

◆ Hurricane Watch – advises that hurricane conditions are possible.

◆ Hurricane Warning – advises that hurricane conditions are expected and that people should take immediate action (e.g. evacuate to high ground or take shelter).

Think about it

Storm surges are often the greatest threat to life and property from a tropical storm. Why do you think this is?

Protection

There are a number of options available to protect people from the hazards associated with tropical storms.

◆ Windows, doors and roofs reinforced to strengthen buildings to withstand strong winds.

◆ Storm drains constructed in urban areas to take away excessive amounts of rainfall and prevent flooding.

◆ Sea walls built to protect key properties from storm surges.

◆ Houses close to the coast constructed on stilts so that a storm surge will pass beneath.

◆ In Bangladesh nearly 2000 cyclone shelters have been built (photo **B**).

B *Cyclone shelter in Bangladesh*

Constructed of strong concrete

Stairs to take people to the safety of the first floor

Shutters over windows

Bicycles used to give warnings to remote communities

Built on stilts in case of floods

Built on raised ground

Shelter can be used as a community centre, school or medical centre for most of the time

Planning

It is unrealistic to stop the tens of millions of people living and working in coastal areas that are at risk from tropical storms. Many people rely upon fishing or tourism to make a living. Even in rich countries like the USA, vast urban developments have been allowed to take place on vulnerable barrier islands off the coast of Florida, for example Miami Beach. South Miami was hit by a powerful hurricane in 1992. However, building developments have still taken place on land at risk from flooding. It's only a matter of time before Miami gets hit again.

Planning to reduce the tropical storm hazard is mostly about raising individual and community awareness. People need to understand the potential dangers and be able to respond. In the USA there is a National Hurricane Preparedness Week (image **C**) which focuses on educating people about potential dangers ahead of the next hurricane season. Families are encouraged to devise their own plan of action should a warning be issued.

C *National Hurricane Preparedness Week (USA)*

D *Bikes carry cyclone warnings to rural communities in Bangladesh*

Bangladesh – a success story

Early warning systems, cyclone shelters (photo **B**) and greater awareness have helped reduce the death toll from tropical cyclones in Bangladesh. The number of deaths has decreased 100-fold over the past 40 years from 500 000 deaths in 1970 to 4234 in 2007.

Tropical cyclones are tracked by the Bangladesh Meteorological Department. Warnings are issued in several languages by radio, television and via social media. In rural areas, even the most remote communities are reached – sometimes by bike (photo **D**)!

ACTIVITIES

1 **a** Use map **A** to describe the characteristics of Typhoon Haiyan at 11 pm on Thursday 7 November.

   b Describe the direction of the predicted track of the typhoon.

   c How many hours was the typhoon expected to take to cross the Philippines?

   d Where was the typhoon expected to make landfall after the Philippines?

2 What are the special design features of the cyclone shelter in photo **B** to reduce the impacts of a storm surge?

Stretch yourself

Do some further research about the work of the National Hurricane Center in Miami.

- How are hurricanes forecast and predictions made?
- What advice is given to people who live in vulnerable areas to help them prepare?

Practice question

Explain why planning and being prepared is the best option for reducing the effects of tropical storms. *(4 marks)*

Weather hazards in the UK

On this spread you will find out how the UK is affected by a variety of weather hazards

What are the UK's weather hazards?

The *weather* is a description of the day-to-day conditions of the atmosphere. We might talk about the temperature, amount of cloud, the strength and direction of the wind or whether it is raining. When we talk about *climate*, this is the average weather over a long period of time. Data are used over a 30-year period to describe the climate of a place.

Weather hazards are extreme weather events. Even though the UK has a moderate climate, it experiences its share of **extreme weather**. Weather is driven towards the UK by south-westerly prevailing winds. Fuelled by the warm and moist conditions of the Atlantic Ocean, strong winds and heavy rain batter the exposed western areas.

Did you know?
Between 30 and 60 people are struck by lightning in the UK each year. Most people survive!

Thunderstorms

In July 2014 dramatic electrical storms resulted in 3000 lightning strikes across southern Britain following a period of hot weather (photo **A**). Torrential rainfall associated with thunderstorms can result in sudden 'flash' flooding as happened in Boscastle in Cornwall in 2004 (photo **B**).

Prolonged rainfall

Persistent rainfall over a long period can lead to river floods. This is common in the UK especially during the late winter and early spring when snowmelt makes the problem worse. During the very wet winter of 2014 flooding was widespread across much of southern England.

Drought and extreme heat

The UK has experienced long spells of dry, hot weather resulting in drought. Rivers can dry up and reservoirs become dangerously low, which affects water supplies and wildlife. Very high temperatures – heatwaves – can be dangerous to frail and elderly people. In 2003 much of Europe suffered the most extreme heatwave for 500 years (photo **C**). Over 20 000 people died, and several countries, including the UK, recorded their highest ever temperatures.

Heavy snow and extreme cold

Long periods of severe winter weather have become less common in recent years, but there are occasions when heavy snow and severe cold can cause great hardship to people particularly in the north of the UK.

A *Lightning above Canary Wharf, London, July 2014*

B *Boscastle flash flood, 2004*

Strong winds

The UK does occasionally get battered by strong winds. Sometimes the remnants of hurricanes travel over the Atlantic from the USA and Caribbean. These can cause disruption to power supplies and damage from fallen trees. Read about the strong winds that hit the UK in February 2014 (extract **D**).

Strong winds bring chaos to UK

There was widespread disruption to road and rail networks, leaving 21 000 people without power, as strong winds continued to batter the UK this week.

Electricity supplies were affected in South Wales, the south-west and the West Midlands. Wind speeds of up to 105 mph were recorded in Aberdaron in north-west Wales, with gusts of 92 mph recorded on the Gower Peninsula, south-west Wales.

The Met Office has warned that coastal areas of the UK could be battered by large waves. Clifton suspension bridge in Bristol was closed briefly for the first time ever because of high winds, and storms have brought down many trees.

C 'Hot enough to fry an egg' – heatwave in 2003

D News report, February 2014

Why does extreme weather occur in the UK?

The UK is rather like a roundabout (map **E**) because it is at the meeting point of several different types of weather from different directions. This explains why we experience such varied weather from week to week and how occasionally we can be affected by extreme weather events.

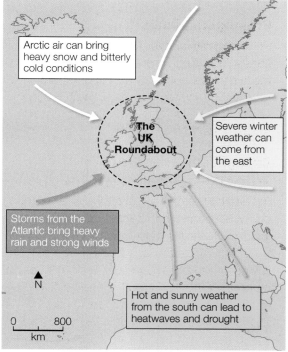

Arctic air can bring heavy snow and bitterly cold conditions

The UK Roundabout

Severe winter weather can come from the east

Storms from the Atlantic bring heavy rain and strong winds

Hot and sunny weather from the south can lead to heatwaves and drought

▲ N

0   800
    km

E The UK's weather roundabout

ACTIVITIES

1 What is the difference between weather and climate?

2 What are the hazards associated with thunderstorms?

3 Use evidence from photo **B** to describe the impacts of the flash flood on the lives of local residents in Boscastle.

4 What is a drought and what impact does it have on the natural world?

5 Read extract **D**. What were the effects of the strong winds in February 2014?

Stretch yourself

The European heatwave of 2003 was a truly extreme weather event.

- How long did the heatwave last and what was the highest temperature?

- What were the impacts of the heatwave on people and the natural world?

Practice question

Describe two types of weather hazard that could affect the UK. *(4 marks)*

Example

On this spread you will find out about flooding on the Somerset Levels in 2014

Where are the Somerset Levels?

Somerset is a county in south-west England. The Somerset Levels and the Somerset Moors form an extensive area of low-lying farmland and wetlands bordered by the Bristol Channel and Quantock Hills to the west and the Mendip Hills to the north (map **A**).

The area is drained by several rivers, most notably the Tone and the Parrett, which flow to the Severn Estuary via Bridgwater. Flooding has occurred naturally here for centuries. As the area has been developed for farming and settlement, many people are now at risk from extreme flood events.

What caused the floods in 2014?

There were several factors that led to extensive flooding of the Somerset Levels.

◆ It was the wettest January since records began in 1910. A succession of depressions (areas of low pressure) driven across the Atlantic Ocean brought a period of wet weather lasting several weeks. About 350 mm of rain fell in January and February, about 100 mm above average.

◆ High tides and storm surges swept water up the rivers from the Bristol Channel. This prevented fresh water reaching the sea and it spilled over the river banks.

◆ Rivers had not been dredged for at least 20 years, and had become clogged with sediment.

What were the impacts of the flood?

Between December 2013 and February 2014, the Somerset Levels hit the national headlines as the area suffered extensive flooding. It was the most severe flooding ever known in the area.

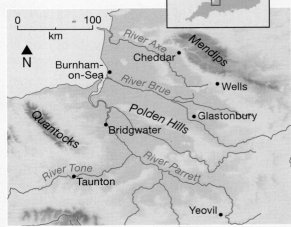

A The Somerset Levels

B Rescuing a resident from Moorland

Social	Economic	Environmental
• Over 600 houses flooded • 16 farms evacuated • Residents evacuated to temporary accommodation for several months • Villages such as Moorland and Muchelney cut off. This affected people's daily lives, e.g. attending school, shopping, etc. • Many people had power supplies cut off	• Somerset County Council estimated the cost of flood damage to be more than £10 million • Over 14 000 ha of agricultural land under water for 3–4 weeks • Over 1000 livestock evacuated • Local roads cut off by floods • Bristol to Taunton railway line closed at Bridgwater	• Floodwaters were heavily contaminated with sewage and other pollutants including oil and chemicals • A huge amount of debris had to be cleared • Stagnant water that had collected for months had to be reoxygenated before being pumped back into the rivers

C The impacts of the Somerset Levels floods

Managing the floods

Immediate responses

As the floodwaters spread out over the Somerset Levels, homeowners coped as best they could. Villagers cut off by the floods used boats to go shopping or attend school. Local community groups and volunteers gave invaluable support.

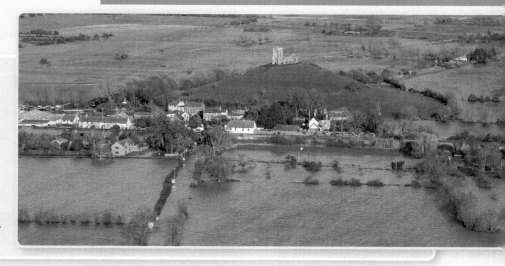

The village of Burrowbridge **D** *almost cut off by the floods*

Longer-term responses

A £20 million Flood Action Plan has been launched by Somerset County Council who will work together with agencies such as the Environment Agency to reduce the risk of future flooding.

◆ In March 2014, 8 km of the Rivers Tone and Parratt were dredged to increase the capacity of the river channel (diagram **E**).

◆ Road levels have been raised in places to maintain communications and enable businesses to continue during future flood events.

◆ Vulnerable communities will have flood defences.

◆ River banks are being raised and strengthened and more pumping stations will be built.

◆ In the longer term – by 2024 – consideration will be given to a tidal barrage at Bridgwater.

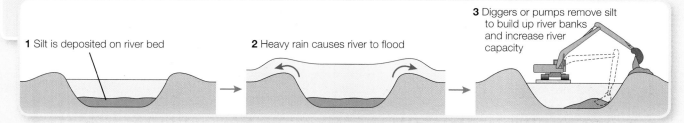

1 Silt is deposited on river bed

2 Heavy rain causes river to flood

3 Diggers or pumps remove silt to build up river banks and increase river capacity

 How dredging works

ACTIVITIES

1 Why do you think the Somerset Levels are prone to flooding (map **A**)?

2 What were the factors contributing to the floods in 2014?

3 **a** Describe the extent of the flooding (photo **D**).

   b Several roads were flooded. What impact did this have on local people?

   c Suggest the impact of the flooding on farmers in the area.

4 Make a copy of diagram **E**. Add labels to describe how dredging can help reduce the flood risk.

Stretch yourself

Imagine you are a local councillor in Somerset. Analyse research plans to construct a tidal barrage at Bridgwater (they can be found on the internet).

What would this scheme involve and how would it reduce the risk of flooding?

Practice question

Using table **C**, evaluate the main impacts of the flooding of the Somerset Levels. *(6 marks)*

Example

On this spread you will use a 1:25 000 map to find out about flooding on the Somerset Levels in 2014

Map **A** is a 1:25 000 map extract of the Somerset Levels a few kilometres south-east of the town of Bridgwater. Photo **B** shows flooding in the village of Moorland (also called Northmoor Green: grid reference 3332). The key for OS maps can be found on page 352.

A *1:25 000 map extract of the Somerset Levels*

© Crown copyright

1 km

Practice question

Suggest the likely social, economic and environmental impacts of the flooding. Use evidence from photo **B** to support your answer. *(4 marks)*

Think about it

How can maps help you to interpret aerial photographs?

B *Flooding in Moorland, Somerset (2014)*

ACTIVITIES

1 Use map **A** to answer the following questions.

a What is the evidence from the map that this area is very flat and low-lying?

b What is the name of the main river?

c Why do you think the area has hundreds of drainage ditches?

d What is the six-figure grid reference of the pumping station?

e Why do you think there is a pumping station at this location?

f In what direction is Burrowbridge from the pumping station?

g To the nearest 100 m, what is the straight line distance from the pumping station to the bridge over the river at Burrowbridge (grid reference 357304)?

h What is the evidence that most of this area is farmland?

i Use evidence from the map to suggest why Moorland is at risk from flooding.

2 Photo **B** shows part of the flooded village of Moorland. Locate the church at the bottom left of the photo. Now locate the church on map **A**. It is at the road junction in the centre of the village.

a What is the six-figure grid reference of the church?

b In what direction is the photo looking?

c What is the name of the farm at the top left of the photo?

d What has been done to try to stop this property from flooding?

e Describe the extent of flooding in the photo.

On this spread you will find out if the UK's weather is becoming more extreme

What is the evidence?

There have been many extreme weather events in the UK throughout history. However, scientists have noticed that these events seem to be occurring more frequently than in the past. Look at the diagram below to read about extreme weather events in the UK since 2000.

2003 Heatwave
The UK recorded its highest ever temperature of 38.5°C in Kent. Over 2000 people died due to the heat, railway tracks buckled and in places the roads melted!

2009 Heavy snow
Parts of south-west and south-east England were affected by heavy snow with 20cm falling in the capital.

2009 Floods
The town of Cockermouth in Cumbria was devastated by floods. Record rainfall amounts fell in November in the Lake District.

2010 Heavy snow
Much of the UK was hit by heavy snowfalls in December. Northern Ireland recorded a record low temperature of −18.7°C at Castlederg.

2008 Floods
Severe flooding occurred in south-west and north-east England with Somerset, Worcestershire and Northumberland badly hit.

2007 Floods
Several people died and many were left homeless by summer floods affecting Hull, Sheffield and Gloucestershire.

2013/14 Floods
Severe flooding occurred across southern England causing the River Thames to burst its banks and vast areas of the Somerset Levels to become inundated. It was England's wettest winter in 250 years.

2015/16 Floods
Severe storms and exceptionally heavy rainfall caused devastating floods to many areas, especially the north of England. Yorkshire and Cumbria (photo **C**) were badly affected. December 2015 was the wettest and warmest month ever recorded in the UK.

 A *Trafalgar Square during the heatwave of 2003*

 B *Snow causes traffic chaos in 2010*

Why might extreme weather events be on the increase?

Recent extreme weather events have also occurred elsewhere in the world. There have been devastating floods in Pakistan (2010), intense heatwaves in Russia (2010) and severe droughts in western USA (2014).

No single extreme weather event can be blamed on climate change. However, scientists believe that a trend over many years could be linked to a warming world.

◆ More energy in the atmosphere could lead to more intense storms.

◆ The atmospheric circulation (see page 22) may be affected, bringing floods to normally dry regions and heatwaves to normally cooler areas.

In 2011 the Intergovernmental Panel on Climate Change concluded that extreme weather would become more common as global warming heats the planet.

Could our weather patterns be getting 'stuck'?

Weather systems cross the UK mainly from west to east, driven by winds from the *jet stream*. The jet stream moves north and south but can 'stick' in one position, resulting in a long period of the same type of weather, such as heavy rain or drought. A large area of high pressure over Northern Europe can block the easterly movement of weather systems and have a similar effect on UK weather.

In 2014 scientists in Germany published a report. It suggested that in recent years weather patterns have become 'stuck' for long periods of time. This has resulted in prolonged periods of high temperatures (heatwaves and droughts) and heavy rain (floods).

These periods seem to have become more frequent in recent years and this could be due to climate change. A warming Arctic, for example, may slow down the atmospheric circulation in the northern hemisphere mid-latitudes resulting in the weather 'sticking' for long periods of time. This could explain the recent heatwaves and floods.

C *Floods in Carlisle, 2015*

D *The Intergovernmental Panel on Climate Change, 2014*

Stretch yourself

Imagine you're a journalist writing an article about extreme weather in the UK since January 2014.

- What happened and what were the impacts?
- How have these events – along with the others since 2000 – been linked with climate change?

ACTIVITIES

1 Draw a timeline to describe the extreme weather events in the UK since 2000. Use text boxes to describe the impacts of the events and illustrate your timeline using photos. Add any recent events.

2 What UK weather records have been broken by extreme weather events since 2000?

3 Newspapers sometimes blame an individual extreme weather event on climate change. Why is this misleading?

Practice question

Suggest why the UK's weather might be becoming more extreme. *(4 marks)*

On this spread you will consider the evidence for climate change from the beginning of the Quaternary period to the present day

It's not as cold as it used to be!

Graph **A** shows the pattern of global temperatures for the last 5.5 million years. This may sound like a long time but remember that the Earth was formed 4600 million years ago!

The graph shows how temperature has changed over time (blue line) compared to today's average temperature (shown by the dashed line at 0 °C). The last 2.6 million years is called the *Quaternary period*. During this geological period temperatures have fluctuated a great deal. Despite these fluctuations the graph shows there has been a gradual cooling during this period.

The cold 'spikes' during the Quaternary period are *glacial periods* when ice covered parts of Europe and North America. The warmer periods in between are called *inter-glacial periods*. Notice that today's average temperature is higher than during almost all of the Quaternary period.

Graph **B** shows that in the last few decades the average global temperature has increased relative to the 1901–2000 average. This has become known as 'global warming', the most recent indication of climate change.

Since 1880 the average global temperature has risen by 0.85 °C. Most of this increase has occurred since the mid-1970s.

Global effects of climate change

Climate change has already had significant effects on global ecosystems and on people's lives.

◆ Many of the world's glaciers and ice caps are shrinking.

◆ Arctic sea ice is less extensive than in the past, affecting wildlife such as polar bears (pages 78–9). However, this may provide opportunities for ships to use the North West Passage in the future.

◆ Low-lying Pacific Islands such as Tuvalu and the Maldives are under threat from sea-level rise.

◆ Sea levels may rise by 1 m by 2100 flooding agricultural land in Bangladesh, Vietnam, India and China

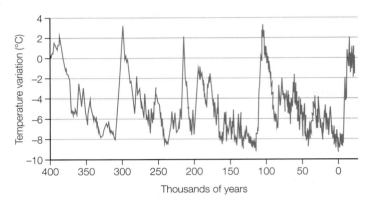

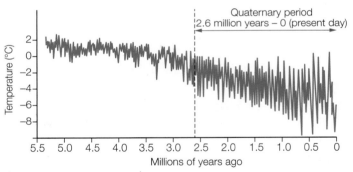

A *Average global temperatures for the last 5.5 million years using information from sediment cores*

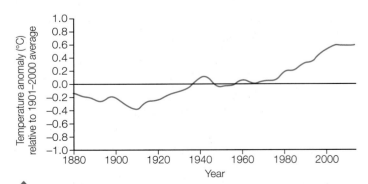

B *Average global temperature (1880–2013) based on recorded temperature records*

What is the evidence for climate change?

Temperature is measured directly using an instrument called a thermometer. Reliable measurements using thermometers go back only about a hundred years. In the UK, for example, reliable weather records began in 1910. So, how do we know what temperatures were in the distant past?

Without the use of thermometers, scientists use indirect data stored as a fossil record. These are found in deep ocean sediments and frozen ice cores.

When layers of sediment or fresh falls of snow become buried they trap and preserve evidence of the global temperature at that time. Scientists can study the oxygen in ocean sediments or water molecules in ice to calculate temperature. They can be accurately dated and this information used to plot graphs such as graph **A**. Ice cores have been used to reconstruct temperature patterns from as long as 400000 years ago (photo **C**).

C *Extracting ice cores from the Antarctic ice sheet*

Things are heating up!

Direct measurements of temperature using thermometers have indicated a clear warming trend (graph **B**). There is other evidence that climate change is taking place.

Shrinking glaciers and melting ice

Glaciers throughout the world are shrinking and retreating. It is estimated that some may disappear completely by 2035. Arctic sea ice has thinned by 65 per cent since 1975 and in 2014 its extent was at an all-time low (photo **D**).

What is the recent evidence for climate change?

Rising sea level

According to the Intergovernmental Panel on Climate Change (IPCC), the average global sea level has risen between 10 and 20cm in the past 100 years. There are two reasons why sea levels have risen.

- When temperatures rise and freshwater ice melts, more water flows to the seas from glaciers and ice caps.
- When ocean water warms it expands in volume – this is called thermal expansion.

Seasonal changes

Studies have suggested that the timing of natural seasonal activities such as tree flowering and bird migration is advancing. A study of bird nesting in the mid-1990s discovered that 65 species nested an average of 9 days earlier than in the 1970s. Could this be evidence of a warming world?

D *Shrinkage of Arctic sea ice, 1979–2012 (yellow line indicates extent in 1979)*

ACTIVITIES

1 Describe the pattern of temperatures during the Quaternary period (graph **A**).

2 a Describe the trend of the average temperature between 1880 and 1940 (graph **B**).

   b Describe the trend in average temperature since 2000.

   c Do you think this graph provides strong evidence for global warming?

3 Briefly describe how ice cores provide scientists with data about past temperatures.

Stretch yourself

Research the shrinking of the world's glaciers and how this is providing evidence of climate change. Find images to show the changes that have taken place in the last few decades. What impact might the melting of glaciers have on people's lives?

Practice question

Study photo **D**. Explain how the shrinkage of Arctic sea ice could be evidence of climate change. *(4 marks)*

On this spread you will find out about the natural causes of climate change

Natural causes of climate change

Scientists believe that there are several natural causes for climate change. These include:

- changes in the Earth's orbit
- variations in heat output from the Sun
- volcanic activity.

Orbital changes

Milutin Milankovitch was a Serbian geophysicist and astronomer. Whilst he was imprisoned during the First World War (1914–18) he studied the Earth's orbit and identified three distinct *cycles* that he believed affected the world's climate. These are known as *Milankovitch cycles* (diagram **A**).

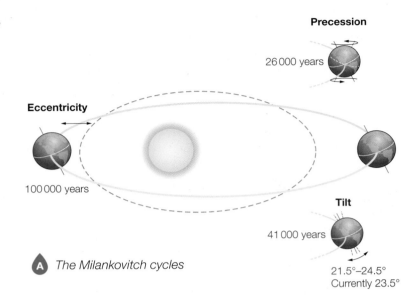

A *The Milankovitch cycles*

Scientists believe that these cycles affect the timings and seasons of the Earth's climate. In particular, the 100 000-year eccentricity cycle coincides closely with the alternating cold (glacial) and warm (inter-glacial) periods in the Quaternary period.

Eccentricity
This describes the path of the Earth as it orbits the Sun. The Earth's orbit is not fixed – it changes from being almost circular to being mildly elliptical. A complete cycle – from circular to elliptical and back to circular again – occurs about every 100 000 years.

Axial tilt
The Earth spins on its axis, causing night and day. The Earth's axis is currently tilted at an angle of 23.5 degrees. However, over a period of about 41 000 years, the tilt of the Earth's axis moves back and forth between two extremes – 21.5 degrees and 24.5 degrees. You can see this on diagram **A**.

The Milankovitch cycles

Precession
This describes a natural 'wobble' rather like a spinning top. A complete wobble cycle takes about 26 000 years. The Earth's wobble accounts for certain regions of the world – such as northern Norway – experiencing very long days and very long nights at certain times of the year.

Solar activity

Scientists have identified cyclical changes in solar energy output linked to the presence of *sunspots*. A sunspot is a dark patch that appears from time to time on the surface of the Sun (photo **B**). The number of sunspots increases from a minimum to a maximum and then back to a minimum over a period of about 11 years. This 11-year period is called the *sunspot cycle*.

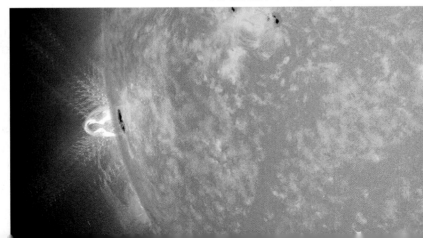

B *Sunspot and solar flare on the surface of the Sun, 2014*

- When sunspot activity is at a *maximum*, the Sun gives off more heat. Large explosions occur on the surface of the sun resulting in solar flares.

- When sunspot activity is at a *minimum* the solar output is reduced. This can lead to lower temperatures on Earth.

For example, very few sunspots were observed between the years 1645 and 1715. This coincided with the coldest period during the so-called 'Little Ice Age', when Europe experienced a much colder climate with severe winters (image **C**).

C A 'Frost Fair' on the River Thames during the 'Little Ice Age'

Volcanic activity

Violent volcanic eruptions blast huge quantities of ash, gases and liquids into the atmosphere.

- Volcanic ash can block out the Sun, reducing temperatures on the Earth. This tends to be a short-term impact.

- The fine droplets that result from the conversion of sulphur dioxide to sulphuric acid act like tiny mirrors reflecting radiation from the Sun. This can last a lot longer and can affect the climate for many years.

The cooling of the lower atmosphere and reduction of surface temperatures is called a *volcanic winter*.

Eruption of Mount Tambora, 1815

In 1815 there was a massive volcanic eruption of Mount Tambora in Indonesia (image **D**). It was the most powerful eruption in the world for 1600 years! Ash and sulphuric acid caused average global temperatures to fall by 0.4 °C–0.7 °C and 1816 became known as 'The year without a summer'.

Across the world harvests failed. There were major food shortages throughout North America and Western Europe, including the UK. Food prices rose sharply and there were riots and looting in European cities. It was the worst famine in Europe in the nineteenth century, resulting in an estimated 200 000 deaths.

D Artist's impression of the eruption of Mount Tambora in 1815

Stretch yourself

Carry out some research about 1816, 'The year without a summer'.
Find more information about the impacts of the eruption of Mount Tambora. Could this happen again in the future?

Practice question

Use the example of Mount Tambora to explain how and why volcanic activity can affect global climate. *(4 marks)*

ACTIVITIES

1 Use diagram **A** to answer the following questions.

  a Which of the Milankovich cycles takes 41 000 years to complete?

  b Explain the eccentricity cycle.

  c What is the evidence that the eccentricity cycle has affected global climates?

  d Describe the precession cycle.

2 Describe how sunspot activity can have an effect on global climates.

What are the human causes of climate change?

On this spread you will find out about the human causes of climate change

Human causes of climate change

Many scientists believe that human activities are at least partly to blame for the rapid rise in temperatures – known as global warming – since the 1970s. To understand how this is possible you need to understand a natural feature of the atmosphere called the *greenhouse effect*.

What is the greenhouse effect?

You probably know that a greenhouse is a small building entirely made of glass and used by gardeners to create warm conditions to grow plants. So how does it work?

Glass allows radiation (heat) from the Sun to enter the greenhouse (diagram **A**). However, this heat cannot escape through the glass. As a result, the greenhouse becomes warmer than the air outside and is ideal for growing tomatoes and vegetables which need constant warm conditions.

Like a greenhouse, the atmosphere allows most of the heat from the Sun (short-wave radiation) to pass straight through it to warm up the Earth's surface (diagram **B**). However, when the Earth gives off heat in the form of long-wave radiation, some gases such as carbon dioxide (CO_2) and methane are able to absorb it. These gases are called *greenhouse gases*.

In the same way that glass traps heat inside a greenhouse, the greenhouse effect keeps the Earth warm. Without this 'blanketing' effect it would be far too cold for life to exist on Earth.

> **Think about it**
>
> Think about your own **carbon footprint**. How do you as an individual contribute to the production of greenhouse gases in your everyday life?

A *The greenhouse effect*

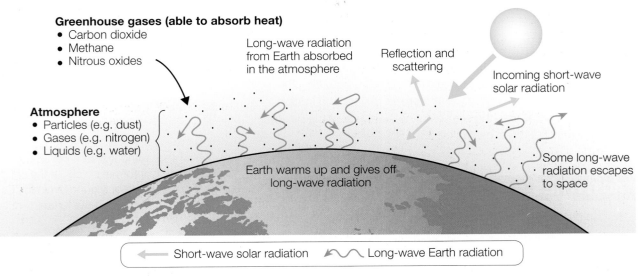

Greenhouse gases (able to absorb heat)
- Carbon dioxide
- Methane
- Nitrous oxides

Long-wave radiation from Earth absorbed in the atmosphere

Reflection and scattering

Incoming short-wave solar radiation

Atmosphere
- Particles (e.g. dust)
- Gases (e.g. nitrogen)
- Liquids (e.g. water)

Earth warms up and gives off long-wave radiation

Some long-wave radiation escapes to space

Short-wave solar radiation   Long-wave Earth radiation

B *How the greenhouse effect works*

The human impact

In recent years, the amounts of greenhouse gases in the atmosphere have increased. Scientists believe that this is due to human activities (diagram **C**).

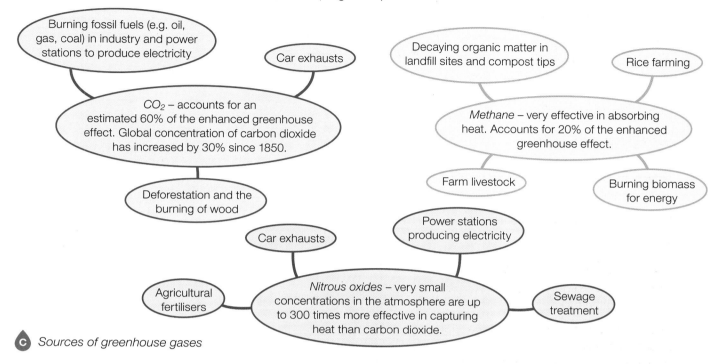

Burning fossil fuels (e.g. oil, gas, coal) in industry and power stations to produce electricity

Car exhausts

Decaying organic matter in landfill sites and compost tips

Rice farming

CO_2 – accounts for an estimated 60% of the enhanced greenhouse effect. Global concentration of carbon dioxide has increased by 30% since 1850.

Methane – very effective in absorbing heat. Accounts for 20% of the enhanced greenhouse effect.

Deforestation and the burning of wood

Farm livestock

Burning biomass for energy

Car exhausts

Power stations producing electricity

Agricultural fertilisers

Nitrous oxides – very small concentrations in the atmosphere are up to 300 times more effective in capturing heat than carbon dioxide.

Sewage treatment

C *Sources of greenhouse gases*

Graph **D** shows the recorded changes in carbon dioxide since the 1960s. The trend of this graph is identical to that of average global temperatures. Many scientists believe that this provides clear evidence that human activities are affecting global climates.

It is the increased effectiveness of the greenhouse effect – the so-called *enhanced greenhouse effect* – that scientists believe is causing recent global warming. For the first time in history, human activities appear to be affecting the atmosphere with potentially dramatic effects on the world's climate. By the end of the century average global temperatures could rise by 1.8–4 °C. This could lead to a rise in sea level of 28–43 cm.

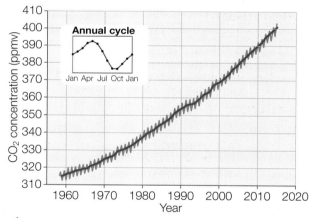

D *Increase in CO_2 obtained from direct readings at the Mauna Loa Observatory, Hawaii*

ACTIVITIES

1 Make a large copy of diagram **B**. Add the main sources of greenhouse gases (diagram **C**) using simple sketches.

2 **a** Describe the trend of CO_2 concentration in the atmosphere (graph **D**).

   b Can you suggest why CO_2 in the atmosphere increases in the winter but decreases in the summer? (Hint: think about plants.)

   c Does the graph support the suggestion that human activities may be contributing to global warming? Explain your answer.

3 Explain the enhanced greenhouse effect.

Stretch yourself

Research more information about the sources of greenhouse gases resulting from human activities. Focus your research on carbon emissions and find out which countries are the highest contributors.

Practice question

Outline two reasons why human activities effect the concentration of CO_2 in the atmosphere. *(4 marks)*

On this spread you will find out how the causes of climate change can be managed (mitigated)

How can climate change be managed?

Alternative energy sources

The burning of **fossil fuels** (coal, oil and gas) to produce electricity, fuel vehicles and power industry contributes 87 per cent of all human-produced CO_2 emissions. The rest comes from land use changes – mostly deforestation (9 per cent) and industrial processes like making cement (4 per cent).

To help reduce carbon emissions many countries are turning to alternative sources of energy such as:

◆ hydro-electricity

◆ nuclear power

◆ solar, wind, and tides.

These do not emit large amounts of CO_2. Some are also renewable and will last into the future. Nuclear power uses uranium to generate electricity but does not emit CO_2 as a by-product.

The UK aims to produce 15 per cent of its energy from renewable sources by 2020. There has been investment in renewable energy projects like wind power. Power companies are encouraged to use renewable sources. A new nuclear reactor is being built at Hinkley Point in Somerset (photo **A**).

A Artist's impression of the new Hinkley Point nuclear reactor

Carbon capture

Coal is the most polluting of all fossil fuels. China gets 80 per cent of its electricity from burning coal, India 70 per cent and the USA 50 per cent. How can coal continue to be used in a less damaging way?

Carbon capture and storage (CCS) uses technology to capture CO_2 produced from the use of fossil fuels in electricity generation and industrial processes. It is possible to capture up to 90 per cent of the CO_2 that would otherwise enter the atmosphere.

Diagram **B** shows how carbon capture works. Once captured, the carbon gas is compressed and transported by pipeline to an injection well. It is injected as a liquid into the ground to be stored in suitable geological reservoirs.

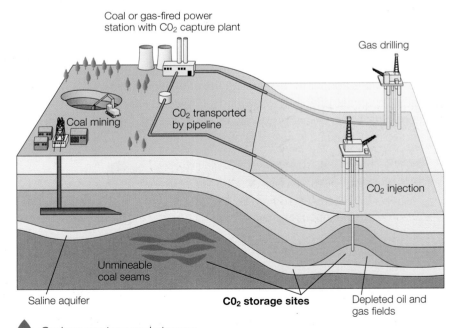

B Carbon capture and storage

Planting trees

Trees act as carbon sinks, removing CO_2 from the atmosphere by the process of *photosynthesis*. They also release moisture into the atmosphere. This has a cooling effect by producing more cloud, reducing incoming solar radiation.

Tree planting is well established in many parts of the world. Plantation forests can absorb CO_2 at a faster rate than natural forests and can do so effectively for up to 50 years.

International agreements

Climate change is a global issue and requires global solutions. Carbon emissions spread across the world and affect everyone (figure **C**).

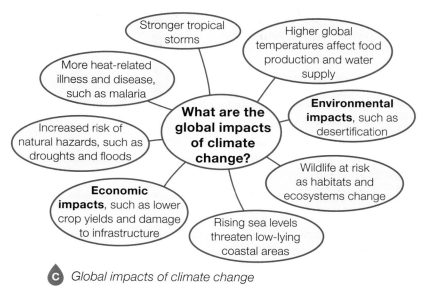

What are the global impacts of climate change?

- Stronger tropical storms
- Higher global temperatures affect food production and water supply
- More heat-related illness and disease, such as malaria
- **Environmental impacts**, such as desertification
- Increased risk of natural hazards, such as droughts and floods
- Wildlife at risk as habitats and ecosystems change
- **Economic impacts**, such as lower crop yields and damage to infrastructure
- Rising sea levels threaten low-lying coastal areas

c *Global impacts of climate change*

2005

The Kyoto Protocol – the first international treaty – became law. Over 170 countries agreed to reduce carbon emissions by an average of 5.2 per cent below their 1990 levels by 2012. Of the major greenhouse gas emitters, only the USA and Australia refused to sign the treaty.

2009

World leaders met in Copenhagen to consider international agreements on tackling climate change beyond 2012. The outcome was the Copenhagen Accord. It pledged to reduce emissions with financial support for developing nations to help them cope with the impacts of climate change. But there was no legally binding agreement.

2015

Paris Agreement 2015 – 195 countries adopted the first ever universal and legally binding global climate deal.

- To peak greenhouse gas emissions as soon as possible and achieve a balance between sources and sinks of greenhouse gases in the second half of this century.
- To keep global temperature increase below 2°C and limited to 1.5°C above pre-industrial levels.
- To review progress every five years.
- US$100 billion a year to support climate change initiatives in developing countries by 2020, with further finance in the future.

There have been criticisms that many of these agreements are 'promises' or aims and not firm commitments.

ACTIVITIES

1. How can alternative sources of energy address the problem of carbon dioxide emissions?
2. Make a copy of diagram **B**. Use detailed annotations to describe how carbon capture and storage works.
3. Why are forests described as 'carbon sinks'?
4. Do you think international agreements will help to solve the problems associated with climate change? Explain your answer.

Stretch yourself

Find out more about carbon capture and storage.

- What are the benefits of this mitigation option?
- What problems and issues need to be overcome for it to be widely used in the future?

Maths skills

Draw a pie chart to show changes to human-produced CO_2 emissions: burning fossil fuels, land use and industrial processes. (Don't forget to multiply the percentages by 3.6 to convert them into degrees.)

Practice question

'International agreements are critical in the challenge to reduce global carbon emissions.' Use evidence to support this statement. *(6 marks)*

On this spread you will find out how climate change can be managed by adapting to changes

How can we adapt to climate change?

Scientists believe that climate change will have a huge impact on agricultural systems across the world.

◆ Patterns of rainfall and temperature will change.

◆ Extreme weather events such as heatwaves, droughts and floods will become more common.

◆ The distribution of pests and diseases will change.

Farmers will need to adapt to these changes.

Agricultural adaptation in low latitudes

Scientists think that the greatest changes to agriculture will occur in low latitudes. Southern Africa's maize crop could fall by 30 per cent by 2030 and the production of rice in South Asia could fall by 10 per cent.

There are several adaptations that can be made (photo **A**).

Agricultural adaptation in middle latitudes

A warmer climate in Europe and North America could lead to an increase in production of certain crops such as wheat. In the UK, Mediterranean crops such as vines (photo **B**) and olives may thrive.

Introducing drought-resistant strains of crops

Educating farmers in water harvesting techniques

Shade trees can be planted to protect seedlings from strong sunshine

New irrigation systems

New cropping patterns can be introduced, e.g. changing planting/sowing dates

A *Irrigating crops in the Gambia*

 B *Vineyards such as this one in West Sussex may become more widespread in the UK*

Managing water supply

Climate change is already causing more severe and more frequent droughts and floods. Unreliable rainfall and periods of water shortage require careful management. Future climate change will affect the current patterns of water supply, impacting on the quantity and quality of our water. It is the most vulnerable, particularly in rural parts of poorer countries, who are likely to be affected the most.

Managing water supply in the Himalayas

Millions of people in Asia depend on rivers fed by snow and glacial melt for their domestic and agricultural water supply. In the Himalayas most of the 16 000 glaciers are receding rapidly due to global warming. This threatens the long-term security of water supply in the region.

Photo **C** shows an artificial glacier project that will supply water to villages in Ladakh, India. Water is collected in winter through a system of diversion canals and embankments and it freezes. When the 'glacier' melts in spring it will provide water for the local villages.

C *Creating artificial glaciers to provide water for villages*

Reducing risk from rising sea levels

Did you know that average sea levels have risen by 20 cm since 1900? By 2100 sea levels are expected to rise by a further 26–82 cm. This will flood important agricultural land in countries such as Bangladesh, India and Vietnam.

As sea levels rise, rates of coastal erosion will increase. Fresh water supplies will become contaminated by saltwater and coastal areas will be prone to damage from storm surges.

Managing rising sea levels in the Maldives

The Maldives are a group of tiny islands in the Indian Ocean some 500 km south-west of India. The highest point on the islands is just 2.4 m. Some climate models suggest that the islands may be uninhabitable by 2030 and submerged by 2070.

The 380 000 inhabitants have a very uncertain future as sea levels rise.

Construction of sea walls – a 3 m sea wall is being constructed around the capital Male with sandbags used elsewhere (as in this photo)

Building houses that are raised off the ground on stilts

Restoration of coastal mangrove forests – their tangled roots trap sediment and offer protection from storm waves

How can the Maldives **D** *manage sea-level rise?*

Ultimately the entire population could be relocated to Sri Lanka or India

Construction of artificial islands up to 3 m high so that people most at risk could be relocated

ACTIVITIES

1 How can farmers adapt to the possible impacts of climate change?

2 **a** What do you think the people in photo **C** are doing?

   b How does this system of water harvesting work?

   c Why is it important that local communities in remote areas start taking action to secure their water supply?

3 How might rising sea levels affect coastal communities?

4 How is the Maldives managing sea-level rise?

Stretch yourself

Research more information about water harvesting techniques. Find out, for example, how water droplets in fog can be harvested from the air to support people living in the Atacama Desert in Chile.

Practice question

Choose *either* the risk of reduced water supply *or* rising sea levels. For the issue chosen, describe examples of strategies used to manage them. *(6 marks)*

Section B The living world

Amazon rainforest in north west Brazil

Unit 1 Living with the physical environment is about physical processes and systems, how they change, and how people interact with them at a range of scales and in a range of places. It is split into three sections.

Section B The living world includes:

- ecosystems
- tropical rainforests
- hot deserts
- cold environments.

You have to study ecosystems and tropical rainforests.
You will then study *either* hot deserts or cold environments.

What if...

1 the sun stopped shining?

2 all the world's rainforests were cut down?

3 the world turned to desert?

4 the Antarctic was fully industrialised?

Your key skills

To be a good geographer, you need to develop important geographical skills – in this section you will learn the following skills:

- Drawing labelled maps and diagrams.

- Drawing a climate graph.

- Literacy – writing a news report.

- Finding evidence from photos.

- Describing patterns from maps and data.

- Using numerical data.

- Carrying out personal research.

Your key words

As you go through the chapters in this section, make sure you know and understand the key words shown in bold. Definitions are provided in the Glossary on pages 346–9. To be a good geographer you need to use good subject terminology.

Your exam

Section B makes up part of Paper 1 – a one and a half hour written exam worth 35 per cent of your GCSE.

5.1 Introducing a small-scale ecosystem

On this spread you will find out about the components of a small-scale ecosystem in the UK

Example

What is an ecosystem?

An **ecosystem** is a natural system made up of plants, animals and the environment. There are often complex interrelationships (links) between the living and non-living components of an ecosystem. *Biotic* components are the living features of an ecosystem such as plants and fish. *Abiotic* components are non-living environmental factors such as climate (temperature and rainfall), soil, water temperature and light.

Ecosystems can be identified at different scales:

◆ a local small-scale ecosystem can be a pond (photo **A**), hedgerow or woodland

◆ a global-scale ecosystem can be a tropical rainforest or deciduous woodland. These global ecosystems are called biomes.

A freshwater pond ecosystem

Freshwater ponds provide a variety of habitats (homes) for plants and animals. There are big variations in the amount of light, water and oxygen available in different parts of a pond.

Diagram **B** shows how different habitats suit certain plants, insects and animals.

Here are some terms that you need to understand.

A *Each of these environments is an important habitat and forms part of the pond ecosystem*

Plants like reeds grow in the water around the edge of the pond.

On the banks grow grasses, bushes and trees.

At the edges of the pond, the water is shallow and there will be plants like water lilies.

On the surface are ducks and small insects such as water boatman.

At the centre the water is deeper and there will be fish.

Term	Definition
Producers	Producers convert energy from the environment (mainly sunlight) into sugars (glucose). The most obvious producers are plants that convert energy from the Sun by photosynthesis.
Consumers	Consumers get energy from the sugars produced by the producers. A pond snail is a good example of a consumer because it eats plants.
Decomposers	Decomposers break down plant and animal material and return the nutrients to the soil. Bacteria and fungi are good examples of decomposers.
Food chain	A food chain shows the direct links (hence the term 'chain') between producers and consumers in the form of a simple line (diagram **C**).
Food web	A food web shows all the connections between producers and consumers in a rather more complex way (hence the term 'web' rather than 'chain') (diagram **D**).
Nutrient cycling	Nutrients are foods that are used by plants or animals to grow. There are two main sources of nutrients: • rainwater washes chemicals out of the atmosphere • weathered rock releases nutrients into the soil. When plants or animals die, the decomposers help to recycle the nutrients making them available once again for the growth of plants or animals. This is the nutrient cycle.

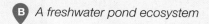 *A freshwater pond ecosystem*

Energy from the Sun

Pond margin – plenty of oxygen and light here. Plenty of shelter for the plants and insects, for small animals to eat.

Pond surface – plenty of oxygen and light here. Animals breathe through their gills, lungs or skin.

Above the pond surface – birds and animals breathe oxygen. Food is found in or on the water, or in the margins.

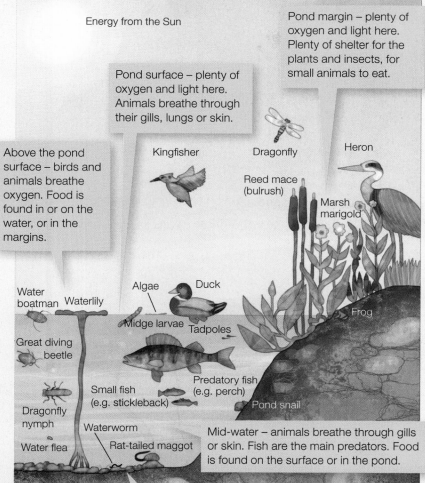

Kingfisher

Dragonfly

Heron

Reed mace (bulrush)

Marsh marigold

Water boatman   Waterlily

Algae   Duck

Midge larvae   Tadpoles

Frog

Great diving beetle

Small fish (e.g. stickleback)

Predatory fish (e.g. perch)

Pond snail

Dragonfly nymph

Waterworm

Water flea

Rat-tailed maggot

Mid-water – animals breathe through gills or skin. Fish are the main predators. Food is found on the surface or in the pond.

Pond bottom – little oxygen or light. Plenty of shelter (rotting plants and stones) and food. Decomposers and scavengers live here.

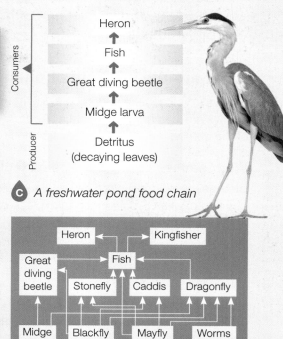

Heron
↑
Fish
↑
Great diving beetle
↑
Midge larva
↑
Detritus (decaying leaves)

Consumers

Producer

C *A freshwater pond food chain*

Heron ← → Kingfisher

Great diving beetle   Fish

Stonefly   Caddis   Dragonfly

Midge larva   Blackfly   Mayfly   Worms

Detritus   Algae and microscopic plants

D *Freshwater pond food web*

Species	Energy source
Frog tadpole	Microscopic plants, algae, midge larvae
Algae	Sunlight
Sticklebacks	Tadpoles, young fish, water fleas, beetles
Heron	Fish, frogs and tadpoles, larger insects
Perch	Small fish (e.g. sticklebacks), beetles, water fleas

E *Freshwater pond species and energy sources*

ACTIVITIES

1 a Describe the pond ecosystem in photo **A**.
   b Identify a producer.
   c Why are ducks good examples of consumers?
   d Imagine that the pond became polluted. How would this impact on the ecosystem?

2 a Where do most decomposers live in the pond ecosystem?
   b Why are decomposers important in nutrient recycling?

3 Look at table **E**. The organisms in the table form a food chain, but they are not in the correct order! Place them in the correct order – you have to decide who is eating who!

Practice question

Explain and describe the features of a small-scale ecosystem in the UK. *(4 marks)*

Stretch yourself

Find a food web diagram for a different small-scale ecosystem in the UK, such as a hedgerow or deciduous woodland.

• Identify the producers in your food web.

• Use a colour or highlighter to show a food chain within the food web.

On this spread you will find out how changes to the ecosystem affect its components

What are the impacts of change on an ecosystem?

Ecosystems can take hundreds if not thousands of years to develop. If an ecosystem is to be sustainable it needs to be in balance. If there is a change to one of the components it may well have knock-on effects for the rest of the ecosystem.

What causes change to ecosystems?

Changes to an ecosystem can occur naturally or result from human activities. Change can take place on different scales:

◆ global-scale changes, such as climate change

◆ local-scale changes, such as changes to a habitat – for example, when a hedge is removed.

Natural changes

Ecosystems can adapt to slow natural changes with few harmful effects. But rapid changes can have serious impacts. Extreme weather events like droughts can be devastating to ponds and lakes. They could dry up in places, which changes the edge-of-pond environment (photo **A**). Plants will dry out and die. Fish, starved of oxygen, might not survive.

Changes due to human activities

Human activity can have many impacts on ecosystems (diagram **B**). Once a component has been changed it can have serious knock-on effects on the ecosystem.

A *Pond in Brighton affected by drought, after a period of dry weather*

DANGER DEEP WATER

B *The impact of human changes on small-scale ecosystems*

Agricultural fertilisers can lead to eutrophication: nitrates increase growth of algae, which will deplete oxygen and fish may die.

Ponds may be drained to use for farming. Aquatic plants will die, as will fish and other pond life.

Woods cut down, destroying habitats for birds and affecting the nutrient cycle.

Hedgerows removed to increase size of fields. Habitats will be destroyed, altering the plant/animal balance.

How can changes affect the pond ecosystem?

Look back to diagram **B** on page 53. What if the pond owner added some perch, a predator, to the ecosystem?

> The perch will eat more of the smaller fish and small animals, like frogs.

> This will reduce the amount of food for creatures further up the food chain, like herons.

> With fewer frogs, there will be an increase of creatures below frogs in the food chain, like slugs.

Avington Park lake, Winchester, Hampshire

Avington Park is a country estate close to Winchester in Hampshire. The lake in the grounds of the estate is of historical and ecological importance. Lack of maintenance in recent years resulted in the accumulation of silt and the growth of vegetation. This created an excellent habitat for birds, but the impressive view of the lake from the house had been lost.

Restoration of the lake was carried out in 2014 (photo **C**). The aim was to restore the lake as part of the landscape, and to preserve and improve its function as a habitat for birds. Restoration involved desilting and redefining the lake and creating new waterside habitats to attract nesting birds and waterfowl. Following its restoration, the lake can again be seen from the house, and has become a healthy ecosystem for a diverse range of wildlife.

 Avington Park lake restoration

ACTIVITIES

1. **a** What evidence can you find in photo **A** that the pond has been affected by drought?

   b Suggest the effects of the drought on the pond margin. How could this affect the pond ecosystem?

   c If the pond dried up completely, what effect would this have on the ecosystem?

2. **a** Select one change in diagram **B**. Describe how it could affect the ecosystem.

   b Imagine the landowner cut down all the vegetation at the side of the pond to create a wooden deck for fishing. How might this affect the ecosystem?

3. **a** What were the changes that caused the Avington Park lake's poor condition?

   b How has the lake been restored?

Stretch yourself

Frogs are an important part of the pond ecosystem. Imagine that disease wipes out all the frogs in a pond. Find out how this would affect the ecosystem in the short term and the long term.

Practice question

Using a named example, explain how change can have short-term and long-term effects on an ecosystem. *(6 marks)*

Think about it

Consider a pond near to where you live or close to school. To what extent is the pond a thriving and healthy ecosystem?

On this spread you will find out about the distribution and characteristics of global ecosystems

The distribution of global ecosystems

Large-scale ecosystems are known as **global ecosystems** (or biomes). These are defined mainly by the dominant type of vegetation that grows in the region, such as tropical rainforest or **tundra**.

Global ecosystems form broad belts across the world from west to east, parallel to the lines of latitude (map **A**). This is because the climate and characteristics of ecosystems are determined by global atmospheric circulation (pages 22–3).

Variations in these west-to-east belts of vegetation are due to factors such as:

◆ ocean currents   ◆ winds   ◆ the distribution of land and sea.

These factors produce small variations in temperature and moisture which in turn affect the ecosystems. For example, the Mediterranean region – with its dry, hot summers and warm, wet winters – has its own global ecosystem.

A Global ecosystems

Key
- Tundra
- Coniferous forest
- Temperate deciduous forest
- Temperate grassland
- Mediterranean
- Desert
- Tropical rainforest
- Tropical grassland (savanna)
- Other biomes (e.g. polar, ice, mountains)

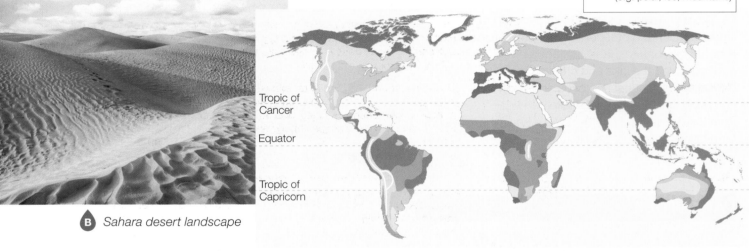

Tropic of Cancer

Equator

Tropic of Capricorn

B Sahara desert landscape

C Coniferous forest, Russia

D Tropical savanna, Kenya

E Alpine tundra, Peru

The characteristics of global ecosystems

Global ecosystem	Location	Characteristics
Tropical rainforest	Close to the Equator	High temperatures and heavy rainfall associated with equatorial low pressure belt creates ideal conditions for plants to grow. Rainforests cover 6 per cent of the Earth's land surface. More than half the world's species of plants and animals live in this global ecosystem. A quarter of all medicines come from rainforest plants.
Desert	Roughly 30° north and south of the Equator	Deserts cover one fifth of the world's land surface. Hot deserts are associated with the sub-tropical high pressure belts. Sinking air stops clouds from forming, resulting in high daytime temperatures, low night-time temperatures and low rainfall. Plants and animals have to be well adapted to survive in these conditions.
Polar	Arctic/Antarctic	Cold air sinks at the north and south Poles, resulting in very low temperatures and dry conditions. The main polar regions are Antarctica and Greenland where temperatures can fall below –50ºC.
Deciduous and coniferous forests	Roughly 50–60° north of the Equator	Deciduous trees shed their leaves in winter to retain moisture. Coniferous trees are cone-bearing evergreens, retaining their leaves to maximise photosynthesis during the brief summer months. The UK's natural vegetation is deciduous forest. Further north, for example in Canada and Scandinavia, coniferous forests dominate as they are better suited to colder climates.
Temperate grassland	Roughly 30–40° north and south of the Equator. Inland away from coasts, with hot summers and cold winters.	This includes the vast areas of grassland in North America (prairies) and Eastern Europe (steppes). These areas experience warm, dry summers and cold winters. Grasses can tolerate these conditions and this land is mainly used for grazing animals.
Mediterranean	Roughly 40–45° north of the Equator. Also isolated locations south of the Equator (South Africa, Western Australia)	Countries around the Mediterranean enjoy hot, sunny and dry summers, with mild winters. This is due to the pressure belts migrating slightly north and south during the year. Mediterranean vegetation includes olive trees and fruit trees, such as lemons and oranges. Other parts of the world have a similar climate, for example California (USA), South Africa and parts of Australia.
Tropical grassland (savanna)	Between 15–30° north and south of the Equator.	The tropical climate in these low latitudes is characterised by distinct wet and dry seasons. The dry season can be very hot and wild fires can break out. Violent thunderstorms can occur during the wet season. Large herds of animals graze on these grasslands, along with predators such as lions and leopards.
Tundra	From the Arctic Circle to about 60–70° north (e.g. Canada, Northern Europe). There are only very small areas of tundra in the southern hemisphere due to the lack of land at these latitudes.	Tundra is characterised by low-growing plants adapted to retain heat and moisture in the cold, windy and dry conditions. These regions are found in northern Canada and across Northern Europe. It is a fragile ecosystem, easily damaged by humans and threatened by developments such as oil exploitation and tourism. Animals such as reindeer are adapted to survive the cold.

ACTIVITIES

1 Describe the pattern of global ecosystems (biomes) in North America.

2 Why do most global ecosystems form broad latitudinal belts across the world?

3 Why is the Mediterranean popular with northern Europeans in both the summer and the winter?

Stretch yourself

Do some research to compare deciduous and coniferous forests. What are the characteristics of each ecosystem and how have plants and animals adapted?

Practice question

Describe the global pattern of the tundra ecosystem.
(4 marks)

6 Tropical rainforests

6.1 Environmental characteristics of rainforests

On this spread you will find out about the environmental characteristics of tropical rainforests

If you were to enter a tropical rainforest you would need a torch and strong shoes, as it is dark and damp. It is also very noisy with the clicks, howls and whistles of insects and animals. You would have difficulty moving about because the vegetation is very lush and dense (photo **A**).

◆ The trees grow very tall, often up to 45 m high.

◆ There is a great variety of wildlife – often up to 100 species in a single hectare!

Where are tropical rainforests found?

Tropical rainforests are found in a broad belt through the Tropics in:

◆ Central and South America
◆ South East Asia
◆ central Africa
◆ northern Australia.

(You can see these areas on map **A** on page 56.)

What is the climate like?

Tropical rainforests thrive in warm and wet conditions. The equatorial zone where they are found is characterised by high rainfall (over 2000 mm a year) and high temperatures (averaging about 27 °C) throughout the year.

Look at table **B**. It provides climatic information for the weather station at Manaus, in the Amazonian rainforest in Brazil.

◆ *The temperature is high and constant throughout the year.* This is because the powerful Sun is overhead for most of the time.

◆ *The rainfall is high.* This is because the global atmospheric circulation causes an area of low pressure to form at the Equator. The rising air creates clouds and triggers heavy rain.

◆ *Rainfall varies throughout the year*, with a distinct wet season lasting about six months. This is due to a period of intense rainfall when the equatorial low pressure area is directly overhead.

A *Rainforest vegetation*

Maths skills

Use the data in table **B** to draw a climate graph for Manaus showing temperature and rainfall.

● Shade the gap between the average minimum and maximum temperature lines. This is the *temperature range*.

● How will you show humidity and sunshine information on your graph?

Month	Temperature (°C)		Rainfall (mm)	Relative humidity	Sunshine (average hours per day)
	Max.	Min.			
January	31	24	249	89	4
February	31	24	231	89	4
March	31	24	262	89	4
April	31	24	221	90	4
May	31	24	170	89	5
June	31	24	84	87	7
July	32	24	58	87	8
August	33	24	38	85	8
September	33	24	46	84	8
October	33	24	107	85	7
November	33	24	142	86	6
December	32	24	203	88	5

B *Climate data for Manaus, Brazil*

What are the soils like?

Tropical rainforest soils are surprisingly infertile. Most nutrients are found at the surface, where dead leaves decompose rapidly in the hot and humid conditions. Many trees and plants have shallow roots to absorb these nutrients. Fungi growing on the roots transfer nutrients straight from the air. This is a good example of *nutrient cycling*.

Heavy rainfall can quickly dissolve and carry away nutrients. This is called *leaching*. It leaves behind an infertile red, iron-rich soil called *latosol*.

What plants and animals are there?

Tropical rainforests support the largest number of species of any biome. Over half of all plant and animal species on the planet live on just 7 per cent of the land surface. Tropical rainforests have a huge **biodiversity**, providing habitats for an enormous range of species.

◆ Birds live in the canopy (branches) feeding on nectar from flowers.

◆ Mammals, like monkeys and sloths, are well adapted to living in the trees.

◆ Animals like deer and rodents live on the forest floor.

How have plants adapted to rainforests?

A tropical rainforest is made up of layers (diagram **C**). The majority of plant and animal species are found in the canopy where there is most light. In contrast, the forest floor is dark and A rainforest is a very fragile ecosystem. Plants and animals (biotic factors), along with fungi and bacteria on the forest floor, enjoy a close but fragile relationship with the abiotic factors such as soils, temperature and moisture. Small changes to biotic or abiotic factors, such as deforestation or water pollution, can have serious knock-on effects on the entire ecosystem.

Water drips off leaves

Fast-growing trees such as kapok out-compete other trees to reach sunlight – such trees are called emergents

Many leaves have flexible bases so that they can turn to face the Sun

Many leaves have a 'drip tip' to allow the heavy rain to drip off the leaf

Top canopy

Middle canopy

Lower tree canopy

Height above ground (m)

Shrub layer and ground layer

Thin, smooth bark on trees to allow water to flow down easily

Buttresses – massive ridges help support the base of the tall trees and help transport water. May also help oxygen/carbon dioxide exchange by increasing the surface area

Lianas – woody creepers rooted to the ground but carried by trees into the canopy where they have their leaves and flowers

Plants called epiphytes can live on branches high in the canopy to seek sunlight – they obtain nutrients from water and air rather than soil

 C *Stratification and vegetation adaptations in a tropical rainforest*

ACTIVITIES

1 a Draw the areas of tropical rainforest (map **A**, page 56) onto a blank outline map of the world. Use an atlas to label the countries or regions (e.g. Ecuador, the Amazon, etc.) where tropical rainforests are found.

b Write a paragraph describing the distribution of tropical rainforests. Link their distribution to global atmospheric circulation (diagram **B** on page 22).

2 Select *three* species of plants or trees from diagram **C**. Describe how each has special adaptations so that it can thrive in this environment.

Case study

On this spread you will find out about the causes of deforestation in Malaysia

About Malaysia

- Malaysia is a country in South East Asia.

- It is made up of Peninsular Malaysia and East Malaysia, which is part of the island of Borneo

- The natural vegetation in Malaysia is tropical rainforest.

- 67 per cent of Malaysia's land is covered by rainforest.

A The location of Malaysia

Deforestation in Malaysia

'Orang-utan' means 'person of the forest'. They are losing their natural habitat. As natural rainforest in Malaysia is destroyed, many young orang-utans are killed or orphaned (photo **B**).

Deforestation is the cutting down of trees, often on a very large scale. The timber is a highly valued export. Deforestation means the land can be used for other profit-making enterprises, like cattle ranching, commercial farming, the production of rubber and palm oil.

The rate of deforestation in Malaysia is increasing faster than in any tropical country in the world. Between 2000 and 2013, Malaysia's total forest loss was an area larger than Denmark!

B A young orphaned orang-utan

Did you know?

Between 1990 and 2004 orang-utans in Borneo lost habitat twice the size of Wales.

ACTIVITIES

1 Complete a spider diagram to show the causes of deforestation in Malaysia. Use photos to illustrate your diagram.

2 Write a two-minute news report about deforestation in Malaysia. Focus on where it is, why it's important and how it has been destroyed.

Stretch yourself

Investigate commercial oil palm farming in Malaysia.

1 What is oil palm used for?

2 How is rainforest cleared to make way for this type of farming?

3 What damage is done to habitats and the natural environment?

Practice question

Photo **D** shows a hydroelectric dam in Sarawak, Malaysia. Evaluate *two* possible environmental impacts of developments like the Bakun Dam. *(4 marks)*

What are the threats to Malaysia's rainforests?

Logging

Malaysia became the world's largest exporter of tropical wood in the 1980s. Clear felling, where all trees are chopped down in an area, was common. This led to the total destruction of forest habitats.

Recently, clear felling has largely been replaced by **selective logging** (photo **C**), where only fully-grown trees are cut down. Trees that have important ecological value are left unharmed.

Road building

Roads are constructed to provide access to mining areas, new settlements and energy projects.

Logging requires road construction to bring in machinery and take away the timber.

  Road construction and logging in Sarawak, East Malaysia

Energy development

In 2011, after five decades of delays, the controversial Bakun Dam in Sarawak started to generate electricity

The Bakun Dam (205 m) is Asia's highest dam outside China.

Several more dams are planned to boost Malaysia's electricity supplies.

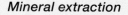 The Bakun Dam

The dam supplies energy for industrialised Peninsular Malaysia.

The dam's reservoir flooded over 700 km² of forests and farmland.

E Slash and burn

Mineral extraction

Mining (mainly tin and smelting) is common in Peninsular Malaysia. Rainforest has been cleared for mining and road construction. Drilling for oil and gas has recently started on Borneo.

Population pressure

In the past, poor urban people were encouraged by the government to move into the countryside from the rapidly growing cities. This is called transmigration. Between 1956 and the 1980s, about 15 000 hectares of rainforest was felled for the settlers. Many then set up plantations.

Commercial farming

Malaysia is the largest exporter of palm oil in the world. During the 1970s, large areas of land were converted to palm oil plantations. Plantation owners receive 10-year tax incentives, so increasing amounts of land have been converted to plantations.

Subsistence farming

Tribal people living in the rainforest practise subsistence farming. Traditionally, local communities would hunt and gather food from the forest and grow some food crops in cleared pockets of forest. This type of farming is small scale and sustainable.

One method of clearing land is 'slash and burn' (photo **E**). This involves the use of fire to clear the land. The burning creates valuable nutrients that help plants to grow. These fires can grow out of control, destroying large areas of forest.

On this spread you will find out about the impacts of deforestation in Malaysia

Case study

Impacts of deforestation in Malaysia

Photo **C** shows a rainforest landscape devastated by deforestation. Imagine all the habitats that have been destroyed. Notice that the trees have been reduced to stumps. The hillslopes have been stripped of vegetation, exposing the soil to erosion by rain and wind. Chart **A** shows the impact of deforestation on the area covered by Malaysia's rainforest.

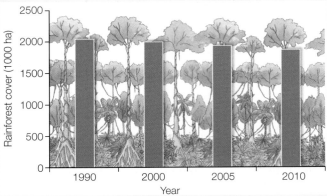

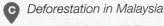

A *The size of Malaysia's rainforest, 1990–2010*

Soil erosion

Soil takes thousands of years to form – but it can be stripped away in a matter of hours. Removal of soil by wind and rain is called **soil erosion**. The roots of trees and plants bind the soil together. So deforestation means that soil can easily become loose and erode away.

Loss of biodiversity

Biodiversity is a measure of the variety of plants and animals in a particular ecosystem. Rainforests are the most biodiverse ecosystem in the world. Deforestation destroys the ecosystem and the many habitats that exist on the ground and in the trees. This reduces the biodiversity.

> *Biodiversity in the Main Range, Peninsular Malaysia*
>
> The Main Range is an upland region stretching for 500 km along the backbone of Peninsular Malaysia. This region is important because:
>
> ◆ it is the largest area of continuous forest left in Peninsular Malaysia (photo **B**)
>
> ◆ the forests are particularly rich in their biodiversity, with over 600 species
>
> ◆ the highland forests are home to over 25 per cent of all plant species found in Malaysia
>
> ◆ there are still many undiscovered plants that have medicinal qualities that could provide cures for diseases.

B *Main Range Mountains, Peninsular Malaysia*

C *Deforestation in Malaysia*

Contribution to climate change

Deforestation can have an impact on local and global climates. During photosynthesis, trees absorb CO_2 and emit oxygen. CO_2 is a greenhouse gas that is partly responsible for global warming. By absorbing CO_2, trees store the carbon and help to reduce the rate of global warming.

So, deforestation can affect climate because:

◆ trees give off moisture by the process of transpiration; deforestation reduces the moisture in the air resulting in a drier climate

◆ the process of evaporation uses up heat and cools the air; if trees are cut down, this cooling ceases and temperatures rise.

Economic development

Deforestation in many parts of the world is driven by profit. However, whilst deforestation may result in short-term economic gains, it may lead to long-term losses.

 Mining in the Malaysian rainforest

Economic gains

◆ Development of land for mining, farming and energy will lead to jobs both directly (construction, farming) and indirectly (supply and support industries)

◆ Companies will pay taxes to the government which can be used to improve public services, such as education and water supply

◆ Improved transport **infrastructure** opens up new areas for industrial development and tourism

◆ Products such as oil palm and rubber provide raw materials for processing industries

◆ Hydro-electric power will provide cheap and plentiful energy

◆ Minerals such as gold are very valuable

Economic losses

◆ Pollution of water sources and an increasingly dry climate may result in water shortages.

◆ Fires can cause harmful pollution. They can burn out of control, destroying vast areas of valuable forest.

◆ Rising temperatures could devastate some forms of farming such as growing tea, fruit and flowers

◆ Plants that could bring huge medical benefits and high profits may become extinct

◆ Climate change could have economic costs as people have to adapt to living in a warmer world

◆ The number of tourists attracted by rainforests could decrease

ACTIVITIES

1 Study photo **C**.

   a Describe the environment in the foreground.

   b Are there signs of slash and burn farming? What are they?

   c How might soil erosion become a problem in the future?

   d How do you think this forest clearance will have affected the species living in the forest?

2 Describe and explain the effects of deforestation on climate change.

3 Make a poster, with a photo showing deforestation in Malaysia. Add a series of colour-coded text boxes explaining possible economic gains and losses.

Stretch yourself

Write a report about the Main Range in Peninsular Malaysia.
- What are the characteristics of its biodiversity?
- Why is it considered to be a special area?

Illustrate your account with labelled photos.

Practice question

Explain, with reference to an example, why it is important to retain biodiversity. *(4 marks)*

On this spread you will find out about rates of deforestation and why rainforests need to be protected

What are the rates of deforestation?

Tropical rainforests are perhaps the most endangered ecosystem on Earth. Every two seconds an area of rainforest the size of a football field (about one hectare) is being destroyed. That's over 1500 hectares an hour!

◆ Tropical rainforests once covered over 15.5 million km^2. The figure is now just over 6.2 million km^2.

◆ An area of rainforest the size of China has been lost.

Look at graph **A**. The fastest rates of deforestation are in Brazil and Indonesia. These countries account for over 40 per cent of the world's deforestation. But deforestation in Brazil is decreasing, and in Indonesia it is increasing.

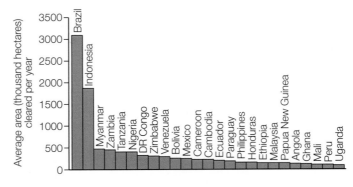

A Rates of tropical deforestation, 2010–15

Deforestation in Brazil

The rates of deforestation in Brazil have varied a lot since 1988 (graph **B**). Historically, small-scale farming was largely responsible for deforestation in Brazil.

However, most deforestation now involves large landowners and big companies. Most rainforest has been cleared for cattle ranching.

Since 2004, the rate of deforestation in Brazil has fallen by nearly 80 per cent to the lowest levels on record. There are several reasons for this:

◆ the Brazilian government has cracked down on illegal deforestation

◆ Brazil is leading the world in conservation – over half of the Amazon is now protected

◆ Brazil is committed to reducing carbon emissions to tackle climate change

◆ consumer pressure not to use products from deforested areas has led to a decline in cattle ranching.

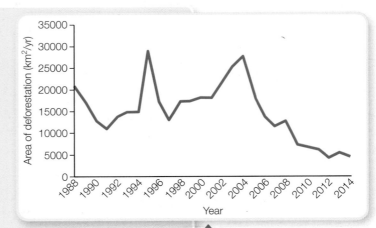

 B Deforestation in the Brazilian Amazon, 1988–2014

Maths skills

Use the information on the right to draw a pie chart. (Remember: To calculate degrees, multiply each percentage by 3.6.)
• Shade each segment a different colour.
• Write the name of the country and the percentage alongside each segment.

Share of tropical deforestation, 2010–15 (%)

Country	%	Country	%	Country	%
Brazil	27	Tanzania	4	Zimbabwe	3
Indonesia	17	Nigeria	4	Venezuela	3
Myanmar	4	DR Congo	3	Other tropical countries	31
Zambia	4				

Why should tropical rainforests be protected?

There are several reasons why tropical rainforests should be protected from further deforestation.

Biodiversity
Tropical rainforests contain half of all the plants and animals in the world. They are home to thousands of different species. Some plants may become extinct before they have even been discovered!

Medicine
Around 25 per cent of all medicines come from rainforest plants. More than 2000 tropical forest plants have anti-cancer properties.

Climate change
Rainforests absorb and store carbon dioxide, a gas that is partly responsible for climate change.

Resources
Tropical rainforest trees provide valuable hardwoods as well as nuts, fruit and rubber.

Climate
Known as the 'lungs of the world', 28 per cent of the world's oxygen comes from the rainforests. They prevent the climate from becoming too hot and dry.

Water
Rainforests are important sources for clean water – 20 per cent of the world's fresh water comes from the Amazon Basin.

People
Indigenous tribes live in harmony in the world's rainforests making use of the forest's resources without causing any long-term harm.

The Achuar people in the Peruvian Amazon

The Achuar are a primitive tribe of about 11 000 people. They live in small communities and rely on the resources of the rainforest for their buildings, food and fuel (photo **C**). They treat the rainforest with respect as their lives depend upon it.

There are rich reserves of oil in this region. Oil companies want permission to explore and drill for oil. If this happens the Achuar will lose some of their traditional lands, and may see their environment damaged by oil pollution.

The Achuar are resistant to oil exploration. They have had success in defending their land. In 2012, the oil company Talisman Energy stopped their oil exploration in the region.

C *A traditional Achuar hut*

ACTIVITIES

1 Study graph **A**. Why do you think that most deforestation happens in Brazil and Indonesia?

2 Study graph **B**.

   a Describe the trend in deforestation since 2005.

   b Why do you think the rate of deforestation has decreased recently?

3 Outline reasons why rainforests should be protected.

Stretch yourself

Find out about the Achuar people.

- What effect is deforestation having on the tribe?
- What are the main threats to their continued life in the Peruvian rainforest?

Practice question

'The rainforest is more valuable when left intact than when destroyed.' Using a case study, use examples to support or challenge this view. *(9 marks)*

On this spread you will find out about different strategies for managing rainforests sustainably

How can rainforests be managed sustainably?

To protect the world's tropical rainforests they need to be managed **sustainably**. There are two main reasons for this:

◆ to ensure that rainforests remain a lasting resource for future generations

◆ to allow valuable rainforest resources to be used without causing long-term damage to the environment.

Tribes like the Achuar (see page 65) have been managing rainforests sustainably. It is the large companies, wealthy landowners and illegal loggers whose drive for profit can result in unsustainable practices.

Selective logging and replanting

The most damaging form of deforestation is *clear felling*. All trees, big and small, are chopped down in the area being cleared. This completely destroys the ecosystem.

◆ **2 years before felling:** Pre-felling study to identify what is there.

◆ **1 year before felling:** Trees marked for felling. Arrows painted on trees to indicate direction of felling to avoid damaging other valuable trees.

◆ **Felling:** Trees felled by licence-holders.

◆ **3–6 months after felling:** Survey to check what has been felled. Prosecution may result from illegal felling.

◆ **2 years after felling:** Treatment plan drawn up to restore forest.

◆ **5–10 years after felling:** Remedial and regeneration work by state forestry officials. Replacement trees planted.

◆ **30–40 years after felling:** Cycle begins again.

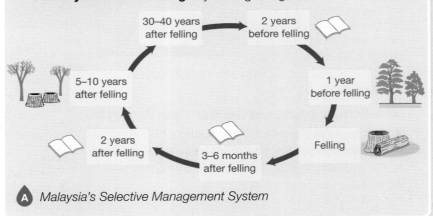

A *Malaysia's Selective Management System*

A more sustainable approach to logging involves *selective logging* (diagram **A**). Managed carefully, this technique – which was introduced in Malaysia in 1977 – is completely sustainable.

Conservation and education

Rainforest can be preserved in conservation areas, such as national parks or nature reserves. These areas can be used for education, scientific research and tourism.

Recently, large international businesses have supported conservation projects in exchange for carrying out scientific research or the provision of raw materials.

B *Tonka beans*

Givaudan

Givaudan is a Swiss perfume company. It works with Conservation International, and aims to protect 148 000 hectares of rainforest in the Caura Basin, Venezuela. Local Aripao people are encouraged to harvest and market tonka beans (photo **B**), which have a caramel-like smell. A warehouse where beans can be dried and stored was built in 2012. This improves their quality and increases their value.

Ecotourism

Countries like Costa Rica, Belize and Malaysia have promoted their forests for **ecotourism**. Ecotourism aims to introduce people to the natural world, to benefit local communities and protect the environment for the future. Through income generated by ecotourism, local people and governments benefit from retaining and protecting their rainforest trees. This is a more sustainable option than cutting them down for short-term profit.

International agreements

Rainforests are now understood to be of global importance. They absorb carbon dioxide from the atmosphere, releasing oxygen and maintaining levels of humidity. International agreements have been made to help protect rainforests.

Hardwood forestry

The Forest Stewardship Council (FSC) is an international organisation that promotes sustainable forestry. Products sourced from sustainably managed forests carry the FSC label.

The FSC tries to educate manufacturers and consumers about the need to buy sustainable hardwood like mahogany. It aims to reduce demand for the rare and valuable hardwoods.

Debt reduction

Some countries have borrowed money to fund developments. To pay off these debts some have raised money from massive deforestation programmes. Recently, some donor countries and organisations have reduced debts in return for agreement that rainforests will not be deforested. This has become known as 'debt-for-nature swapping'.

C *Tourist accommodation in an eco-lodge*

Carbon sinks

In 2008, the Gola Forest on Sierra Leone's southern border with Liberia became a protected national park (photo **D**).

The park plays a significant role in reducing global warming. It acts as a carbon sink by absorbing carbon dioxide from the air.

D *Scientists in Sierra Leone's Gola Forest*

ACTIVITIES

1 Why is Malaysia's Selective Management System a good example of sustainable management?

2 How can encouraging people to make commercial use of forest products like the tonka bean (photo **B**) help conserve rainforests?

3 How can we as consumers help to conserve rainforests?

Stretch yourself

Find out more about the Gola Forest National Park and its rainforest conservation projects. What do you think its donors get out of the deal? Is it a 'win-win' situation?

Practice question

Describe and explain *two* benefits of international cooperation in sustainably managing tropical rainforests.
(4 marks)

On this spread you will find out about the characteristics of hot deserts

What are hot deserts like?

A desert is an area that receives less than 250 mm of rainfall per year. The resulting dryness or aridity is the main factor controlling life in the desert (photo **A**). Did you know that there are both hot and cold deserts? Parts of the Arctic and Antarctic are as dry as some of the desert areas in Africa.

Where are hot deserts found?

Map **B** shows the locations of the world's hot deserts. Notice that they are mostly found in dry continental interiors, away from coasts, in a belt at approximately 30°N and 30°S. There are some coastal deserts too, for example, the Atacama Desert in South America.

What is the climate like?

The location of the hot deserts can be largely explained by the global atmospheric circulation (see page 22). At these latitudes air that has risen at the Equator descends forming a persistent belt of high pressure. This explains the lack of cloud and rain and the very high daytime temperatures. It also explains why, with the lack of cloud cover, temperatures can plummet to below freezing at night during the winter. Graph **C** describes the climate of In Salah, a weather station in the middle of the Sahara Desert in Algeria.

A *Jordan – a hot desert environment*

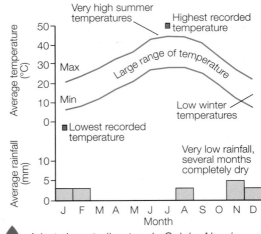

C *A hot desert climate – In Salah, Algeria*

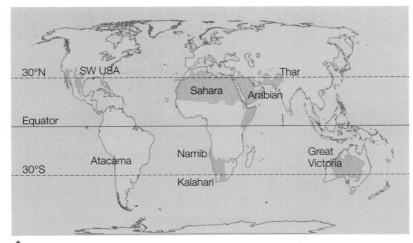

B *Location of the world's hot deserts*

What are desert soils like?

Desert soils tend to be sandy or stony, with little organic matter due to the general lack of leafy vegetation. They are dry but can soak up water rapidly after rainfall. Evaporation draws salts to the surface, often leaving a white powder on the ground. Desert soils are not very fertile.

What plants and animals are found in a hot desert?

Hot deserts are home to a surprising diversity of plants, animals and birds. In all but the driest areas, plants and animals find ways of surviving the hostile conditions. Plants tend to have very thin leaves or spines to reduce water loss and some have very long roots to reach deep underground water. For example, the cactus is a common desert plant.

How have plants and animals adapted?

With the lack of water, vegetation is low growing and sparse. Plants have developed several adaptations to enable them to cope with the environment (diagram **D**).

Many rodents are nocturnal, surviving the extremely high daytime temperatures by living in burrows underground and only venturing out during the cooler nights (photo **E**). Snakes and lizards retain water by having a waterproof skin and producing only tiny amounts of urine. Camels – the 'ships of the desert' – are well adapted to cope with many days without water.

E *A desert jerboa*

Think about it

Most people who live in the Sahara Desert are nomads who move from place to place. How does this lifestyle suit living in a hot desert?

Some plants have horizontal root systems, just below the surface

Seeds can stay dormant for years, but can germinate quickly when it rains

Some plants store water in their roots, stems, leaves or fruit (these are called succulents)

Small leaves, spines, glossy and waxy leaves all help reduce water loss

D *Plant adaptations to hot desert climate*

Some plants have long taproots (7–10 metres deep) to reach groundwater

---

ACTIVITIES

1 a Describe the characteristics of the hot desert shown in photo **A**.

  b Why is this landscape a hostile environment for plants and animals?

2 a Make a copy of map **B** to show the location of the world's hot deserts. Use an atlas or the internet to find the names of the hot deserts and label them on your map.

  b Describe the pattern of hot deserts.

  c Explain the cloudless conditions shown in photo **A**.

  d How do these cloudless conditions account for the high daytime temperatures and low night-time temperatures?

3 a Describe the pattern of temperatures for In Salah (graph **C**).

  b The total annual rainfall in London is about 600 mm. How does this compare with the rainfall for In Salah?

  c How do plants cope with very low rainfall (diagram **D**).

Stretch yourself

Find out about plants called succulents and how they have special adaptations for living in hot deserts.

Practice question

Use diagram **D** to explain how plants have adapted to the hostile conditions in hot deserts. *(4 marks)*

Opportunities for development in hot deserts

On this spread you will find out about how people use hot desert environments

Where is the Thar Desert?

The Thar Desert is one of the major hot deserts of the world. It stretches across north-west India and into Pakistan (map **A**). The desert covers an area of some 200 000 km² mostly in the Indian state of Rajasthan. It is the most densely populated desert in the world.

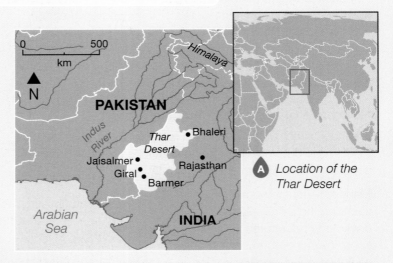

 Location of the Thar Desert

Did you know?

The Thar Desert is just slightly smaller than the whole of the UK!

The landscape is mainly sandy hills with extensive mobile sand dunes and clumps of thorn forest vegetation – a mixture of small trees, shrubs and grasses

Rainfall is low, between 100 and 240 mm per year, and summer temperatures in July can reach 53°C

Soils are sandy and not very fertile, with little organic matter to enrich them. They drain very quickly so there is little surface water.

What are the opportunities for development?

Despite the hostile conditions, the Thar Desert offers a number of opportunities for human activity and economic development.

B *The desert environment near Jaisalmer*

Mineral extraction

The desert region has valuable reserves of minerals which are used all over India and exported across the world. The most important minerals are:

◆ gypsum (used in making plaster for the construction industry and in making cement)

◆ feldspar (used to make ceramics)

◆ phospherite (used for making fertiliser)

◆ kaolin (used as a whitener in paper).

There are also valuable reserves of stone in the region. At Jaisalmer the Sanu limestone is the main source of limestone for India's steel industry. Valuable reserves of marble are quarried near Jodhpur, to be used in the construction industry.

Tourism

In recent years the Thar Desert, with its beautiful landscapes, has become a popular tourist destination. Tens of thousands visit the desert each year, many from neighbouring Pakistan.

Desert safaris on camels, based at Jaisalmer, have become particularly popular with foreigners as well as wealthy Indians from elsewhere in the country. An annual Desert Festival held each winter is a popular attraction. Local people benefit by providing food and accommodation and by acting as guides or rearing and looking after the camels.

Energy

The Thar Desert is a rich energy source.

◆ *Coal* – there are extensive lignite coal deposits in parts of the Thar Desert and a thermal energy plant has been constructed at Giral (map **A**).

◆ *Oil* – a large oilfield has been discovered in the Barmer district which could transform the local economy.

◆ *Wind* – recently there has been a focus on developing wind power, a renewable form of energy. The Jaisalmer Wind Park was constructed in 2001 (photo **C**). This is India's largest wind farm.

◆ *Solar* – with its sunny, cloudless skies, the Thar Desert offers ideal conditions for solar power generation. At Bhaleri solar power is used in water treatment.

C   *The Jaisalmer Wind Park*

Farming

Most of the people living in the desert are involved in subsistence farming. They survive in the hot and dry conditions by grazing animals on the grassy areas and cultivating vegetables and fruit trees.

Commercial farming, which has grown in recent decades, has been made possible by irrigation. The construction of the Indira Ghandi Canal in 1958 (see page 73) has revolutionised farming and crops such as wheat and cotton now thrive in an area that used to be scrubby desert (photo **D**). Other crops grown under irrigation include pulses, sesame, maize and mustard.

D ▶ *Growing wheat on irrigated land in the desert*

ACTIVITIES

1 a Describe the natural environment of the Thar Desert in photo **B**.

   b What are the challenges of this environment for the local people?

   c What are the animals in the photo?

   d Why is this traditional form of farming appropriate to the environment?

2 a Why is the Thar Desert a good location for a large wind farm?

   b Why do you think the Indian government is keen to develop wind and solar energy in the Thar Desert?

3 How has irrigation led to improvements in farming?

4 Design a poster describing how people can make use of the Thar Desert.

Stretch yourself

Find out more about the Carbon Neutral Company's Thar Desert Wind Farm.

● What is the potential for renewable energy projects in the Thar Desert?

● Why is it important to develop 'carbon neutral' energy in the future?

Practice question

Explain how hot deserts like the Thar Desert can provide opportunities for development. *(6 marks)*

On this spread you will find out about the challenges of development in hot desert environments

Challenges for development in the Thar Desert

Extreme temperatures

The Thar Desert suffers from extremely high temperatures (graph **A**), sometimes exceeding 50°C in the summer. This presents challenges for people, animals and plants living in this environment.

◆ Working outside in the heat of the day can be very hard, especially for farmers.

◆ High rates of evaporation lead to water shortages which affect people as well as plants and animals.

◆ Plants and animals have to adapt to survive in the extreme heat. Some animals are nocturnal, hibernating in the cooler ground during the daytime. Livestock, such as cattle and goats, need shade to protect them from the intense sun.

Water supply

Why are there water shortages?

Water supply has become a serious issue in the Thar Desert. As the population has grown and farming and industry have developed, demand for water has increased. Water in this region is a scarce resource.

The desert has low annual rainfall, high temperatures and strong winds. This causes high rates of evaporation.

What are the sources of water?

There are several sources of water in the Thar Desert.

◆ Traditionally, drinking water for people and animals is stored in ponds, some of which are natural (*tobas* – photo **B**) and others are man-made (*johads*).

◆ There are a few rivers and streams that flow through the desert, such as the River Luni which feeds a marshy area called the Rann. But these are intermittent, and flow only after rainfall. Most settlements are found alongside these rivers.

◆ Some water can be obtained from underground sources (aquifers) using wells but this water is salty and not very good quality.

Altitude 60m   Average temperature 27.3°C   Total rainfall 313mm

A *Climate graph for the Thar Desert*

B *Collecting water in the desert*

The Indira Gandhi Canal

The main form of irrigation in the desert is the Indira Gandhi (Rajasthan) Canal (photo **C**). This source of fresh water has transformed an extensive area of the desert and has revolutionised farming.

Commercial farming, growing crops such as wheat and cotton, now flourishes in an area that used to be scrub desert.

Two of the main areas to benefit from the canal are centred on the cities of Jodhpur and Jaisalmer where over 3500 km² of land is under irrigation.

The canal provides drinking water to many people in the desert.

Constructed in 1958 the canal has a total length of 650 km.

C *The Indira Ghandi Canal*

Accessibility

Due to the very extreme weather and the presence of vast barren areas there is a very limited road network across the Thar Desert. The high temperatures can cause the tarmac to melt and the strong winds often blow sand over the roads.

Many places are accessible only by camel, which is a traditional form of transport in the region. Public transport often involves seriously overladen buses (photo **D**).

ACTIVITIES

1 **a** What is the source of water shown in photo **B**?

   b What do you think the quality of the water is like? Explain your answer.

   c The rivers and streams in the desert are 'intermittent'. What does this mean and why is it a problem for water supply?

2 Why do you think the Rann is a protected area in the Thar Desert?

3 How has the Indira Ghandi brought benefits to the region?

4 Why do the high temperatures and the limited road network in the Thar Desert present challenges for development?

D *Traditional transport in the Thar Desert*

Practice question

Suggest two reasons why irrigation is important for future human development of the Thar Desert. *(4 marks)*

Stretch yourself

Carry out some research about the Indira Ghandi Canal.

- Search for a detailed map to show its route through the desert.

- Investigate how it has 'revolutionised' farming in the region. Illustrate your work with captioned or annotated photos.

- Do you think another canal should be constructed and, if so, where?

On this spread you will find out about the causes of desertification

What is desertification?

Desertification happens where land is gradually turned into a desert (photo **A**), usually on the edges of an existing desert. This can occur when land is **overgrazed** by livestock or stripped of vegetation by people collecting firewood. Once exposed to the weather, it will crack and break up. It will then be eroded by wind and water.

Where is desertification a problem?

Most of the areas at risk from desertification are on the borders of existing deserts, for example the Sahara Desert in Africa (map **B**). An estimated one billion people live in the areas at risk. Desertification affects rich countries as well as poorer ones. It is a significant problem in parts of the USA, Europe (especially Spain) and Australia.

What causes desertification?

Desertification can be caused by natural events, such as droughts, as well as poor land management. The areas close to deserts are ecologically very fragile. Slight changes in temperature and rainfall associated with climate change can have serious impacts. This makes these semi-desert areas even more prone to overgrazing or over-cultivation.

In Australia over 40 per cent of the 5 million km² of desert and semi-desert land has been affected by desertification. This is caused mainly by the pressure of grazing on fragile land affected by drought.

A Oryx grazing alongside the Namib Desert, Namibia

Did you know?

A heavy storm can erode soil that has taken thousands of years to form. Half of the topsoil on Earth has been lost in the last 150 years!

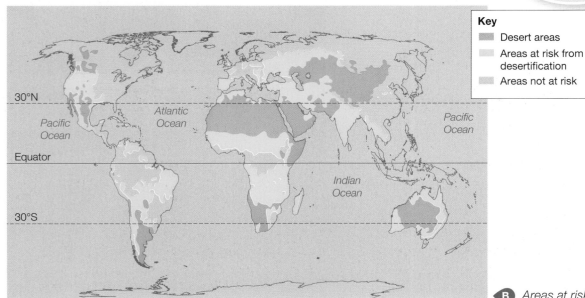

Key
- Desert areas
- Areas at risk from desertification
- Areas not at risk

30°N

Pacific Ocean

Atlantic Ocean

Pacific Ocean

Equator

30°S

Indian Ocean

B Areas at risk of desertification

In some desert regions, such as the Sahel on the southern fringes of the Sahara Desert, *climate change* is resulting in drier conditions and unreliable rainfall. On average it now rains less than it did 50 years ago.

Population pressure can result in land close to existing deserts being overgrazed. This means that there are too many animals to be supported by the limited vegetation. When the vegetation has been destroyed the land will turn to desert.

Soil erosion is often linked to desertification. When vegetation has been destroyed the soil is exposed to the wind and the rain making it vulnerable to erosion.

Over-cultivation resulting from the need to produce more food can lead to the soil becoming exhausted. It will turn to dust and become infertile.

Population growth is also increasing the demand for *fuelwood*. Trees are stripped of their branches and eventually die.

C *Causes of desertification*

Desertification in the Badia, Jordan

The Badia is a dry rocky desert in eastern Jordan. Its average annual rainfall is less than 150 mm and summer temperatures exceed 40 °C. The lack of water in this region is a major problem affecting the people who live there.

Much of the land has been traditionally grazed by the nomadic Bedouin who herd sheep, goats and camels on the rough shrubby grassland. An influx of sheep from Iraq following the 1991 Gulf War led to overgrazing and desertification.

Desertification made the land unproductive and people moved away from the area. Without vegetation, soil erosion became a major problem too.

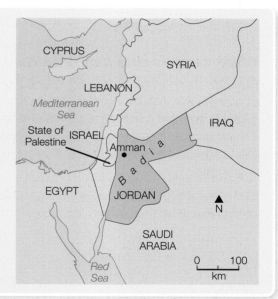

D *Location of the Badia*

ACTIVITIES

1 **a** Describe the landscape in photo **A**.
   b Can you see any evidence that this area is suffering from desertification?

2 Use map **B** to describe the pattern of areas 'at risk' from desertification.

3 **a** How does population pressure increase the risk of land becoming desert (photo **C**)?
   b What human activities can lead to soil erosion?
   c Is soil erosion an environmental disaster?

4 What caused desertification to occur in the Badia region in Jordan?

Stretch yourself

Investigate desertification in a HIC, such as the USA, Spain or Australia.

• What are the impacts and causes?

• What forms of management are being used to address the issue?

Practice question

'Desertification is largely caused by poor land management.' Use evidence to discuss this statement. *(6 marks)*

On this spread you will find out about how desertification can be reduced in hot desert environments

Holding back the desert

Land at risk of desertification needs to be managed sustainably so that people can live and prosper without damaging the environment.

Water and soil management

Commercial farming in hot deserts often involves **irrigation**. Water from underground sources or from rivers and canals can be sprayed onto crops or used to flood fields. But too much irrigation can cause problems leading to a process called *salinisation* (diagram **A**). The high rate of evaporation in hot deserts leads to a build-up of salts on the surface. This reduces soil fertility and kills plants.

Water management is at the centre of attempts to combat desertification in Australia. Local farmers are encouraged to use the following methods to prevent soil erosion.

◆ Ponding banks – areas of land enclosed by low walls to store water.

◆ Contour traps – embankments built along the contours of slopes to prevent soil from being washed down during heavy rainfall.

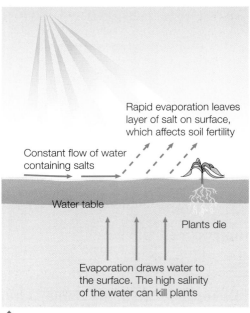

Rapid evaporation leaves layer of salt on surface, which affects soil fertility

Constant flow of water containing salts

Water table

Plants die

Evaporation draws water to the surface. The high salinity of the water can kill plants

A *The process of salinisation*

Water and soil management in the Badia, Jordan

The Tal Rimah Rangeland Rehabilitation Project aims to reverse the desertification caused by overgrazing in the 1990s.

◆ Local people have built low stone walls to stop water running down slopes after heavy rainfall.

◆ This water is used to irrigate newly-planted *Atriplex* shrubs that are well adapted to semi-desert environments (photo **B**).

◆ *Atriplex* hold the soil together and provide grazing for sheep and goats.

◆ As soil conditions improve, plants have started to grow, attracting birds and butterflies to the area.

B Atriplex *shrubs growing in shallow ditches in the Badia*

National parks

In some parts of the world, hot desert areas at risk of desertification have been protected by making them into national parks.

◆ The Desert National Park in the Thar Desert, India was created in 1992 to protect some 3000 km² of desert and reduce the risk of desertification.

◆ The Zion National Park (photo **C**) is one of four desert national parks in the USA. It was established in 1919 to protect a desert canyon near Las Vegas.

C *Zion National Park*

Tree planting in the Thar Desert, India

Tree planting is an important way of reducing erosion. Tree roots bind the soil together and the leaves and branches provide shade, grazing for animals and fuelwood.

The most important tree in the Thar Desert is the *Prosopis cineraria* (photo **D**). It is well adapted to the hostile desert conditions and has many uses.

The *Prosopis cineraria* provides:
- plenty of foliage and seed pods for animals to eat
- good quality firewood
- strong wood for building material
- shade and moist growing conditions for plants
- roots to stabilise sand dunes.

D Prosopis cineraria

E *Building a wall with 'magic stones'*

Appropriate technology

Many people living on the edges of deserts are poor. **Appropriate technology** involves using methods and materials that are appropriate to their level of development. They may not have access to expensive machinery. Sustainable approaches have to be practical and appropriate.

'Magic stones' in Burkina Faso, West Africa

In rural parts of Burkina Faso lines of stones have been used to reduce soil erosion. Using basic tools and trucks to transport the stones, local people have built low stone walls between 0.5 m and 1.5 m high along the contours of slopes (photo **E**). When rain washes down the hillside, the walls trap water and soil. This has helped to increase crops by up to 50 per cent and reduce desertification.

ACTIVITIES

1 What is salinisation and how is it caused (diagram **A**)?

2 **a** Why do you think the *Atriplex* shrubs in photo **B** are planted in shallow trenches?

 b How have these plants benefited local people and the natural environment?

3 What are the sustainable qualities of the *Prosopis cineraria* tree (figure **D**)?

4 **a** How do 'magic stones' use appropriate technology to help solve the problem of desertification?

 b What Australian management scheme is similar to the use of 'magic stones'?

 c Make a simple sketch to show how 'magic stones' work.

Stretch yourself

Carry out your own research about *one* of the four desert national parks in the USA:
- Grand Canyon
- Zion
- Arches
- Canyonlands.

How are the authorities in your chosen park managing the environment to reduce desertification?

Practice question

Explain how the effects of salinisation shown in diagram **A** can lead to desertification and how they can be reduced. *(6 marks)*

On this spread you will find out about the characteristics of cold environments

What are cold environments?

Cold environments experience temperatures that are at or below zero degrees Celcius for long periods of time. In the most extreme cold environments, such as the Antarctic, temperatures will be below zero throughout the year. Less extreme cold environments, such as northern Canada and parts of Iceland, simply experience very cold winters.

Key
polar tundra alpine environment

Greenland ice sheet

Antarctic ice sheet

A *The location of the world's cold environments*

Characteristics of cold environments

POLAR

Climate
- Winter temperatures often fall below −50 °C
- These areas have low precipitation (snow) totals.

Soils
Permanently covered by ice so soils are permanently frozen.

Plants
Some plants such as mosses and lichens are found on the fringes of the ice.

Animals
- Polar bears are well adapted to the polar environment. To retain heat they have thick fur, an insulating layer of fat, with a black nose and foot pads to absorb sunshine.
- In the Antarctic, penguins lay their eggs on land and bring up their young before returning to the ocean.

TUNDRA

Climate
- This climate is less extreme. Winter temperatures may drop to −20 °C.
- The brief summers can be quite warm.
- Amounts of precipitation – mainly snow – can be high in coastal regions.

Soils
- Soils are frozen (**permafrost**) but in summer will melt closer to the surface.
- Soils are generally infertile. Water draining through soils removes nutrients.
- Soils become waterlogged because water is trapped by permafrost.

Plants
- Low-growing flowering plants such as bearberry, Arctic moss and tufted saxifrage.
- Low bushes and small trees may grow in warmer regions.

Animals
- With more food options and a less extreme climate, several animals live here, including the Arctic fox and Arctic hare.
- Birds such as ptarmigans and insects such as midges and mosquitoes are abundant in the summer.

How does vegetation adapt to cold environments?

Few plants, if any, are found in polar regions, but a wide variety of plants live in tundra environments. This is because they have evolved a number of special adaptations to cope with the low temperatures, strong winds and dry conditions. For example:

◆ flowering and seed formation happens in a short time so that reproduction can take place during the short summers

◆ plants are low-growing and cushion-like to protect and insulate them from the strong dry winds

◆ hairy stems help to keep plants warm

◆ thin and waxy leaves reduce water loss.

Did you know?
In the Antarctic, temperatures will plunge to –60°C or below in the winter!

How does the bearberry adapt to cold environments?

The bearberry is a plant with red berries and bright green waxy leaves. It is one of the tundra's most abundant plants. The bearberry thrives in the tundra environment because it has evolved a number of adaptations (photo **C**).

Very low-growing (5–15 cm off the ground) to enable it to survive the strong winds

Stems have a thick bark for stability in the windy conditions

Small leathery leaves help retain water in this dry environment

Hairy stems help to retain heat and keep the plant safe from very low temperatures

Bright red berries are eaten by birds and owls and this helps to distribute the seeds

C *The bearberry plant*

Month	Temperature (°C)		Rainfall (mm)
	Maximum	**Minimum**	
January	-7	-11	26
February	-7	-14	25
March	-9	-15	24
April	-5	-12	15
May	-1	-5	20
June	4	1	19
July	7	4	25
August	6	3	40
September	3	0	36
October	-1	-5	39
November	-3	-8	37
December	-6	-10	31

B *Climate data for Spitzbergen, Svalbard*

Maths skills

Use table **B** to help you answer the following questions.
- What is the average annual precipitation on Spitzbergen?
- How does this compare to London's average annual precipitation (600 mm)?
- Write a few short sentences describing the climate of Spitzbergen.

Stretch yourself

Carry out some research online to answer the following questions:
- How are polar bears and penguins able to survive in polar environments?
- What adaptations do animals have in tundra environments?

ACTIVITIES

1 a Draw the extent of the polar and tundra environments (map **A**) onto a blank world map outline.

   b Label the main countries and regions that have these environments.

2 a Use table **B** to add notes describing the climate and soils of these two cold environments.

   b Describe how vegetation has adapted to cold environments.

Practice question

Explain the features of plants and soils in the tundra environment. *(4 marks)*

On this spread you will find out about the opportunities for development in Svalbard

Case study

Where is Svalbard?

Svalbard is a Norwegian territory in the Arctic Ocean and the most northerly permanently inhabited group of islands in the world.

Much of Svalbard experiences a polar climate with 60 per cent of the land covered by glaciers. The rest of the land is tundra much of which is frozen. There is no arable farming and there are no trees – it is just too cold! Svalbard has:

◆ five major islands, the largest of which is Spitzbergen

◆ a population of about 2700, most living in the main town of Longyearbyen

◆ more polar bears and snowmobiles than people.

A *Location of Svalbard*

What are the opportunities for development in Svalbard?

Mineral extraction

Svalbard has rich reserves of coal, but mining on Svalbard is a controversial issue. Environmental groups are against it as burning coal is a major source of greenhouse gases. However, coal mining is vital to the economy of Svalbard because:

◆ it is the main economic activity

◆ more than 300 people are employed in the mines and as support staff. In recent years the industry has faced a decline due to lower world coal prices and some jobs have been lost.

In 2014 a new mine opened near Svea (photo **B**). To gain access to the new mine, a road had to be constructed over a glacier!

B *The new Lunckefjell coal mine near Svea*

Energy developments

Some of the coal mined on Svalbard is burned to generate electricity in the Longyearbyen power station (photo **C**). It is Norway's only coal-fired power station and supplies all of Svalbard's energy needs. Environmentalists believe the power station should be closed down and renewable sources should be explored. The most likely future source is **geothermal energy**, tapping into the heat of the earth and using it to generate electricity.

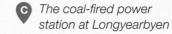

C *The coal-fired power station at Longyearbyen*

Like Iceland, which uses mainly geothermal energy, Svalbard is located close to the Mid-Atlantic Ridge, a constructive plate margin. Here the Earth's crust is thin and hot rocks are close to the surface.

Another future option involves capturing carbon dioxide from burning coal and circulating this instead of water to generate electricity. This is called *carbon capture and storage* (page 46).

Fishing

The cold waters of the Barents Sea south of Svalbard (map **A**) are one of the richest fishing grounds in the world (photo **D**)

Apart from cod, there are an estimated 150 species of fish here, including herring and haddock. These waters are extremely important breeding and nursery grounds for fish stocks and need to be protected from pollution.

Fishing in the Barents Sea is jointly controlled and monitored by Norway and Russia to ensure that fishing is sustainable and the ecosystem is protected.

Did you know?

If you have recently eaten cod and chips, there is a good chance that the cod came from the Barents Sea which has the largest stocks of cod in the world!

Tourism

Tourism in Svalbard has grown in recent years as people seek to explore extreme natural environments.

◆ In 2011, 70 000 people visited Longyearbyen and 30 000 of these were cruise passengers.

◆ The harbour at Longyearbyen has been enlarged to cope with the increase in the number of cruise ships.

◆ Tourism provides around 300 jobs for local people.

◆ Most tourists come from Norway and most visit as part of organised tours.

Tourists visit Svalbard to explore the natural environment – the glaciers, fjords and the wildlife, especially polar bears. Adventure tourism is becoming more popular with activities such as hiking, kayaking and snow mobile safaris (photo **E**). In the winter, tourists visit to experience the amazing Northern Lights.

D *Svalbard fishing trawlers*

ACTIVITIES

1 Describe the location of Svalbard.

2 Photo **B** shows Svalbard's newest coal mine.

   a Do you think coal is extracted from the surface or from below ground? Explain your answer.

   b What evidence suggests this is a difficult environment in which to operate a coal mine?

   c How does coal mining benefit the people of Svalbard?

   d Can you suggest any environmental problems associated with coal mining?

3 How does Svalbard benefit from the abundant fish stocks in the Barents Sea?

E *Adventure tourism in Svalbard*

Stretch yourself

Carry out further research about tourism in Svalbard.

- Why has Svalbard become a popular tourist destination in recent years?
- Why is tourism being promoted?
- How can tourism bring both benefits and problems?

Practice question

Explain why sustainable fishing requires international cooperation. *(2 marks)*

On this spread you will find out about the challenges of development in Svalbard

What are the challenges for development?

There are several challenges that need to be overcome when living and working in such an extreme environment (photo **A**).

Did you know?

The sun sets in Longyearbyen on 25 October and does not rise again for four months – on 8 March!

How can homes be kept warm during the very cold winters?

How do people travel around when roads are covered by snow?

How is water connected to homes and how is it prevented from freezing?

How is sewage and waste water removed from people's homes?

How can roads be constructed and maintained?

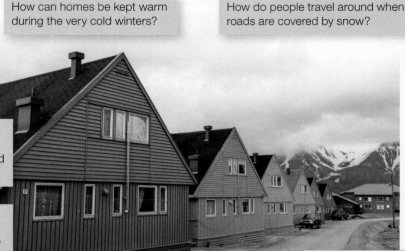

A Challenges of living and working in Svalbard

Extreme temperatures

Even in Longyearbyen winter temperatures can fall below −30 °C. In the northern glacial regions, it can be even colder. Such extreme temperatures make it dangerous to work outside, with a serious risk of frostbite. People have to dress very warmly (diagram **B**) and this can make outdoor work very slow and difficult. Imagine trying to build a house or work on the roads wearing several layers of thick clothes and gloves!

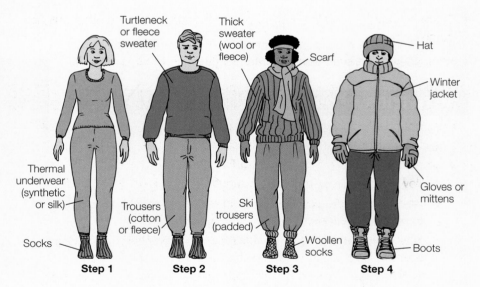

Turtleneck or fleece sweater

Thick sweater (wool or fleece)

Scarf

Hat

Winter jacket

Thermal underwear (synthetic or silk)

Trousers (cotton or fleece)

Ski trousers (padded)

Gloves or mittens

Socks

Woollen socks

Boots

Step 1　Step 2　Step 3　Step 4

B Dressing for extreme cold

Construction

There are many forms of construction that have been carried out in Svalbard:

◆ building houses, shops and offices

◆ constructing and maintaining roads

◆ constructing and enlarging harbour facilities for fishing vessels and cruise ships

◆ constructing buildings and mining operations to extract coal.

Working outdoors in extreme temperatures and also in limited light during the winter is very demanding. As a result most construction work is carried out during the brief summer period.

The frozen ground (permafrost) can provide good solid foundations but it has to be protected from melting. If the top surface of the permafrost melts, then it becomes unstable and could lead to buildings and roads cracking or even collapsing.

Most roads are dirt or gravel roads raised up above the ground surface (photo **C**). These roads are relatively cheap and easy to maintain from year to year.

Services

Services include water, electricity and sanitation. These are very important both to the resident population and also to tourists who expect comfortable living conditions with warm rooms and hot showers.

Unlike the UK most services are provided to individual buildings by overground heated water and sewage pipes (photo **D**). They need to be kept off the ground to prevent them causing any possible thawing of the permafrost and to allow easy maintenance.

Accessibility

Svalbard is located in a remote part of Europe and can only be reached by plane or ship. The islands themselves are inaccessible and almost all transport systems are restricted to the immediate area around Longyearbyen.

◆ There is one airport close to Longyearbyen capable of handling international flights from Norway and Russia. Smaller aircraft can be used to connect to other islands.

◆ There is 50 km of road in Longyearbyen. No roads serve the other outlying communities.

◆ Most people use snowmobiles, particularly in the winter (photo **E**).

C *Wheeled dog sled on a dirt road near Longyearbyen*

D *Overground service pipes*

E *Snowmobiles parked in Longyearbyen*

ACTIVITIES

1 a What are the challenges facing people living in Longyearbyen (photo **A**)?

b Can you think of any other challenges suggested by the photo?

2 Diagram **B** comes from the Svalbard Tourist Board.

a How many layers of clothing are suggested for the body and legs?

b Why is 'layering' a good idea to keep warm?

c What is suggested to keep your feet warm?

d Why is it important to instruct tourists on how to dress when visiting Svalbard?

e How might dressing for extreme cold make it difficult to work outside?

3 Why are services such as heated water and sewage connected to houses by overground pipes?

4 Suggest why snowmobiles are the most commonly used form of transport in the winter (photo **E**).

Stretch yourself

Imagine that you were going to visit Svalbard on a school trip in September. Find out what weather conditions you might expect and research appropriate clothes to cope with these conditions. Consider what footwear would be best suited for walking on rocks and snow.

Practice question

Explain how cold environments like Svalbard can provide challenges for development. *(6 marks)*

On this spread you will find out how cold environments are at risk from economic activities and why these wilderness areas need protecting

Why are cold environments fragile?

Cold environments are extremely fragile and they can be easily damaged by human activities. Tundra vegetation (photo **A**) takes a very long time to become established. It is a very delicate ecosystem that can be easily disturbed.

Relatively minor developments – such as constructing a footpath – can have serious long-term effects.

Tundra is a **fragile environment** and can take a very long time to recover from any damage that is done by human activity.

Off-road vehicle damage in Alaska

Off-road vehicle driving is a popular tourist activity in the Alaskan wilderness. People may be hunting wild animals or simply want to travel to a remote area for walking or fishing.

Most off-road driving takes place in the summer when the snow has melted. Warmer temperatures melt the upper surface of the soil making it extremely soggy. Photo **B** was taken in Alaska's Wrangell-St Elias National Park. Notice how a single vehicle can leave deep tyre tracks through the swampy tundra. As other vehicles skirt the muddy pools, damage extends over a large area.

It will take many years – possibly decades – for this land to recover from damage that may have taken just a few minutes to cause.

A *Autumn tundra landscape, northern Canada*

B *Damage by off-road vehicles in Alaska*

Trees killed by the oil spill

Electricity pylons will have resulted in tree clearance and environmental damage

Risk of fire, either started deliberately or by a lightning strike

Oil has leaked from this broken pipeline

River has become polluted and is now totally lifeless

River edge habitats polluted and destroyed – the vegetation may never recover

C *Oil-polluted river in Siberia, Russia*

How can cold environments be harmed by economic development?

Cold environments have rich reserves of oil, gas and other precious minerals such as gold. Oil and gas in particular are in high demand as a source of energy and countries are keen to exploit their resources for the economic benefits. To extract these resources, roads have to be constructed through forests and across the tundra and supply bases built. Housing for hundreds of workers also needs to be constructed. All this can have a huge impact on the environment.

Photo **C** is an example of one of the worst environmental disasters that can happen in cold environments – an oil spill. The damage to rivers and other natural ecosystems resulting from oil spills is long-lasting. It is hard to imagine this landscape ever recovering.

Why do cold environments need to be protected?

Many indigenous people live a traditional life here. For example, the Inuit who live in Arctic Alaska, Canada and Greenland depend on the wildlife and survive by hunting and fishing.

Cold environments are home to many birds, animals and plants, such as penguins, polar bears, the Arctic fox and many species of tundra vegetation.

Unpolluted and unspoilt, cold environments are important outdoor laboratories for scientific research such as the effects of climate change.

Their beauty and potential for adventure activities attracts tourists who bring huge benefits to countries such as Norway (Svalbard), Iceland and Alaska.

They provide opportunities for forestry and fishing.

D *The small town of Qeqertarsuaq, Greenland*

ACTIVITIES

1 **a** Describe the vegetation in photo **A**.
   b Are there any signs of human activity?
   c How might people use and damage this environment?
   d Why is it important that this environment is protected?

2 Imagine photo **B** is to be used in an environmental campaign against off-road driving. Write a short account (no more than 100 words) to run alongside the photo describing the harmful effects of off-road driving in a tundra environment.

3 What could be done to stop oil leaks such as the one shown in photo **C**?

4 Use the information in figure **D** along with your own research to design an information poster to explain why cold environments need protecting. Use captioned photos to illustrate your poster.

Stretch yourself

Find other examples of how human activities have damaged fragile cold environments.
Find at least *two* photographs that illustrate the problems caused and add labels to describe what has happened and what the impact has been.

Practice question

Outline three possible environmental impacts of economic development on cold environments. *(6 marks)*

On this spread you will find out about strategies to reduce the risks to cold environments

How can the risk to cold environments be reduced?

Cold environments offer many opportunities for economic development. To ensure that they do not suffer any long-term damage, they need to be managed sustainably. This can be done through:

◆ the use of technology

◆ action by governments

◆ the work of conservation groups.

 A Route of trans-Alaskan pipeline

The use of technology – the trans-Alaskan pipeline

In 1969 oil was discovered at Prudhoe Bay on the north coast of Alaska (map **A**). Winter sea ice in the Arctic Ocean prevented oil being transported by tanker so an alternative way had to be found. In 1974 the trans-Alaskan pipeline was opened. It enabled oil to be transported the 1300 km from Prudhoe Bay to the port of Valdez (photo **B**). Technology has been used to reduce its impact on the environment (table **C**).

Problem	Technological solution
The pipeline crosses rivers and mountains such as the Brookes Ranges	Pumping stations keep the oil moving. The pipeline passes beneath rivers to minimise the impact on the landscape.
Oil from the ground is very hot (49°C) which helps it to flow, but could melt the permafrost	The pipeline is raised and insulated to retain heat and prevent it melting the permafrost.
Possible cracks caused by earthquakes can cause oil leaks	The pipeline is supported and can slide if earthquakes happen. The flow of oil stops automatically if there is a leak.
Large herds of animals such as caribou migrate across the route of the pipeline.	The pipeline is raised to allow caribou to migrate underneath.

C Using technology to overcome problems in constructing the trans-Alaskan pipeline

B The trans-Alaskan pipeline

Action by governments – Alaska, USA

The United States government has been involved in the protection of Alaska ever since oil was discovered there in the 1960s.

◆ The National Environmental Policy Act, ensuring that companies involved with the extraction and transportation of oil protect the natural environment and recognise the rights of native people.

◆ The creation of the Western Arctic Reserve (map **A**) – a 9 million-hectare protected wilderness managed by the Department of the Interior. Home to thousands of caribou, millions of migratory birds, musk ox, wolves and even polar bears. Drilling for oil is kept away from sensitive areas.

◆ The National Oceanic and Atmospheric Administration (NOAA) oversees sustainable fisheries in Alaska and protects marine habitats.

International agreements – the Antarctic Treaty

In 1959 the Antarctic Treaty was signed by countries with territorial claims to Antarctica. Its main aim is to protect the natural environment of the largest wilderness on Earth (photo **D**). Despite the discovery of valuable minerals, the Antarctic Treaty has been successful in preventing economic development. The Treaty:

◆ recognises the importance of the continent for scientific research, particularly into climate change

◆ controls tourism and keeps disturbance to a minimum.

D *Antarctica – the world's last great wilderness*

Conservation groups – WWF in Canada

The World Wildlife Fund (WWF) is a conservation group that helps to protect Arctic environments in Canada. It provides scientific information, expertise and resources. The WWF:

◆ works with local communities to manage critical ecosystems, for example the Beaufort Sea

◆ supports scientific research to help protect important species such as polar bears, narwhal and Greenland shark

◆ works with oil companies, local Inuit organisations and government regulators to plan for a sustainable future for the Arctic.

> ### Think about it
>
> *Polar bears live along the northern coast of Alaska and western Canada. Their population has declined by 40% in just 10 years. What are the possible reasons?*

Should cold environments be protected as wilderness areas?

A wilderness is a wild and unspoiled area unaffected by human activity. There are few true **wilderness areas** left – mainly rainforests, deserts or cold environments (photo **D**). There are strong arguments for and against protecting wilderness areas from economic development (table **E**).

Arguments in favour	Arguments against
• Wilderness areas are fragile and are easily damaged by economic activities.	• Cold environments are rich in resources, such as oil, precious minerals, fish and timber.
• Untouched natural environments form important outdoor laboratories for scientific research.	• Over 4 million people already live in the Arctic in balance with the environment.
• Rare plants and animals will be protected.	• Technology now allows cold environments to be exploited with less impact.

E *Should cold environments be protected as wilderness areas?*

ACTIVITIES

1 Why was the trans-Alaskan pipeline built?

2 Why is it important to prevent permafrost melting?

3 How is the US government helping to protect cold environments in Alaska?

4 Suggest why the Antarctic Treaty is such an important international agreement.

Stretch yourself

Work in small groups. Read table **E**. Do you think cold environments should be treated as wilderness areas, and protected from economic development? Give your reasons.

Practice question

Using a case study, explain how different strategies can help reduce environmental damage in cold environments. *(9 marks)*

The estuary of the River Mawddach, Bamouth, mid-Wales, with Cadair Idris visible top right (see page 138)

Unit 1 Living with the physical environment is about physical processes and systems, how they change, and how people interact with them at a range of scales and in a range of places. It is split into three sections.

Section C Physical landscapes in the UK includes:

- an introduction to UK physical landscapes
- coastal landscapes in the UK
- river landscapes in the UK
- glacial landscapes in the UK.

You have to study the introduction to the UK physical landscapes. You must also study two of the other three types of landscapes.

> **What if...**
>
> **1** the UK landscapes were made entirely of granite?
>
> **2** all our cliffs collapsed?
>
> **3** the UK had droughts every year?
>
> **4** the world entered a new Ice Age?

Your key skills

To be a good geographer, you need to develop important geographical skills – in this section you will learn the following skills:

- Drawing cross-sections.
- Drawing labelled sketches and diagrams.
- Drawing sketches from photos.
- Using and describing information in photos.
- Using OS and atlas maps.
- Literacy – describing landforms and processes.

Your key words

As you go through the chapters in this section, make sure you know and understand the key words shown in bold. Definitions are provided in the Glossary on pages 346–9. To be a good geographer you need to use good subject terminology.

Your exam

Section C makes up part of Paper 1 – a one and a half-hour written exam worth 35 per cent of your GCSE.

9.1 | The UK's relief and landscapes

On this spread you will find out about the relief and landscapes found in the UK

What is relief?

Relief is a term used by geographers to describe the physical features of the landscape. This includes:

◆ height above sea level
◆ steepness of slopes
◆ shapes of landscape features.

The relief of an area is determined mainly by its *geology* – the rocks that form the landscape. Tough, resistant rocks such as granite and slate form some of the UK's most dramatic mountain ranges such as those in Arran, Scotland (photo **A**). Weaker rocks such as clays and limestone often form low-lying plains and gently rolling landscapes (photo **B**).

A *Dramatic mountains in Arran, Scotland*

The UK's landscapes

A landscape is an area whose character is the result of the action and interaction of natural and human factors. In the UK a wide variety of rock types are responsible for creating our varied landscapes. Map **C** is from an atlas and shows the relief of the UK. The key uses different colours to show land heights and sea depths.

The UK's river systems

You can see on map **C** that the UK has a very extensive river system. Most rivers have their source in the mountain ranges or hills and flow to the sea (can you see any that don't?).

B *The Cotswold Hills*

ACTIVITIES

Use map **C** to answer Activities 1–3.
1 **a** What is the name and height of the highest mountain in the UK? In what mountain range is it located?
   b What is the name and height of the highest mountain in Wales?
   c Describe the relief of your home area or that of your school.
   d Compare the location and relief of Arran and the Cotswold Hills (photos **A** and **B**).
   e What is special about the relief of the Fens in eastern England?
2 Describe the pattern of upland areas in the UK.
3 **a** Describe the course of the River Severn from its source to its mouth.
   b Locate the river that is closest to your home or school. Describe its course.
4 As a class complete a wall display of photos from around the UK showing the different landscapes and river systems. If possible, locate each photo on a large map of the UK.
5 Alphabet Run! Can you find an upland area, mountain or river for each letter of the alphabet?

Stretch yourself

Use the spot heights and the height values in the key for map **C** to try drawing a cross-section sketch across the UK. For example, from Snowdon in North Wales to the Norfolk Broads in East Anglia; or from Snowdon to Romney Marsh in Kent, to include include the Cotswolds (photo **B**). Take time to choose an appropriate vertical scale.

Practice question

Explain how different types of rock determine the UK's landscapes. *(4 marks)*

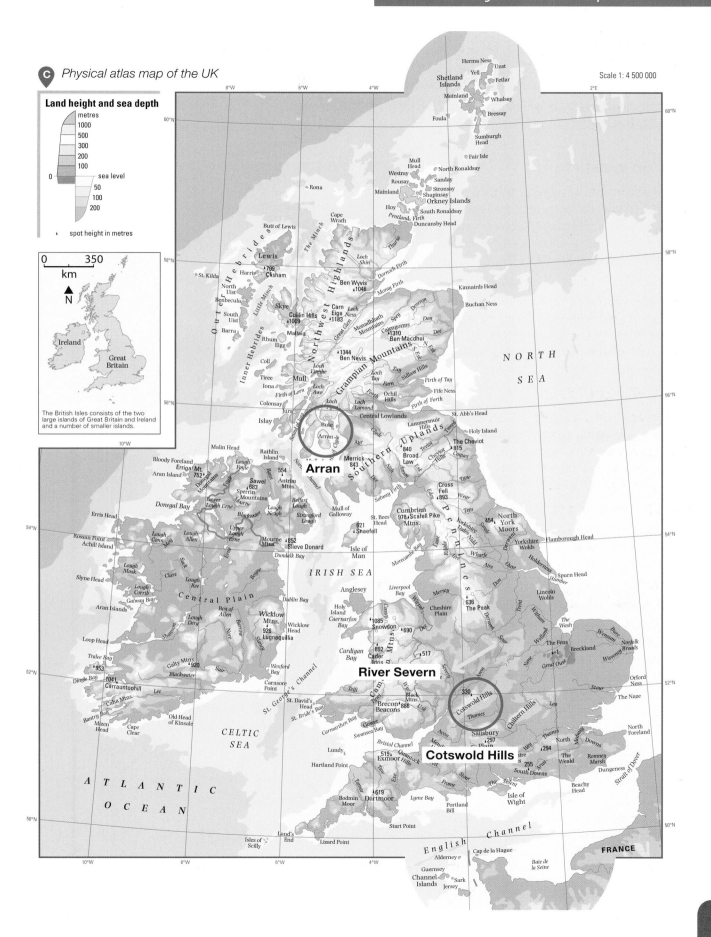

C *Physical atlas map of the UK*

Scale 1: 4 500 000

Land height and sea depth

metres
1000
500
300
200
100
0 — sea level
50
100
200

▲ spot height in metres

0 — 350
km
▲ N

Ireland   Great Britain

The British Isles consists of the two large islands of Great Britain and Ireland and a number of smaller islands.

Arran

River Severn

Cotswold Hills

On this spread you will find out about the formation and characteristics of waves

How do waves form?

Waves are formed by the wind blowing over the sea. Friction with the surface of the water causes ripples to form and these develop into waves. The distance the wind blows across the water is called the *fetch*. The longer the *fetch*, the more powerful the wave.

Waves can also be formed more dramatically when earthquakes or volcanic eruptions shake the seabed. These waves are called *tsunami*. In March 2011 a wall of water up to 40 m high crashed into the Japanese coast north of Tokyo destroying several coastal settlements and killing over 20 000 people (photo **B**).

What happens when waves reach the coast?

In the open sea, despite the wavy surface, there is little horizontal movement of water. Only when the waves approach the shore is there forward movement of water as waves break and surge up the beach (diagram **C**).

The seabed interrupts the circular movement of the water. As the water becomes shallower, the circular motion becomes more elliptical. This causes the crest of the wave to rise up and eventually to collapse onto the beach. The water that rushes up the beach is called the *swash*. The water that flows back towards the sea is called the backwash.

A *Surfing at Newquay, Cornwall*

B *Tsunami waves hit the coast of Japan*

 Waves approaching the coast

Circular orbit in open water

Friction with the seabed distorts the circular orbital motion

Top of wave moves faster

Increasingly elliptical orbit

Wave begins to break

Water from previous wave returns

Water rushes up the beach

Shelving seabed (beach)

Wave types

It is possible to identify two types of wave at the coast.

Constructive waves

These are low waves that surge up the beach and 'spill' with a powerful swash (diagram **D**). They carry and deposit large amounts of sand and pebbles and 'construct' the beach making it more extensive. Surfers prefer constructive waves because they give longer rides (photo **A**)! These waves are formed by storms often hundreds of kilometres away.

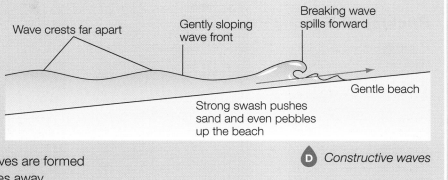

Wave crests far apart

Gently sloping wave front

Breaking wave spills forward

Gentle beach

Strong swash pushes sand and even pebbles up the beach

D *Constructive waves*

Destructive waves

These are formed by local storms close to the coast, and they can 'destroy' the beach – hence their name. They are closely spaced and often interfere with each other producing a chaotic swirling mass of water. They become high and steep before plunging down onto the beach (diagram **E**). There is little forward motion (swash) when a destructive wave breaks but a powerful backwash. This explains the removal of sand and pebbles and the gradual destruction of the beach.

Waves close together

Steep wave front

Breaking wave plunges downwards

Steep beach

Strong backwash pulls sand and even pebbles out to sea

E *Destructive waves*

ACTIVITIES

1. **a** Copy diagram **C** and draw an arrow to show the direction of the waves.

   b Add the labels *swash* and *backwash* in the correct places.

   c What causes the waves to rise up and break on the beach?

   d When waves break on a sandy or pebbly beach the amount of backwash is often less than the amount of swash. Why do you think this is?

   e Larger pebbles are found at the top of the beach with smaller ones near the bottom. Use your answer to **d** to suggest reasons why.

2. Why do surfers prefer constructive waves to destructive waves?

3. Outline the characteristics of constructive and destructive waves. Complete a copy of the table below.

Wave characteristic	Constructive wave	Destructive wave
Wave height		
Wave length		
Type of wave break (spilling or plunging)		
Strength of swash		
Strength of backwash		
Net beach sediment (gain or loss)		

Stretch yourself

Carry out some research about the tsunami waves that struck Japan in March 2011.

- Why were the waves so high and so powerful?

- What were the impacts on people and human activities?

- What effect did the waves have on the physical geography of the coast of Japan?

Practice question

Compare the characteristics of constructive and destructive waves. *(4 marks)*

On this spread you will find out about processes of weathering and mass movement at the coast

Rockfall at Beachy Head, 2001

Photo **A** shows a dramatic rockfall that happened at Beachy Head in East Sussex. During the wet winter of 2000 the chalk rock became saturated with water. The water froze during the winter. In April 2001 this caused a rockfall – a huge slab of chalk broke away and collapsed into the sea. Processes like this combine with the action of the waves to shape the coastline.

What causes cliffs to collapse?

Cliffs collapse because of different types of weathering. This is the weakening or decay of rocks in their original place on, or close to, the ground surface. It is mostly caused by weather factors such as rainfall and changes in temperature.

A   *Rockfall at Beachy Head, Sussex*

There are three types of weathering:

◆ **Mechanical (physical) weathering** – the disintegration (break-up) of rocks. Where this happens, piles of rock fragments called *scree* can be found at the foot of cliffs.

◆ **Chemical weathering** – caused by chemical changes. Rainwater, which is slightly acidic, very slowly dissolves certain types of rocks and minerals.

◆ *Biological weathering* – due to the actions of flora and fauna. Plant roots grow in cracks in the rocks. Animals such as rabbits burrow into weak rocks such as sands.

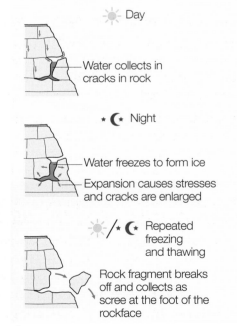

Day

Water collects in cracks in rock

Night

Water freezes to form ice

Expansion causes stresses and cracks are enlarged

Repeated freezing and thawing

Rock fragment breaks off and collects as scree at the foot of the rockface

B   *The process of freeze-thaw weathering*

C   *Landslip at Holbeck Hall, Scarborough*

Weathering process	Description
Freeze-thaw (mechanical)	Look at diagram **B**. • Water collects in cracks or holes (pores) in the rock. • At night this water freezes and expands and makes cracks in the rock bigger. • When the temperature rises and the ice thaws, water will seep deeper into the rock. • After repeated freezing and thawing, fragments of rock may break off and fall to the foot of the cliff (*scree*).
Salt weathering (mechanical)	• Seawater contains salt. When the water evaporates it leaves behind salt crystals. • In cracks and holes these salt crystals grow and expand. • This puts pressure on the rocks and flakes may eventually break off.
Carbonation (chemical)	• Rainwater absorbs CO_2 from the air and becomes slightly acidic. • Contact with alkaline rocks such as chalk and limestone produces a chemical reaction causing the rocks to slowly dissolve.

What are the processes of mass movement?

Mass movement is the downward movement or **sliding** of material under the influence of gravity. In 1993, 60 m of cliff slipped onto the beach near Scarborough in North Yorkshire taking with it part of the Holbeck Hall Hotel (photo **C**). The hotel was left on the cliff edge and had to be demolished.

Diagram **D** describes some of the common types of mass movement found at the coast. Both mass movement and weathering provide an input of material to the coastal system. Much of this material is carried away by waves and deposited further along the coast.

D *Types of mass movement at the coast*

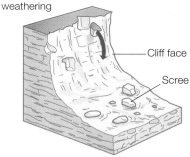

a Rockfall – fragments of rock break away from the cliff face, often due to freeze-thaw weathering

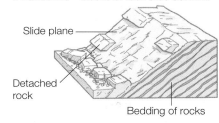

b Landslide – blocks of rock slide downhill

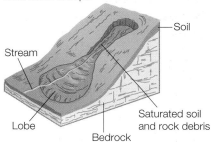

c Mudflow – saturated soil and weak rock flows down a slope

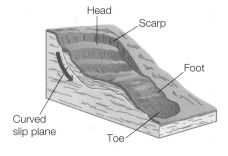

d Rotational slip – slump of saturated soil and weak rock along a curved surface

ACTIVITIES

1 a Draw a simple sketch of the coastline in photo **A**. Label the rockfall, the chalk cliffs and the rocky beach.

 b Do you think freeze-thaw is active here?

 c What is scree? Label this feature on your sketch.

 d How might rockfalls be a hazard to people?

2 Make a copy of diagram **B** and add detailed annotations to describe the process of freeze- thaw weathering.

3 Describe the process of mass movement in photo **C** and suggest the causes.

Stretch yourself

Investigate the Beachy Head rockfall in 2001.

• Which weathering and mass movement processes were responsible?

• What impact did the rockfalls have on the shape of the coast?

• Find out how and why the Belle Tout lighthouse had to be moved.

Practice question

Describe the effects of weathering and mass movement on a cliffed coastline. *(6 marks)*

On this spread you will find out about the processes of erosion and deposition

Coastal erosion

Erosion involves the removal of material and the shaping of landforms. There are several different processes of coastal erosion.

 *Processes of erosion*

Solution
The dissolving of soluble chemicals in rocks, e.g. limestone.

Corrasion
Fragments of rock are picked up and hurled by the sea at a cliff. The rocks act like tools scraping and gouging to erode the rock.

Hydraulic power
This is the power of the waves as they smash onto a cliff. Trapped air is forced into holes and cracks in the rock eventually causing the rock to break apart. The explosive force of trapped air operating in a crack is called *cavitation*.

Abrasion
This is the 'sandpapering' effect of pebbles grinding over a rocky platform often causing it to become smooth.

Attrition
Rock fragments carried by the sea knock against one another causing them to become smaller and more rounded.

Coastal transportation

Sediment of different sizes can be transported in four different ways. (diagram **B**):

- ◆ **solution**
- ◆ **suspension**
- ◆ **saltation**
- ◆ **traction**.

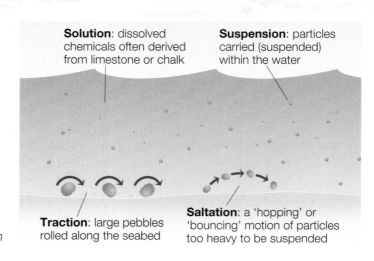

Solution: dissolved chemicals often derived from limestone or chalk

Suspension: particles carried (suspended) within the water

Traction: large pebbles rolled along the seabed

Saltation: a 'hopping' or 'bouncing' motion of particles too heavy to be suspended

B *Types of coastal transportation*

Longshore drift

The movement of sediment on a beach depends on the direction that waves approach the coast (diagram **C**). Where waves approach 'head on', sediment is simply moved up and down the beach. But if waves approach at an angle, sediment will be moved *along* the beach in a 'zigzag' pattern. This is called **longshore drift**.

Longshore drift is responsible for a number of important coastal landforms including beaches and spits (pages 100–1).

Coastal deposition

Coastal **deposition** takes place in areas where the flow of water slows down. Waves lose energy in sheltered bays and where water is protected by spits or bars (see page 101). Here sediment can no longer be carried or moved and is therefore deposited. This explains why beaches are found in bays, where the energy of the waves is reduced. This is called *wave refraction* (diagram **D**).

Mudflats and saltmarshes are often found in sheltered estuaries behind spits where there is very little flow of water.

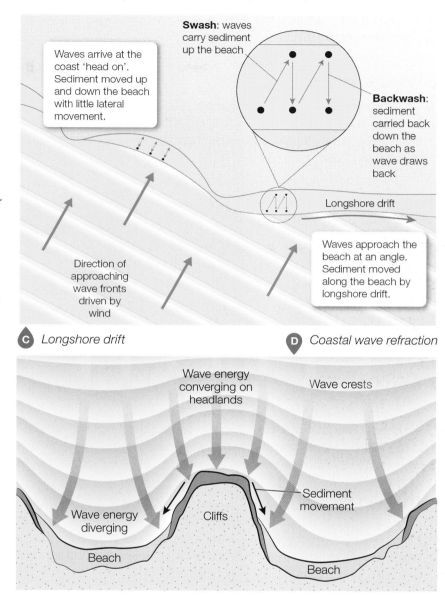

Swash: waves carry sediment up the beach

Waves arrive at the coast 'head on'. Sediment moved up and down the beach with little lateral movement.

Backwash: sediment carried back down the beach as wave draws back

Direction of approaching wave fronts driven by wind

Longshore drift

Waves approach the beach at an angle. Sediment moved along the beach by longshore drift.

C *Longshore drift*

D *Coastal wave refraction*

Wave energy converging on headlands

Wave crests

Wave energy diverging

Cliffs

Sediment movement

Beach

Beach

ACTIVITIES

1 **a** Draw an annotated diagram similar to **B** to show the processes of *erosion*. Show a wave breaking against the foot of a cliff.

   b Add detailed labels to describe the *five* processes of erosion.

2 **a** What is meant by the term 'longshore drift' (diagram **C**)?

   b Why does this only occur on some beaches?

   c Draw a diagram to show the process of longshore drift. Add labels to describe what is happening.

   d Imagine you are doing a fieldwork investigation for evidence of longshore drift along a stretch of coast. What evidence would you look for and why?

Stretch yourself

Find out more about the coastal locations where deposition occurs.

- Focus on a stretch of coastline near to your school or one that you have visited.

- Use maps and satellite images to zoom in on locations where deposition has happened. Describe the material that has been deposited and suggest reasons why.

Practice question

What factors affect the processes operating along a stretch of coastline? *(6 marks)*

On this spread you will find out about the characteristics and formation of coastal landforms

What is a landform?

You will come across the term 'landform' all the time in physical geography. A landform is a feature of the landscape that has been formed or sculpted by processes of:

◆ erosion ◆ transportation ◆ deposition.

What factors influence coastal landforms?

Some rocks are tougher and more resistant than others. Rocks such as granite, limestone and chalk form impressive cliffs and headlands because they are more resistant to erosion. Softer rocks, clays and sands are more easily eroded to form bays or low-lying stretches of coastline.

Geological structure includes the way that layers of rocks are folded or tilted. This can be an important factor in the shape of cliffs. **Faults** are cracks in rocks. Enormous tectonic pressures can cause rocks to 'snap' rather than fold (bend) and movement (or *displacement*) happens on either side of the fault. Faults form lines of weakness in rocks, easily carved out by the sea.

Ⓐ *The formation of headlands and bays*

a

Less resistant (softer) clay

Resistant (harder) sandstone

Waves

Clay

Resistant (harder) chalk or limestone

Clay

b

Less resistant rock worn away to leave a bay

Resistant (harder) rock left as a headland

Waves

Sheltered bay – sand is deposited

Headland

Bay

Headlands and bays

Different types of rock at the coastline will be eroded at different rates. Weaker bands of rock (such as clay) erode more easily to form *bays*. As the bays are sheltered, deposition takes place and a sandy beach forms (diagram **A**).

The tougher, more resistant bands of rock (such as limestone or sandstone) are eroded much more slowly. They stick out into the sea to form **headlands**. Erosion dominates in these high-energy environments, which explains why there are no beaches. Most *erosional* landforms are found at headlands.

Ⓑ *Wave-cut platform and beach near Beachy Head*

Cliffs and wave-cut platforms

When waves break against a **cliff**, erosion close to the high tide line will wear away the cliff to form a wave-cut notch. Over a long period of time – usually hundreds of years – the notch will get deeper and deeper, undercutting the cliff. Eventually the overlying cliff can no longer support its own weight and it collapses.

Through a continual sequence of wave-cut notch formation and cliff collapse, the cliff will gradually retreat. In its place will be a gently sloping rocky platform called a **wave-cut platform** (photo **B**). A wave-cut platform is typically quite smooth due to the process of abrasion. However, in some places it may be scarred with rock pools.

Caves, arches and stacks

Lines of weakness in a headland, such *as faults*, are particularly vulnerable to erosion. The energy of the waves wears away the rock along a line of weakness to form a **cave** (diagram **C**). Over time, erosion may lead to two back-to-back caves breaking through a headland to form an **arch**. Gradually the arch is enlarged by erosion at the base and by weathering processes (such as freeze-thaw) acting on the roof. Eventually the roof will be worn away and collapse to form an isolated pillar of rock known as a **stack**.

Remember!

- A cliff, a river meander or a delta *are* all landforms.
- A process such as longshore drift is *not* a landform.
- A geological feature such as a joint in a rock outcrop is *not* a landform.

If you are in any doubt, check with your teacher!

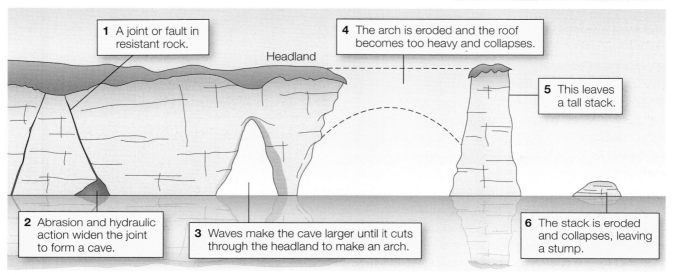

1 A joint or fault in resistant rock.

Headland

4 The arch is eroded and the roof becomes too heavy and collapses.

5 This leaves a tall stack.

2 Abrasion and hydraulic action widen the joint to form a cave.

3 Waves make the cave larger until it cuts through the headland to make an arch.

6 The stack is eroded and collapses, leaving a stump.

C How caves, stacks and arches are formed

A chalk arch, Bwa Gwyn, Anglesey

Bwa Gwyn (Photo **D**) is an impressive arch formed by erosion in an outcrop of white quartzitic rock on the Anglesey coastline. In the past, Bwa Gwyn was quarried for china clay. Today you can still see the grindstone used to extract the clay on the top of the rocks. It is a stunning climb across the rocks, but it can be dangerous because of unstable cliffs.

D Bwa Gwyn arch, Anglesey

Stretch yourself

Find an example of a coastline with headlands and bays. This could be a stretch of coastline near to where you live or one that you have visited.

- Search for a map or satellite photo and add labels to describe the main features.
- Find out about the different types of rock.

Practice question

Use one distinctive coastal landform to illustrate the erosive power of the sea. *(6 marks)*

ACTIVITIES

1 Draw a sequence of diagrams to show the formation of headlands and bays. To test your understanding, draw your coast facing a different direction to diagram **A**.

2 Draw a sequence of labelled diagrams to show how a cliff is undercut by the sea and then collapses to form a wave-cut platform. Use your labels to explain the processes and landforms.

3 Use a sequence of diagrams to explain the formation of a stack (diagram **C**).

On this spread you will find out about the characteristics and formation of coastal deposition landforms

Beaches

Beaches are deposits of sand and shingle (pebbles) at the coast. Sandy beaches are mainly found in sheltered bays (photo **A**). The waves entering the bay are *constructive* waves (see page 93). They have a strong swash and build up the beach.

Not all beaches are made of sand. Much of the south coast of England has pebble beaches. These high-energy environments wash away the finer sand and leave behind the larger pebbles. These come from nearby eroded cliffs or are deposited onshore from vast accumulations out to sea.

Diagram **B** shows the profile of a typical sandy beach. Notice the clear ridges called *berms*. One of these marks the high tide line where seaweed and rubbish get washed up onto the beach.

Sand dunes

At the back of the beach in photo **A**, sand deposited on the beach has been blown inland by onshore winds to form *dunes*. Diagram **C** shows how dunes change in form and appearance the further inland.

A *A sandy beach at Studland Bay, Dorset*

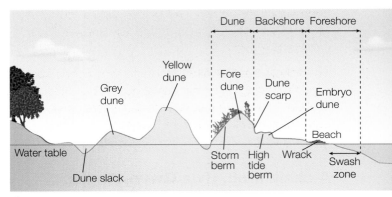

B *Cross section through beach and sand dunes*

> Embryo dunes form around deposited obstacles such as pieces of wood or rocks.

↓

> These develop and become stabilised by vegetation to form *fore dunes* and tall *yellow dunes*. Marram grass is adapted to the windy, exposed conditions and has long roots to find water. These roots help bind the sand together and stabilise the dunes.

↓

> In time, rotting vegetation adds organic matter to the sand making it more fertile. A much greater range of plants colonise these 'back' dunes.

↓

> Wind can form depressions in the sand called *dune slacks*, in which ponds may form.

C *Development of sand dunes*

Saltmarshes and mudflats

Recurved end

Hurst Castle Spit

D *Hurst Castle Spit, Hampshire*

Spits

A **spit** is a long, narrow finger of sand or shingle jutting out into the sea from the land (photo **D**).

Spits form on coasts where there is significant longshore drift. If the coastline changes orientation and bends sharply, sediment is then deposited out to sea (diagram **E**). As it builds up, it starts to form an extension from the land. This process continues with the spit gradually growing further out into the sea. Strong winds or tidal currents can cause the end of the spit to become curved to form a feature called a *recurved end* (photo **D**). There may be a number of recurved ends marking previous positions of the spit.

In the sheltered water behind the spit, deposits of mud have built up. An extensive saltmarsh has formed as vegetation has started to grow in the emerging muddy islands. Saltmarshes are extremely important wildlife habitats and over-wintering grounds for migrating birds.

Bars

Longshore drift may cause a spit to grow right across a bay, trapping a freshwater lake (or *lagoon*) behind it. This feature is called a **bar** (photo **F**).

An offshore bar forms further out to sea. Waves approaching a gently sloping coast deposit sediment due to friction with the seabed. The build-up of sediment offshore causes waves to break at some distance from the coast.

In the UK some offshore bars have been driven onshore by rising sea levels following ice melt at the end of the last glacial period some 8000 years ago. This type of feature is called a *barrier beach*. Chesil Beach in Dorset is one of the best examples of this feature in the UK.

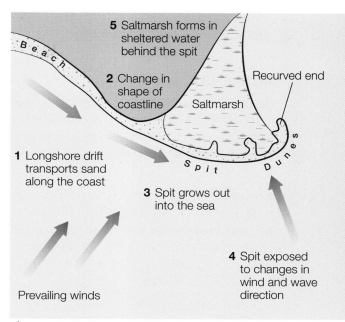

 The formation of a spit

 Bar at Slapton Ley, Devon

ACTIVITIES

1 Describe the processes responsible for the formation of the beach and the sand dunes in photo **A**.

2 Draw a sketch of Hurst Castle Spit. Add labels to describe the characteristic features and the processes responsible for the spit's formation.

3 Describe the characteristics and possible formation of the bar in photo **F**.

Practice question

How do the processes of deposition lead to the formation of distinctive landforms? *(6 marks)*

Stretch yourself

Investigate the characteristics and formation of sand dunes.
- Why do they only form in certain places on the coast?
- Research 'sand dune succession' to find out the sequence of events in the formation of sand dunes.
- What are the characteristics of marram grass and why does it thrive on sand dunes?

On this spread you will find out about coastal erosion and deposition landforms at Swanage, Dorset

Example

Where is Swanage?

Swanage is a seaside town in Dorset on the south coast of England. It is located in a sheltered bay and has a broad sandy beach (photo **B**). This is a classic stretch of coastline with many impressive landforms of coastal erosion and deposition.

Different rock types and geological structure are important in the formation of this coastline. The rocks have been folded and tilted so that bands of different rock types reach the coast. Headlands and bays form where there are alternating bands of more resistant (harder) and less resistant (softer) rocks (map **A**).

Did you know?
The coast around Swanage is part of what is known as the Jurassic Coast. The 154 km stretch of coast in East Devon and Dorset was made a World Heritage Site in 2001 because of its geological importance. Jurassic is the name of the geological period when the rocks on the coast were formed – 145 to 200 million years ago!

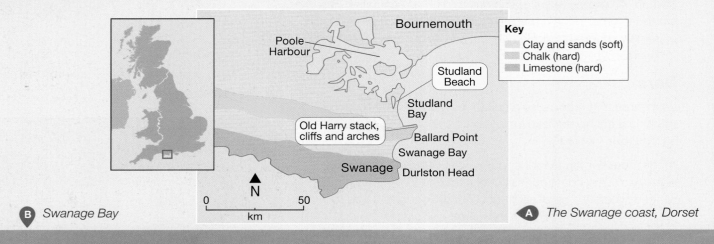

Key
Clay and sands (soft)
Chalk (hard)
Limestone (hard)

Bournemouth
Poole Harbour
Studland Beach
Studland Bay
Old Harry stack, cliffs and arches
Ballard Point
Swanage Bay
Swanage
Durlston Head
N
0   50
km

A The Swanage coast, Dorset

B Swanage Bay

This indented coastline is called a *discordant coastline*. On the south coast there is only one type of rock – limestone. This forms a relatively straight section of coast and is called a *concordant coastline*.

To the north of Swanage is Poole Harbour, one of the UK's largest natural harbours. A great deal of deposition has taken place in this large sheltered bay. You can see two spits at the mouth of the harbour, one on the south side and one on the north.

At Studland there are lagoons, saltmarshes and sand dunes. This area is well known for its wildlife. Photo **C** shows part of the beach and sand dunes at Studland.

c *The beach and sand dunes at Studland*

ACTIVITIES

1 **a** What rock forms the headland at Durlston Head (map **A**)?

   b What type of rock forms Swanage Bay?

   c Explain why headlands and bays have formed along this stretch of coastline.

2 **a** Explain why sediment that has been deposited on the beach in photo **B** (see diagram **D** on page 97).

   b Why has a beach formed in Swanage Bay?

   c Why do you think the beach is popular with visitors?

   d What is the evidence in this photo that the tide is going out?

3 **a** Suggest reasons why sand dunes have formed at the back of the beach in photo **C**.

   b What is the name of the grass growing on the sand dunes?

   c How does this grass help to stabilise the dunes?

   d Describe the characteristics of the barbecue area. Why do you think a specific area has been provided for this purpose?

Stretch yourself

Carry out some further research about Studland.

- What are the main habitats found here?
- Why is it an important area for wildlife?
- Why is it a popular place for visitors?
- Find out how the area is being managed to minimise the harmful effects of visitors.

Practice question

Using evidence from the photos, evaluate any potential conflict between the different uses of the Dorset coast near Swanage and Studland. *(6 marks)*

On this spread you will use map and photo evidence to study landforms of coastal erosion and deposition at Swanage, Dorset

Example

Using the OS map extract and photo

Map **A** is a 1:50 000 extract from an OS map of the Swanage coast. Look back to diagram **A** on page 102 to see how the different rock types form the headlands and bays.

Locate the chalk headland Ballard Point on the map, to the north of Swanage Bay. Photo **B** is an aerial photo of this stretch of coastline. Notice the impressive white chalk cliffs and the many isolated stacks. In the far distance is a well-known local landform, an isolated stack called Old Harry. Photo **C** is a close-up view of Old Harry.

B Aerial view of the coastline between Ballard Point and the Foreland

Old Harry stack

C The Foreland and Old Harry

ACTIVITIES

1 Study map **A** and photo **B**.

  a The Foreland, Peveril Point and Durlston Head are all examples of what coastal landform?

  b In what grid square is the Foreland?

  c In what direction is the photo looking?

  d On the map, what local name is given to the stacks shown in the photo?

  e Describe the characteristics of the chalk cliffs in the photo.

  f Give the six-figure grid reference of Old Harry.

2 a Locate Swanage Bay on the map. Approximately how wide is the bay from Ballard Point to Peveril Point?

  b What map evidence is there that deposition is occurring in Swanage Bay?

  c How does this deposition help to explain the growth of Swanage as a tourist resort?

Practice question

Use evidence from the OS map of the Swanage coast to suggest how the area's human use has been affected by its physical geography. *(4 marks)*

A 1:50 000 OS map extract of Swanage coast

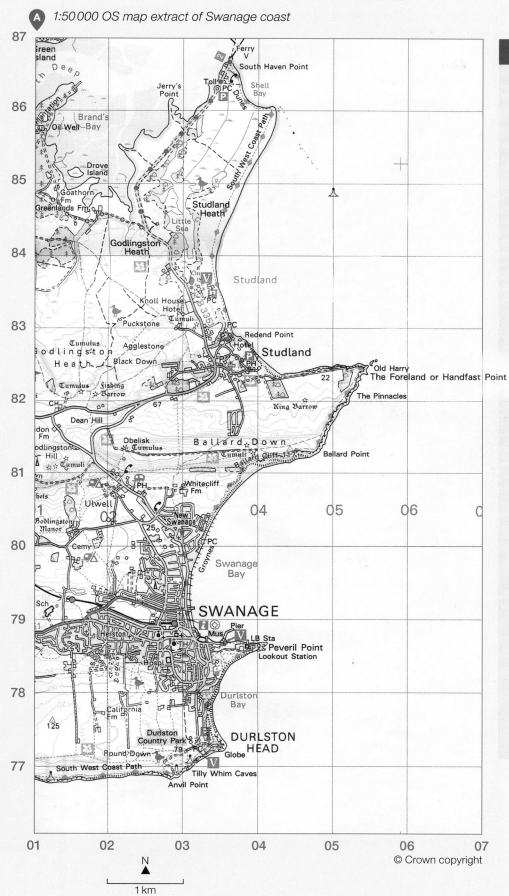

© Crown copyright

N ▲
1 km

ACTIVITIES

3 Study map **A** and photo **C**.

a What are the landforms labelled 1–3 on photo **C**?

b Was the photo taken at high tide or low tide? Explain your answer.

c What additional evidence would you need to confirm that landform 4 is an arch?

d Design an information board to be located on The Foreland to explain the formation of Old Harry.

- Draw a series of annotated diagrams to describe its formation. Refer to the processes of erosion in your annotations.

- Remember that your information board is aimed at the general public so make sure it is clear and attractive.

4 Locate Studland Bay and Studland Heath on map **A**.

a Give the four-figure grid reference for Studland Heath.

b What is the meaning of the blue bird symbol at 033850?

c Describe the different types of natural environment in this area.

d Why is this area popular with visitors?

e Why do you think Studland needs to be managed?

On this spread you will find out how hard engineering can protect coastlines from the effects of physical processes

Why do coasts need to be managed?

Coasts need to be managed to maintain a balance between the forces of nature and the demands of people. People living or working at the coast need to be protected from erosion and flooding. With sea levels expected to rise in the future, coastal defences will become ever more expensive. In some case the increasing costs may outweigh the benefits and coastlines may be left undefended.

What are the coastal management options?

There are three different management strategies for defending the coast.

Hard engineering – using artificial structures such as sea walls to control natural processes

Soft engineering – less intrusive, more environmentally-friendly methods that work with natural processes to protect the coast

Managed retreat – this increasingly popular option enables the controlled retreat of the coastline, often involving allowing the sea to flood over low-lying land

Hard engineering

For centuries people have used hard engineering structures to try to control the actions of the sea and protect property and land. Sea walls, groynes, rock armour and gabions are the most common hard engineering structures used in coastal management.

Sea wall

Description: Concrete or rock barrier against the sea, placed at the foot of cliffs or at the top of a beach. Has a curved face to reflect the waves back into the sea.

Cost: £5000–£10 000 per metre

Advantages:
- Effective at stopping the sea.
- Often has a walkway or promenade for people to walk along.

Disadvantages:
- Can look obtrusive and unnatural.
- Very expensive and high maintenance costs.

 Sea wall at Dawlish, Devon

Groynes

Description: Timber or rock structures built out to sea from the coast. They trap sediment being moved by longshore drift and enlarge the beach. The wider beach acts as a buffer to reduce wave damage.

Cost: Timber groynes £150 000 each (at every 200 m)

Advantages:
- Create a wider beach, which can be popular with tourists.
- Provide useful structures for people interested in fishing.
- Not too expensive.

 Groynes at Eastbourne, Sussex

Disadvantages:
- By interrupting longshore drift they starve beaches further along the coast, often leading to increased rates of erosion elsewhere. The problem is therefore shifted rather than solved.
- Groynes are unnatural and rock groynes in particular can be unattractive.

Nowadays hard engineering approaches are less commonly used because they:

◆ are expensive and involve high maintenance costs

◆ interfere with natural coastal processes and can cause destructive knock-on effects elsewhere – for example, by altering wave patterns erosion can occur further along the coast, leading to new problems such as cliff collapse

◆ look unnatural.

Rock armour

Description: Piles of large boulders dumped at the foot of a cliff. The rocks force waves to break, absorbing their energy and protecting the cliffs. The rocks are usually brought by barge to the coast.

Cost: £200 000 per 100 m

Advantages:
◆ Relatively cheap and easy to maintain.
◆ Can provide interest to the coast.
◆ Often used for fishing.

Disadvantages:
◆ Rocks are usually from other parts of the coastline or even from abroad.
◆ Can be expensive to transport.
◆ Do not fit in with the local geology.
◆ Can be very obtrusive.

C   *Rock armour at Walton on the Naze, Essex*

Gabions

Description: Wire cages filled with rocks that can be built up to support a cliff or provide a buffer against the sea.

Cost: Up to £50 000 per 100 m

Advantages:
◆ Cheap to produce and flexible in the final design.
◆ Can improve drainage of cliffs.
◆ Will eventually become vegetated and merge into the landscape.

Disadvantages:
◆ For a while they look very unattractive.
◆ Cages only last 5–10 years before they rust.

D   *Gabions at Thorpeness, Suffolk*

ACTIVITIES

1   a   Why is a sea wall an example of hard engineering?

    b   What is the purpose of a sea wall?

    c   What are the advantages and disadvantages of a sea wall?

2   Draw a simple diagram to explain how groynes cause a beach to become wider.

3   What are the arguments for and against using gabions as a form of coastal defence?

Stretch yourself

Find out about other options for hard engineering. Consider the following:

• revetments   • offshore breakwaters   • artificial headlands.

What is the cost of construction? Outline the advantages and disadvantages.

Practice question

What are the advantages and disadvantages of hard engineering at the coast? *(6 marks)*

On this spread you will find out how soft engineering can protect coastlines from the effects of physical processes

How does soft engineering protect the coast?

Photo **A** shows **beach nourishment**, one of the most widely used forms of soft engineering. Sand, or in this case shingle, is dredged offshore and transported to the coast by barge. The shingle is then dumped onto the beach and shaped by bulldozers. This is called **reprofiling**. The higher and wider beach now provides greater protection to valuable land and property and creates a natural amenity for tourism and recreation.

Soft engineering approaches such as beach nourishment try to work with natural coastal processes. Photo **B** shows marram grass being replanted to help stabilise sand dunes. This is called **dune regeneration**.

Soft engineering schemes tend to be cheaper than hard engineering although they may require more maintenance. Every few years beaches will need more sand or shingle and sand dunes may need replanting to replace grass that has died or been trampled. However, these schemes are generally more sustainable and are often the preferred option for coastal management today.

A Beach nourishment at Eastbourne, East Sussex

B Sand dune regeneration at Calgary Bay, Mull, Scotland

Beach nourishment

Description: The addition of sand or shingle to an existing beach to make it higher or wider. The sediment is usually obtained offshore locally so that it blends in with the existing beach material. It is usually transported onshore by barge.

Cost: Up to £500 000 per 100 m

Advantages:
- Relatively cheap and easy to maintain.
- Blends in with existing beach.
- Increases tourist potential by creating a bigger beach.

Disadvantages:
- Needs constant maintenance unless structures are built to retain the beach.

Dune regeneration

Description: Sand dunes are effective buffers to the sea but are easily damaged and destroyed by trampling. Marram grass can be planted to stabilise dunes and help them to develop. Fences can be used to keep people off newly-planted areas.

Cost: Cost: £200–£2000 per 100 m

Advantages:
◆ Maintains a natural coastal environment that is popular with people and wildlife.
◆ Relatively cheap.

Disadvantages:
◆ Time-consuming to plant the marram grass and fence areas off.
◆ People don't always respond well to being prohibited from accessing planted areas.
◆ Can be damaged by storms

C *Dune regeneration at Chichester, West Sussex*

Dune fencing

Description: Fences are constructed on a sandy beach along the seaward face of existing dunes to encourage new dune formation. These new dunes help to protect the existing dunes.

Cost: £400–£2000 per 100 m.

Advantages:
◆ Minimal impact on natural systems.
◆ Can control public access to protect other ecosystems.

Disadvantages:
◆ Can be unsightly especially if fences become broken.
◆ Regular maintenance needed especially after storms.

D *Dune fencing at Formby, Merseyside*

ACTIVITIES

1 **a** Describe what is happening in photo **A**.

b Why do you think beach nourishment has been chosen to help defend the coastline at Eastbourne?

c What other forms of coastal defence have been installed here and what is their purpose?

d What are the disadvantages of beach nourishment?

2 **a** Why do you think the area of sand dunes in photo **B** needs to be restored?

b Apart from planting marram grass, what other forms of management will be needed to restore these dunes?

3 Why do you think there is a wide price range for each of the forms of soft engineering?

4 Suggest why *either* hard engineering *or* soft engineering is the best option for defending the coast.

Stretch yourself

Find out more about sand dune regeneration.

• Try to find an example of sand dunes that have had to be regenerated (restored).

• What caused the problems and what solutions have been adopted?

• What are the challenges and opportunities for the future?

Practice question

Identify the differences between hard and soft engineering coastal management strategies.
(4 marks)

Managing coasts – managed retreat

On this spread you will find out how managed retreat can protect coastlines from the effects of physical processes

Managed retreat

Managed retreat is a deliberate policy of allowing the sea to flood or erode an area of relatively low-value land. It is a form of soft engineering as it allows natural processes to take place and does not intervene in the way that hard engineering does.

In the long term, allowing managed retreat is a more sustainable option than spending large sums of money trying to protect the coast with sea walls or groynes. As sea levels continue to rise, managed retreat seems likely to become an increasingly popular choice for managing the coastline.

A *The breach of the sea defences at Medmerry*

Medmerry Managed Retreat, near Chichester, West Sussex

Aerial photo **B** shows a stretch of coastline on the south coast of England near Chichester. This flat, low-lying coast is mainly used for farming and caravan parks. For many years the land was protected by a low sea wall but this is now in need of repair. Building a new sea wall to protect the area against future sea-level rise was a very expensive option.

Given the relatively low value of the land, it was decided to allow the sea to breach the current sea defences (photo **A**) and flood some of the farmland that was previously protected. You can see in the photo how this has happened .

The Medmerry scheme cost £28 million and the controlled breaching of the old sea defences took place in November 2013. In the future, this scheme will:

◆ create a large natural saltmarsh to form a natural buffer to the sea

◆ help to protect the surrounding farmland and caravan parks from flooding

◆ establish a valuable wildlife habitat and encourage visitors to the area.

B *Managed retreat at Medmerry, West Sussex*

You can see on photo **B** that embankments have been constructed inland to give protection to farmland, roads and settlements. This alteration of the coastline is called *coastal realignment*.

Coastal monitoring and adaptation

Much of the coastline of the UK does not require expensive intervention in the form of coastal defences. Land may be low-value farmland, forest or moorland. In many cases these coastal zones can be left alone – this is sometimes called the 'Do Nothing' approach. People living or working in these areas have to adapt by relocating further inland. This might involve moving mobile homes on a holiday park, a path, a fence or a hole on a golf course (photo **C**).

Scientists conduct monitoring of these stretches of coastline. This helps to reduce the possibilities of conflict between managing the coast and the needs and views of local people whose lives are affected. This monitoring involves studying marine processes, mass movement and human activity to ensure safety and to make sure this approach remains the most appropriate. If conditions change, for example the risk of flooding increases and threatens property, then a new approach might be adopted.

Another view

Some experts argue that plans for managed retreat strategies may not take into account the impact on coastal communities. There may be longer-term effects on coastal trade, tourism, infrastructure and businesses, as well as rehousing costs.

C *Manage or adapt?*

ACTIVITIES

1 Why is managed retreat a sustainable option for coastal management?

2 **a** Describe the relief of the area shown in photo **B**.

   b What are the main land uses at X and Y?

   c What is the purpose of the feature at Z?

   d What are the advantages and disadvantages of this scheme?

3 Do you think the stretch of coast in photo **C** should be protected or should people adapt to the natural changes taking place? Justify your answer.

Practice question

Examine why a system of managed retreat may not be a feasible option in some parts of the coast. *(6 marks)*

Stretch yourself

Carry out your own research to find another example of managed retreat.

- What were the pre-existing forms of coastal defence and why has managed retreat now been adopted?
- Assess the advantages and disadvantages of your chosen scheme.
- What are the challenges for the future?

Example

On this spread you will find out about the coastal management schemes at Lyme Regis in Dorset

Lyme Regis is a small coastal town on the south coast of England. It lies at the heart of the World Heritage Site known as the Jurassic Coast. This is one of the most spectacular stretches of coastline in the UK and famous for its fossils. The town is a popular tourist destination. In summer, the population of the town swells from 4000 to 15 000!

What are the issues at Lyme Regis?

Much of the town has been built on unstable cliffs. The coastline is eroding more rapidly than any in Europe due to the powerful waves from the south west. Many properties have been destroyed or damaged, and there has been considerable erosion of the foreshore. The sea walls have been breached many times.

How has the coastline been managed?

The Lyme Regis Environmental Improvement Scheme was set up by West Dorset District Council in the early 1990s. Its aims were to provide long-term coastal protection and reduce the threat of landslips. Engineering works were completed in 2015.

To reduce conflicts between different interest groups, such as property owners, fishermen and environmentalists, there were consultation meetings and the public were kept informed before and during the construction work.

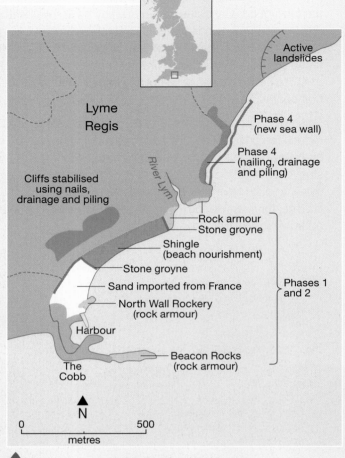

A *Coastal management at Lyme Regis*

Phase 1 Date: 1990s (completed 1995)

- New sea wall and promenade constructed to the east of the mouth of the River Lim.
- In the winter of 2003–2004 a £1.4 million emergency project was completed to stabilise the cliffs. Hundreds of large nails were used to hold the rocks together as well as improving drainage and re-profiling the slope of the beach.

Phase 2 Date: 2005–2007

Extensive improvements made to the sea front costing £22 million. These included:

- construction of new sea walls and promenades
- creation of a wide sand and shingle beach to absorb wave energy and increase use of the shore; shingle dredged from the English Channel and sand imported from France
- extension of rock armour at The Cobb (map **A**) and the eastern end of the sea front, to absorb wave energy and help retain the new beach.

Phase 3 Not undertaken.

The initial plan to help prevent landslips and coastal erosion to the west of The Cobb were shelved. It was decided to leave this stretch of coast alone as the costs outweighed the benefits.

Phase 4 Date: 2013–2015

This final phase focused on the coast east of the town (photo **B**). It cost £20 million and involved:

- constructing a new 390 m sea wall in front of the existing wall (photo **C**) to provide additional protection
- extensive nailing, piling and drainage to provide cliff stabilisation to protect 480 homes.

B *The Jurassic Coast east of Lyme Regis in 2013 before Phase 4 began*

C *Phase 4 coastal defence works at Lyme Regis*

How successful has the management scheme been?

Positive outcomes ✓	Negative outcomes ✗
• The new beaches have increased visitor numbers and seafront businesses are thriving. • The new defences have stood up to recent stormy winters. • The harbour is now better protected, benefiting boat owners and fishermen.	• Increased visitor numbers have led to conflicts with local people who think traffic congestion and litter have increased. • Some people think the new defences have spoilt the natural coastal landscape. • The new sea wall may interfere with coastal processes and affect neighbouring stretches of coastline, causing conflicts elsewhere. • Stabilising cliffs will prevent landslips that may reveal important fossils – a potential conflict.

ACTIVITIES

1 Complete a table listing the different types of hard and soft engineering used at Lyme Regis (map **A**).

2 Photo **C** shows Phase 4 of the coastal defence work at Lyme Regis.

   a Describe what is happening in the photo.

   b What material has been used to construct the sea wall?

   c Suggest some of the issues associated with carrying out this new defence work.

3 Suggest why both hard and soft engineering have been used to protect the coast at Lyme Regis.

4 How has the management of the coast at Lyme Regis reduced possible conflicts between different groups of people?

Stretch yourself

Investigate the management measures at Lyme Regis. You will find plenty of photos and maps online with 'before and after' images and information about the different measures implemented. Consider how successful these measures have been since 2015.

Practice question

To what extent can the coastal management at Lyme Regis be considered a success? *(6 marks)*

11.1 Changes in rivers and their valleys

On this spread you will find out how rivers and their valleys change with distance downstream

What is a drainage basin?

Diagram **A** shows a typical *drainage basin*, an area of land drained by a river and its tributaries. Make sure you are familiar with the key terms on this diagram, as you will need to remember them.

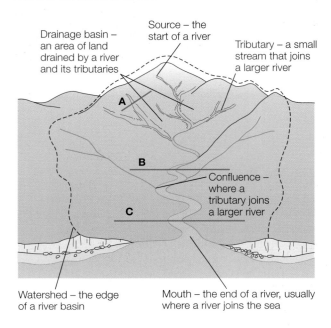

Drainage basin – an area of land drained by a river and its tributaries

Source – the start of a river

Tributary – a small stream that joins a larger river

Confluence – where a tributary joins a larger river

Watershed – the edge of a river basin

Mouth – the end of a river, usually where a river joins the sea

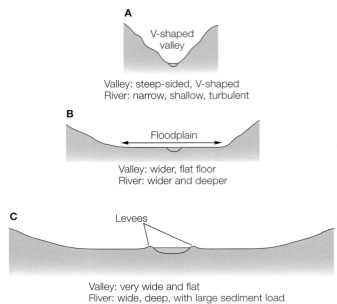

A V-shaped valley

Valley: steep-sided, V-shaped
River: narrow, shallow, turbulent

B Floodplain

Valley: wider, flat floor
River: wider and deeper

C Levees

Valley: very wide and flat
River: wide, deep, with large sediment load

A *Drainage basin*

How does the long profile of a river change downstream?

Imagine that you were on a raft floating down the river in diagram **A**.

◆ In the mountains your speed (velocity) would vary considerably. Where the water is shallow and turbulent there is friction with the bed and banks, slowing the rate of flow. But if you encounter rapids, where the channel narrows and the river becomes deeper, you would move much faster!

◆ Further downstream, the river's channel is much deeper due to the tributaries bringing additional water. Now less water is in contact with the bed and banks and the velocity increases, even though the gradient is less steep than in the mountains. You would now be floating faster!

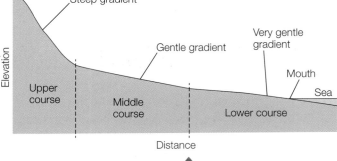

Source

Steep gradient

Very gentle gradient

Gentle gradient

Mouth

Sea

Elevation

Upper course

Middle course

Lower course

Distance

B *The long profile of a river*

If you plotted your journey as a line graph it would look like diagram **B**. This is called the long profile of a river. Notice that the river has a steep gradient in its upper course and a much gentler gradient in its lower course. This concave shape is an ideal profile. In most cases, a river's long profile will vary because, for example, of the river crossing bands of tough and weak rock. A waterfall, for example, creates a step in the long profile of a river.

C The River Tees in County Durham

How does the cross profile of a river and its valley change downstream?

A **cross profile** is an imaginary 'slice' across a river channel and its valley at a particular point. Diagram **A** shows the cross profile of both a river and its valley downstream. The river channel becomes wider and deeper, with the river valley becoming wider and flatter. Its sides are less steep compared with its V-shaped appearance further upstream.

In reality there will be variations in places. For example, river management can alter the shape of a river channel, and different types of rock or human activities such as quarrying can affect the cross profile of a valley.

These changes downstream are due to the amount of water flowing in the river. As tributaries bring water from other parts of the drainage basin the river becomes bigger. With more water and more energy it is able to erode its channel, making it wider and deeper.

The changes to the valley cross profile are mainly due to channel erosion broadening and flattening the base of the valley. Together with weathering and mass movement, these processes make the sides of the valley less steep.

Did you know?

The UK's longest river is the River Severn. It is 354 km in length compared to the River Thames which is 346 km.

ACTIVITIES

1 a Copy the long profile (diagram **B**).

   b Locate the three cross profiles shown in diagram **A** on your diagram. Draw each cross profile and add labels to describe the valley and the river.

   c Describe how a river and its valley change with distance downstream.

2 Photo **C** shows the River Tees in County Durham.

   a Describe the river. Comment on its width, depth and type of flow (turbulent or smooth).

   b Describe the shape of the valley.

   c Suggest where in the long profile of the river this photo was taken. Explain your answer.

Stretch yourself

Investigate the changes in a river close to your home or school. Use a map or photos to show how the river and its valley change with distance downstream.

Practice question

Describe how the shape of a river valley changes downstream. *(4 marks)*

On this spread you will find out how rivers erode, transport and deposit material

What are the processes of erosion?

Photo **A** shows a small river on Exmoor in south west England. There is very little water in the river and very little is happening! It is like this for much of the year. The river is using all its energy to overcome friction and just transport water downstream.

It is only after heavy rainfall that the river has enough energy to erode and enlarge its channel and the river valley. It is possible to identify two types of erosion:

◆ **vertical erosion** (downwards)

◆ **lateral erosion** (sideways).

These combine to cause the downstream changes to the river channel and the river valley described on pages 114–15.

A *An Exmoor river at low flow during the late summer*

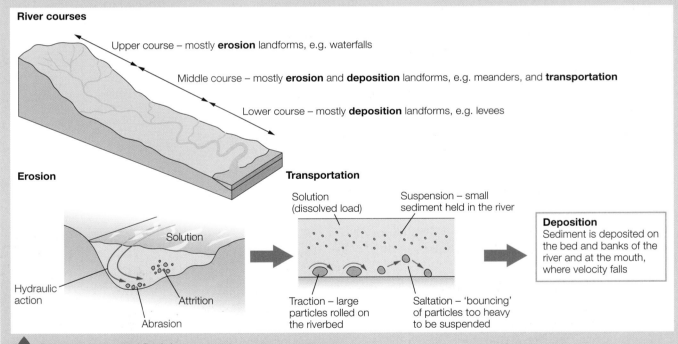

River courses

Upper course – mostly **erosion** landforms, e.g. waterfalls

Middle course – mostly **erosion** and **deposition** landforms, e.g. meanders, and **transportation**

Lower course – mostly **deposition** landforms, e.g. levees

Erosion

Solution

Hydraulic action

Abrasion

Attrition

Transportation

Solution (dissolved load)

Suspension – small sediment held in the river

Traction – large particles rolled on the riverbed

Saltation – 'bouncing' of particles too heavy to be suspended

Deposition
Sediment is deposited on the bed and banks of the river and at the mouth, where velocity falls

B *River courses and fluvial processes*

Diagram **B** shows the four processes of erosion that take place in a river:

◆ **Hydraulic action** – the force of the water hitting the river bed and banks. This is most effective when the water is moving fast and when there is a lot of it.

◆ **Abrasion** – when the load carried by the river repeatedly hits the bed or banks dislodging particles into the flow of the river.

◆ **Attrition** – when stones carried by the river knock against each other, gradually making the stones smaller and more rounded.

◆ **Solution** – when the river flows over limestone or chalk, the rock is slowly dissolved. This is because it is soluble in mildly acidic river water.

What are the processes of transportation?

The material transported by a river is called its *load*. Diagram **B** shows the four main types of *transportation* that occur in a river:

◆ **traction**

◆ **saltation**

◆ **suspension**

◆ **solution**.

The size and total amount of load that can be carried will depend on the river's rate of flow – its *velocity*. After a rainstorm rivers often look very muddy because they are flowing fast and transporting a large amount of sediment (photo **C**). At low flow, when rivers are clear, very little sediment is being transported (photo **A**).

C *A river in high flow*

When does deposition take place?

Deposition occurs when the velocity of a river decreases. It no longer has enough energy to transport its sediment so it is deposited.

◆ Larger rocks tend to be deposited in the upper course of a river. They are only transported for very short distances, mostly by *traction*, during periods of very high flow.

◆ Finer sediment is carried further downstream, mostly held in *suspension*. This material will be deposited on the river bed or banks, where velocity is slowed by *friction*.

◆ A large amount of deposition occurs at the river mouth, where the interaction with tides, along with the very gentle gradient, greatly reduces the river's velocity.

ACTIVITIES

1 **a** What is the evidence in photo **A** that this river is experiencing low flow conditions?

 b Do you think the river is transporting any load? Explain your answer.

 c What evidence is there that erosion and deposition take place in this river?

 d Under what conditions would you expect active erosion to take place?

2 Use diagram **B** to draw a labelled diagram describing the processes of river erosion.

3 How do the size of the sediment and the velocity of the river affect the processes of river transportation?

4 Where and when does deposition take place in a river?

Stretch yourself

1 Investigate how velocity affects the processes of erosion, transportation and deposition.

2 Find out about the Hjulstrom Curve and make a simple copy of the graph. Add annotations to describe what it shows.

Practice question

To what extent is the size and shape of a river valley the result of the work of the river under flood conditions? *(6 marks)*

On this spread you will find out how rivers erode their valleys to form distinctive landforms

What are the distinctive river landforms?

Diagram **A** shows a typical river from source to mouth and its distinctive landforms.

◆ In the river's upper course, erosion dominates to form landforms such as *interlocking spurs*, *waterfalls* and *gorges*.

◆ Further downstream, erosion and deposition combine to form *meanders* and *ox-bow lakes*.

◆ As the river nears the sea, deposition dominates to form a *floodplain*, *levees* and the river estuary. You need to be able to recognise these features and describe how they form.

Of course, not all rivers are 'typical' and it's possible to find landforms of erosion and deposition at various points along the course of a river.

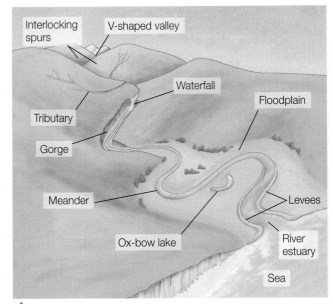

A *River landforms from source to mouth*

Distinctive river erosion landforms

Interlocking spurs

Notice in photo **B** how the Welsh mountain stream weaves its way through the V-shaped valley and around the 'fingers' of land that jut out. These are called *interlocking spurs*. The river is near its source, and is not powerful enough to cut through the 'spurs' of land, so has to flow around them.

B *Blaen Taff Fawr mountain stream, Wales*

Waterfalls

As it makes its way from source to mouth a river often flows over a variety of different rock types. Tougher, more resistant rocks are less easily eroded than weaker rocks and they will form 'steps' in the long profile of a river. These steps form waterfalls (diagram **C**).

Waterfalls are most commonly formed when a river flows over a relatively resistant band of hard rock. When the river plunges over a waterfall it forms a deep and turbulent *plunge pool*. Here the processes of erosion, particularly hydraulic action and abrasion, are active and they combine to undercut the waterfall. Eventually the overhanging rock collapses and the waterfall retreats upstream. Over many years the retreating waterfall will leave behind a steep-sided gorge (diagram **A**).

Waterfalls can also form when a drop in sea level causes a river to cut down into its bed creating a step in the long profile of a river. This step is called a *knick point* and it is marked by the presence of a waterfall. Waterfalls can also be found in glacial hanging valleys (see page 135).

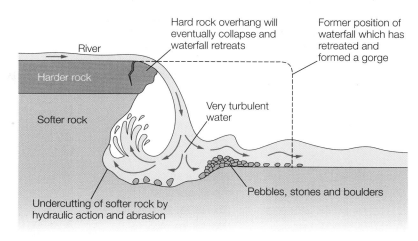

C Formation of a waterfall

Gorges

A gorge is a narrow steep-sided valley that is usually found immediately downstream of a waterfall. It is formed by the gradual retreat of a waterfall over hundreds or even thousands of years (diagram **D**).

Gorges may sometimes form in other ways. At the end of the last glacial period, around 8000 years ago, huge quantities of water from melting glaciers poured off upland areas to form gorges such as Cheddar Gorge in Somerset. More rarely, some gorges form on limestone as a result of the collapse of underground caverns.

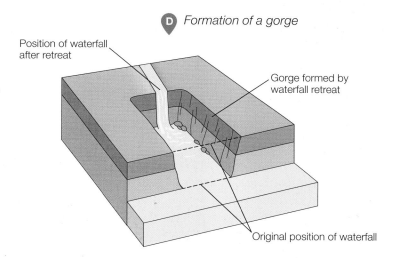

D Formation of a gorge

ACTIVITIES

1 Draw a sketch of the river and its valley in Photo **B**. Label the interlocking spurs and the V-shaped valley. Add labels to describe the valley sides and the river channel. Is it high or low flow?

2 Make a copy of diagram **C**. Add another diagram to show what happens when the overhanging rock collapses.

3 With the aid of diagrams explain how a gorge is formed as a waterfall retreats.

Stretch yourself

1 Search online for a photo to show each of the three landforms described on this spread. They should be examples in the UK.

2 Add detailed labels to describe the main characteristics of each landform. (Don't use a photo of the High Force waterfall on the River Tees, as that appears later in the chapter!)

Practice question

Explain why a waterfall is only a temporary feature on a river's course. *(4 marks)*

On this spread you will find out about river landforms created by deposition and erosion

River landforms

Meanders

Meanders are the wide bends of a river found mainly in lowland areas (photo **A**). They are the most efficient channel for a heavily-laden river as it flows over fine sediment on very gentle slopes. Meanders are constantly changing their shape and position. This is a result of the processes of **lateral (sideways) erosion** and deposition in the river channel.

Diagram **B** shows the main features and processes taking place in a meandering river. The *thalweg* is the line of fastest flow (velocity) within the river. It swings from side to side causing erosion on the *outside* bend and deposition on the *inside* bend. Over time this process of erosion and deposition causes meanders to migrate across the valley floor.

A *Meanders on the River Cuckmere, East Sussex*

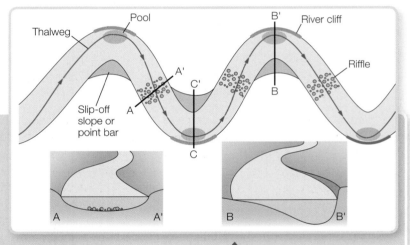

B *Processes and landforms of a meandering river*

Pools and riffles

Meandering streams carrying coarse sediment may develop alternating deep sections (called *pools*) and shallow sections (called *riffles*). Pools are usually found on the outside bends of meanders where, during periods of high flow, the faster flow erodes a deep channel.

Riffles result from the deposition of coarse sediment, also at times of high flow, and are characterised by more turbulent slower-flowing water. During low-flow conditions, however, water tends to flow more slowly through a pool section, depositing fine muddy sediment. Under these low-flow conditions, water may flow slightly faster in a riffle section, accounting for the lack of fine sediment here. This is what you are most likely to see when conducting fieldwork.

Ox-bow lakes

Over time, as meanders migrate across the valley floor, they may start to erode towards each other (diagram **C**). Gradually the neck of the meander narrows until it is completely broken through (usually during a flood) to form a new straighter channel. The old meander loop is cut off by deposition to form an ox-bow lake.

1 The neck of the meander is gradually eroded.

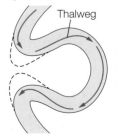

2 Water now takes the shortest (steepest) route.

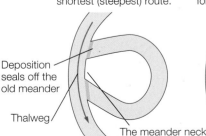

Deposition seals off the old meander

The meander neck is cut through completely

3 Meander is cut off, forming an ox-bow lake.

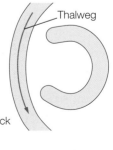

C *Stages in the formation of an ox-bow lake*

River deposition landforms

Floodplains and levees

A floodplain is a wide, flat area of marshy land on either side of a river, and found in the middle and lower courses. Floodplains are made of *alluvium*, a sediment (*silt*) deposited by a river when it floods. Floodplains are used for farming as the soils are very fertile.

There are two processes responsible for the formation of a floodplain (diagram **D**).

◆ Meanders migrate across the floodplain due to lateral erosion. When they reach the edge of the floodplain they erode the valley side (bluff). This explains why floodplains are very wide.

◆ When the river floods it deposits silt, creating a very flat floodplain. Layer upon layer builds up over many years to form a thick deposit of fertile alluvium.

Floodplain
Bluff (the edge of the floodplain)
Meander erodes laterally, widening the floodplain
Levees (raised river banks)
Layers of alluvium deposited during floods
Fine sediment
Coarse sediment
Velocity slows and deposition takes place on river banks
Flooding
Levees

D *The formation of floodplains and levees*

A levee is a raised river bed (*levé* in French means 'rise') found alongside a river in its lower course (diagram **D**). It is formed by flooding over many years. A ridge of sediment is deposited naturally to build up the levee.

During low flow conditions deposition takes place, raising the river bed and reducing the capacity of the channel. When flooding occurs, water flows over the sides of the channel. Here the velocity of the river decreases rapidly leading to deposition of sediment on the river banks. First the coarser sands are deposited and then the finer silt and mud. Gradually after many floods the height of the banks can be raised by as much as two metres.

Estuaries

In the UK most river mouths form wide tidal estuaries, especially in areas where sea levels have risen. Estuaries are *transitional zones* between river and coastal environments and are affected by wave action as well as river processes. The main process operating in estuaries is deposition. During a rising tide river water is unable to be discharged into the sea. The river's velocity falls and sediment is deposited. At low tide these fine deposits form extensive *mudflats*. Over time, mudflats develop into important natural habitats called *saltmarshes*.

ACTIVITIES

1   **a**   Sketch a cross-section of the meander C–C' in diagram **B**.

    b   Draw and label the following: thalweg, deposition, lateral erosion, river cliff, slip-off slope (or point bar).

2   Draw a sequence of labelled diagrams to show how an ox-bow lake forms. Make sure you show the importance of both erosion and deposition in this process.

3   Describe with the aid of a diagram how a levee is formed.

Stretch yourself

Search for an aerial photograph of a floodplain in the UK. Make sure it shows meanders, ox-bow lakes, a floodplain and levees.

• Label these features on your photo.

• Describe how the land is used.

Practice question

The gradient of the River Mississippi drops on average, only 10 cm/km for the last 1000 km of its course to the Gulf of Mexico. Consider how this can result in the river changing course.
(4 marks)

River landforms on the River Tees

On this spread you will find out about the erosion and deposition landforms along a stretch of the River Tees in County Durham, in north east England

Example

Where is the River Tees?

The River Tees is an important river in the north east of England. Its source is high in the Pennine Hills near Cross Fell (height 893 m). From there it flows roughly east for around 128 km to reach the North Sea at Middlesbrough (map **A**). Look back to page 95 to locate the River Tees on the atlas map of the UK.

High Force waterfall and gorge

High Force on the River Tees is one of the UK's most impressive waterfalls. It is located close to Forest-in-Teesdale in the river's upper course. The river drops 20 m as a single sheet of water into the foaming and turbulent plunge pool below. It then continues its course through a spectacular gorge.

The waterfall was formed due to a resistant band of igneous rock (cooled volcanic lava) called *dolerite*, which cuts across the river valley. Unable to erode this tougher band of rock, the river has formed a step in the long profile of the river. This has developed over hundreds of years to form High Force waterfall.

The underlying darker rock with horizontal layers (called *beds*) is the *Carboniferous limestone*. The overlying slightly lighter-coloured rock with vertical joints is the dolerite. As the river plunges over the waterfall, it undercuts the weaker limestone forming an overhang. This eventually collapses and the waterfall gradually retreats upstream to form a gorge. You can see the steep side of the gorge to the left of the waterfall in photo **B**.

Meanders, levees and floodplains near Darlington

Map **C** is a 1:50 000 extract of the River Tees south of Darlington (map **A**). Here the river is flowing from west to east over relatively flat and low-lying land. All along this stretch of the River Tees there are good examples of meanders, levees and floodplains.

Locate grid square 3810 on map **C**. Look closely at the river and notice the embankments, or levees, running alongside the river meander. Notice also the extensive white area of the map alongside the river. The lack of brown contour lines in this area tells us that the land is flat. This is the river's floodplain.

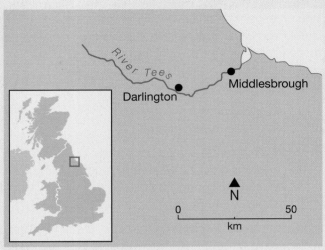

A *The location of the River Tees in north-east England*

B *High Force on the River Tees*

Stretch yourself

Use online mapping to find an aerial satellite photograph of the mouth of the River Tees at Teesport (Middlesbrough).

- Add labels to identify the wide river estuary (what is its width?) and the mudflats.
- Try to identify the industries located alongside the river.
- Why do you think this area has been developed for industries?

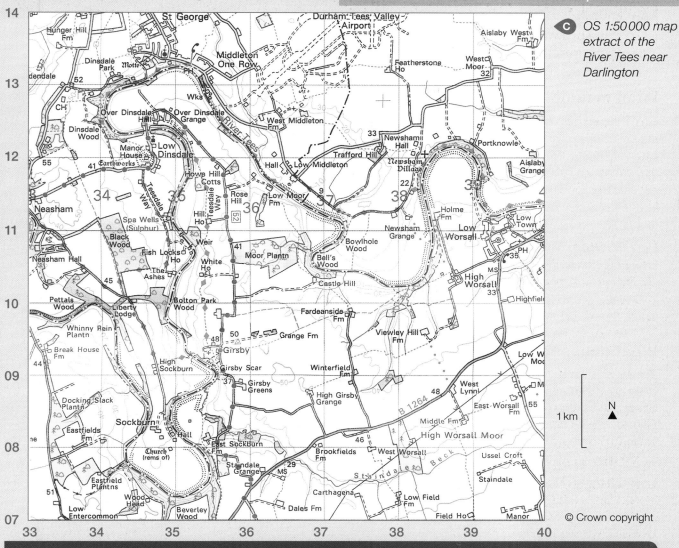

© Crown copyright

ACTIVITIES

1 Draw a sketch of High Force waterfall using photo **B**. Add detailed labels to describe its main characteristics and the fluvial processes that are operating.

2 **a** What is the evidence from photo **B** that the waterfall used to be more extensive in the past?

   b What is the evidence that the waterfall is retreating to leave a gorge?

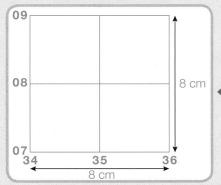

D *Sketch map area for activity 3*

3 Locate Sockburn on map **C** in 3408. Here a sweeping meander passes through four grid squares.

   a Draw an enlarged sketch map of the four grid squares but doubling the scale, so that each square is 4 cm (diagram **D**).

   b Now draw the course of the river and the 20 m contour that runs alongside the river on either side.

   c Draw any levees that have formed on either side of the river.

   d Carefully label the following:
   - the meander
   - the direction of river flow
   - the floodplain
   - levees.

   e How might the course of this meander change in the future? Use a simple sketch to support your answer.

Practice question

To what extent does the River Tees illustrate the features normally associated with a river's course from its source to its mouth? *(6 marks)*

On this spread you will find out about how physical and human factors can increase the risk of flooding

What is flooding?

On 19 November 2009 a remote mountain weather station at Seathwaite in the Lake District recorded an astonishing 314.4 mm of rain in just 24 hours. This was the wettest day ever recorded in the UK. It unleashed a devastating flood that tore through valleys, washing away bridges and inundating the small town of Cockermouth (photo **A**).

Flooding is where land that is not normally underwater becomes inundated. A river **flood** occurs when a river channel can no longer hold the amount of water flowing in it. Water overtops the banks and floods the adjacent land – the floodplain.

What causes flooding?

River floods usually occur after a long period of rainfall, often during the winter. The volume of water steadily increases causing river levels to rise. Eventually the river may overtop its banks to cause a flood.

Sudden floods can occur following torrential storms. These are called *flash floods*. They are more often associated with heavy rainstorms that occur in the summer.

We can identify both physical and human factors that increase *flood risk.*

A   *Cockermouth floods, 2009*

Physical factors

- ◆ **Precipitation** – torrential rainstorms can lead to sudden flash floods as river channels cannot contain the sheer volume of water flowing into them. Steady rainfall over several days can also lead to flooding in lowland river basins.

- ◆ *Geology (rock type) – impermeable rocks* (rocks that do not allow water to pass through them) such as shales and clays encourage water to flow overland and into river channels. This speeds up water flow and makes flooding more likely.

- ◆ *Steep slopes* – in mountain environments steep slopes encourage a rapid transfer of water towards river channels. This increases the risk of flooding.

Human factors (land use)

- ◆ *Urbanisation* – building on a floodplain creates impermeable surfaces such as tarmac roads, concrete driveways and slate roofs. Water is transferred quickly to drains and sewers and then into urban river channels. This rapid movement of water makes flooding more likely.

- ◆ *Deforestation* – much of the water that falls on trees is evaporated or stored temporarily on leaves and branches. Trees also use up water as they grow. When trees are removed much more water is suddenly available and transferred rapidly to river channels, increasing the flood risk.

- ◆ *Agriculture* – in arable farming, soil is left unused and exposed to the elements for periods of time. This can lead to more surface runoff. This is increased if the land is ploughed up and down steep slopes, as water can flow quickly along the furrows.

What is a hydrograph?

The volume of water flowing along a river is its **discharge**. It is measured in *cumecs* – cubic metres per second. A **hydrograph** is a graph that plots river discharge after a storm (graph **B**). It shows how discharge rises after a storm, reaches its peak and then returns to the normal rate of flow.

One of the most important aspects of a hydrograph is the *time lag*. This is the time in hours between the highest rainfall and the highest (peak) discharge. This shows how quickly water is transferred into a river channel and is a key factor in the flood risk. The shorter the time lag the greater the risk of flooding.

B *A flood hydrograph*

What affects the shape of a hydrograph?

The shape of a hydrograph is affected by rainfall and by drainage basin characteristics (table **C**).

C *Factors affecting the shape of a hydrograph*

Drainage basin and precipitation characteristics	'Flashy' hydrograph with a short lag time and high peak	Low, flat hydrograph with a low peak
Basin size	Small basins often lead to a rapid water transfer.	Large basins result in a relatively slow water transfer.
Drainage density	A high density speeds up water transfer.	A low density leads to a slower transfer.
Rock type	Impermeable rocks encourage rapid overland flow.	Permeable rocks encourage a slow transfer by groundwater flow.
Land use	Urbanisation encourages rapid water transfer.	Forests slow down water transfer, because of interception.
Relief	Steep slopes lead to rapid water transfer.	Gentle slopes slow down water transfer.
Soil moisture	Saturated soil results in rapid overland flow.	Dry soil soaks up water and slows down its transfer.
Rainfall intensity	Heavy rain may exceed the infiltration capacity of vegetation, and lead to rapid overland flow.	Light rain will transfer slowly and most will soak into the soil.

ACTIVITIES

1 Describe the effects of the flooding in Cockermouth (photo **A**). Consider the social, economic and environmental impacts.

2 What is the difference between a normal river flood and a flash flood?

3 What features of the urban environment increase the risk of flooding? Give reasons for your answer.

4 What physical and human factors are likely to produce a hydrograph (table **C**) with a short time lag and a high peak?

Stretch yourself

Research online about the Cockermouth flood of 2009.

- What were the main physical and human causes of the flood?
- What were the impacts?
- What has been done since 2009 to reduce the likelihood of future flooding?
- How successful were the post-2009 defences in coping with the extreme rainfall in December 2015?

Practice question

'River flooding is a natural phenomenon.'
To what extent do you consider this statement to be correct? *(6 marks)*

On this spread you will find out about the costs and benefits of hard engineering to manage river flooding

What is hard engineering?

Hard engineering involves using man-made structures to prevent or control natural processes from taking place. This form of flood management is usually very expensive – individual projects can cost several million pounds. But this is the preferred option for protecting expensive property or land, such as housing estates, railways and water treatment works. The costs have to be weighed against the benefits.

◆ *Costs* – the financial cost of the scheme, and any negative impacts on the environment and on people's lives

◆ *Benefits* – financial savings made by preventing flooding, along with any environmental improvements

Diagram **A** shows a drainage basin with hard and soft engineering options.

Dams and reservoirs

Dams and reservoirs are widely used around the world to regulate river flow and reduce the risk of flooding. Most dam projects are multi-purpose, having several functions, for example:

◆ flood prevention   ◆ hydro-electric power generation

◆ irrigation   ◆ recreation.

◆ water supply

Dams can be very effective in regulating water flow. During periods of high rainfall, water can be stored in the reservoir. It can then be released when rainfall is low. But the construction of dams can be very controversial. They cost huge amounts of money and the reservoir often floods large areas of land. Many people may have to be moved from their homes.

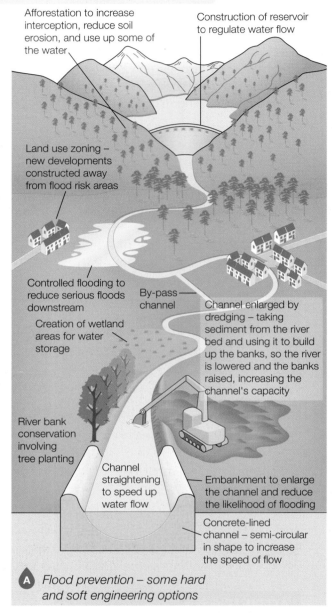

Afforestation to increase interception, reduce soil erosion, and use up some of the water

Construction of reservoir to regulate water flow

Land use zoning – new developments constructed away from flood risk areas

Controlled flooding to reduce serious floods downstream

By-pass channel

Channel enlarged by dredging – taking sediment from the river bed and using it to build up the banks, so the river is lowered and the banks raised, increasing the channel's capacity

Creation of wetland areas for water storage

River bank conservation involving tree planting

Channel straightening to speed up water flow

Embankment to enlarge the channel and reduce the likelihood of flooding

Concrete-lined channel – semi-circular in shape to increase the speed of flow

A *Flood prevention – some hard and soft engineering options*

Clywedog reservoir, Llanidloes, Wales

The Clywedog reservoir (photo **B**) was constructed in the 1960s to help prevent flooding of the River Severn. Its concrete dam is over 70m high and 230m wide and the reservoir stretches for nearly 10km. It has been in continuous use since 1967, filling up in the winter and gradually releasing water in the summer to retain a constant flow. Although some flooding has continued to affect settlements further downstream, Clywedog has undoubtedly prevented catastrophic floods.

B *The Clywedog dam and reservoir*

Channel straightening

River straightening involves cutting through meanders to create a straight channel. This speeds up the flow of water along the river. Whilst river straightening may protect a vulnerable location from flooding, it may increase the flood risk further downstream. The problem is not really solved but shifted somewhere else!

In some places straightened sections of river are lined with concrete. This speeds up the flow and prevents the banks from collapsing, which can cause the channel to silt up. But the concrete channels create a very unattractive and unnatural river environment and can damage wildlife habitats.

Embankments

An embankment is a raised riverbank. Raising the level of a riverbank allows the river channel to hold more water before flooding occurs.

Hard engineering structures involving concrete walls or blocks of stone are frequently used in towns or cities to prevent flooding of valuable property. Sometimes mud dredged from the river may be used. This is cheaper and more sustainable and looks more natural.

Flood relief channels

A flood relief channel is a man-made river channel constructed to by-pass an urban area.

During times of high flow, sluice gates can be opened to allow excess water to flow away into the flood relief channel and reduce the threat of flooding.

The Jubilee River, Maidenhead

In the UK a flood relief channel, named the Jubilee River, has been constructed on the River Thames near Maidenhead in Berkshire (map **C**). The 11 km channel was opened in 2002. It cost £110 million to construct and with a length of nearly 12 km is the longest man-made channel in the UK. As well as reducing the risk of flooding for over 3000 properties, the Jubilee River has had a positive impact on the environment by creating new wetlands. It is also popular for recreational activities such as walking and fishing.

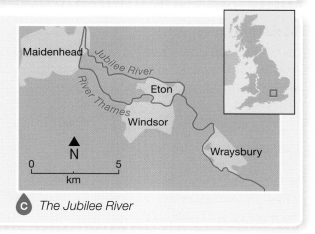

C *The Jubilee River*

ACTIVITIES

1 Draw a diagram in the style of diagram **A** to illustrate the different types of hard engineering described on this spread.

2 Consider the costs (disadvantages) and benefits (advantages) of dams and reservoirs such as at Clywedog.

3 Construct a summary table to describe the costs and benefits of the following hard engineering options:
   - channel straightening
   - embankments
   - flood relief channels.

Stretch yourself

Search online for more information about the Jubilee River flood relief channel.
- Why was it built? (Had there been some serious floods in the past?)
- What have been the environmental and social benefits of the flood relief channel?
- Try to assess the costs and benefits of the Jubilee River.

Practice question

To what extent are hard engineering schemes sustainable? *(6 marks)*

On this spread you will find out about the costs and benefits of managing river flooding using soft engineering

What is soft engineering?

Soft engineering involves working with natural river processes to manage the flood risk. Unlike hard engineering it does not involve building artificial structures or trying to stop natural processes. It aims to reduce and slow the movement of water into a river channel to help prevent flooding. In common with all forms of management there are costs (disadvantages) and benefits (advantages).

Planting trees to establish a woodland or forest is called *afforestation*. Trees obstruct the flow of water and slow down the transfer to river channels. Water is soaked up by the trees or evaporated from leaves and branches. Tree planting is relatively cheap and has environmental benefits.

Wetlands and flood storage areas

Wetland environments on river floodplains are very efficient in storing water (photo **A**). Wetlands are deliberately allowed to flood to form flood storage areas. Water can be stored to reduce the risk of flooding further downstream.

A *Flood storage area, near Rye, East Sussex*

Floodplain zoning

Floodplain zoning restricts different land uses to certain locations on the floodplain (diagram **B**). Areas close to the river and at risk from flooding can be kept clear of high-value land uses such as housing and industry. Instead these areas can be used for pasture, parkland or playing fields. Floodplain zoning can reduce overall losses caused by flood damage. But it can be difficult to implement on floodplains that have already been developed and can cause land prices to fall.

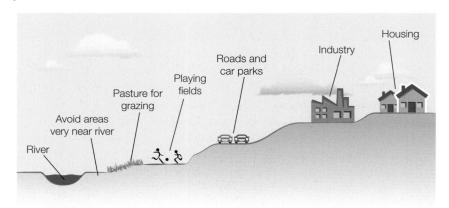

B *Floodplain zoning*

River restoration

Where the course of a river has been changed artificially, river restoration can return it to its original course . River restoration uses the natural processes and features of a river, such as meanders and wetlands to slow down river flow and reduce the likelihood of a major flood downstream (photo **C**).

C *Restoration of the River Glaven, Norfolk*

Preparing for floods

Rivers and river basins are monitored remotely using satellites and computer technology. Instruments are used to measure rainfall and to check river levels. Computer models can then be used to predict discharges and identify areas at risk from flooding.

In England and Wales the Environment Agency issues **flood warnings** if flooding is likely. Warnings are sent to the emergency services and the public using social media, text and email. There are three levels of warning:

◆ *Flood Watch* – flooding of low-lying land and roads is expected. People should be prepared and watch river levels.

◆ *Flood Warning* – there is a threat to homes and businesses. People should move items of value to upper floors and to turn off electricity and water.

◆ *Severe Flood Warning* – extreme danger to life and property is expected. People should stay in an upper level of their home or leave the property.

The Environment Agency makes maps identifying areas at risk from flooding. People living in these areas are encouraged to plan for floods. This might include:

◆ planning what to do if there is a flood warning (e.g. moving valuable items upstairs)

◆ using flood gates to prevent floodwater from damaging property (photo **D**)

◆ using sandbags to keep floodwater away from buildings.

Local authorities and emergency services use these maps to plan responses to floods. For example, installing temporary flood barriers, evacuating people, closing roads and securing buildings and services.

Flood prediction is based on probability and one of the 'costs' is that places can become blighted by being 'at risk' from flooding. This can cause property values to drop and insurance premiums to increase.

Think about it

Is your town, city or village at risk from flooding? What defences are in place to protect the area from floods?

D *Flood gate protecting property from the rising River Severn, Deerhurst, Gloucestershire*

ACTIVITIES

1 What is the purpose of a flood storage area (photo **A**).

2 **a** What is the evidence in photo **C** that this river channel and its floodplain have been modified?

   b Suggest *three* reasons why these changes may lead to a reduction in the flood risk further downstream.

3 Suggest why some river engineers and local people prefer soft rather than hard engineering schemes.

Stretch yourself

Imagine a builder has submitted a planning application to build new houses on the area labelled 'Playing fields' on diagram **B**. Explain why, as the planner considering the proposal, you have rejected the scheme. Propose a better option.

Practice question

Use an example of one soft engineering river flood management strategy to show how it has a limited effect on the environment. *(4 marks)*

Managing floods at Banbury

On this spread you will find out about the flood management scheme in Banbury

Where is Banbury?

Banbury is located in the Cotswold Hills about 50 km north of Oxford (map **A**). The town has a population of around 45 000 people. Much of the town is on the floodplain of the River Cherwell, a tributary of the River Thames.

How has Banbury been affected by flooding?

Banbury has a history of devastating floods. In 1998, flooding led to the closure of the town's railway station, shut local roads and caused £12.5 million of damage (photo **B**). More than 150 homes and businesses were affected. In 2007, the town was hit again by floods that extended over much of central and western England. Many more homes and businesses were affected as the river burst its banks after very heavy rain.

What has been done to reduce the risk of flooding?

In 2012, Banbury's new flood defence scheme was completed. A 2.9 km earth embankment was built parallel to the M40 motorway to create a flood storage area (map **C**). The embankment has a maximum height of 4.5 m. It is capable of holding around 3 million cubic metres of water – that's 1200 Olympic-size swimming pools!

The flood storage area is located mainly on the natural floodplain of the River Cherwell. It collects rainwater that otherwise would have swelled the river and caused it to burst its banks.

Photo **D** shows one of the two flow control structures in the embankment. The specially designed aperture (opening) controls the rate of flow downstream towards Banbury. Any excess water backs up behind the structure, filling up the reservoir rather than continuing towards Banbury. The design avoids the need to open and close flood gates. Map **C** shows how this works.

A *The location of Banbury*

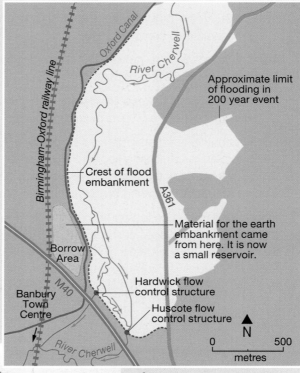

C *Main features of the Banbury Flood Storage Reservoir*

B *Banbury station flooded in 1998*

Additional flood defence measures that are part of the scheme include:

◆ raising the A361 road in the flood storage area (map **C**) plus improvements to drainage beneath the road to prevent flooding

◆ new earth embankments and floodwalls to protect property and businesses, such as the motorsport business Prodrive

◆ a new pumping station to transfer excess rainwater into the river below the town

◆ the creation of a new Biodiversity Action Plan (BAP) habitat with ponds, trees and hedgerows to absorb and store excess water.

What have been the social, economic and environmental costs and benefits?

The table below outlines some of the main costs and benefits associated with the project.

D Flow control structure looking upstream towards the embankment

Social	Economic	Environmental
• The raised A361 route into Banbury will be open during a flood, to avoid disrupting people's lives. • Quality of life for local people is improved with new footpaths and green areas. • Reduced levels of anxiety and depression through fear of flooding.	• The cost of the scheme was about £18.5 million. • Donors included Environment Agency and Cherwell District Council. • By protecting 441 houses and 73 commercial properties, the benefits are estimated to be over £100 million.	• Around 100 000 tonnes of earth were required to build the embankment. This was extracted from nearby, creating a small reservoir (map **C**). • A new Biodiversity Action Plan habitat has been created with ponds, trees and hedgerows. • Part of the floodplain will be deliberately allowed to flood if river levels are high.

ACTIVITIES

1 Use photo **B** to help describe the problems associated with flooding in Banbury.

2 Make a large copy of map **C**. Find photos to show some of the main features of the flood storage area. Add labels or captions to describe what they show.

3 Describe how the flood storage area works and how it is designed to prevent flooding in Banbury.

Practice question

Use the example of Banbury to show how the flood defence scheme benefits both the local people and the environment. *(6 marks)*

Stretch yourself

Work in pairs to list additional social, economic and environmental issues related to the flood defence scheme. Consider the costs and benefits. Research the 'Banbury Flood Alleviation Scheme' to help get you started.

On this spread you will find out about the processes operating in cold glacial environments

What is a glacial environment?

It's hard to believe that the UK once looked like the place in photo **A**! During the last ice age snow and ice covered much of the landscape. Temperatures barely rose above freezing even in the summer! Huge glaciers radiated from the north and west and carved deep glacial valleys and troughs. Further south and east the land was permanently frozen with some meltwater rivers. Map **B** shows the areas that were covered by ice during the last ice age.

Weathering processes in glacial environments

The main weathering process in a cold environment is **freeze-thaw** (see pages 94–5). The amount of liquid water is limited due to the freezing temperatures. In summer water flows into cracks in the rocks and freezes hard the following winter. There is freezing and thawing taking place every day of the year, but the most intense freezing is seasonal.

Freeze-thaw is an important process in glacial environments because;

◆ it helps to shape jagged glacial mountain landscapes.

◆ rocks become weakened by freeze-thaw making it easier for them to be eroded by glaciers.

◆ piles of large angular rocks, called *scree*, collect at the foot of mountains – these become powerful erosion tools when trapped under moving glaciers.

The processes of glacial erosion

Glaciers usually move very slowly – only a few centimetres a year – but they are responsible for an incredible amount of landscape erosion. There are two main types of glacial erosion: **abrasion** and **plucking** (diagram **C**).

◆ **Abrasion** is the 'sandpaper' effect caused by the weight of the ice scouring the valley floor. Abrasion leaves a smooth, polished surface. Scratches (called striations) caused by large rocks beneath the ice can often be seen.

◆ **Plucking** is when meltwater beneath a glacier freezes and bonds the base of glacier to the rocky surface below, rather like glue. As the glacier moves, any loose fragments of rock will be 'plucked' away – like extracting loose teeth! This process of plucking leaves behind a jagged rocky surface.

A *Glacial landscape in British Columbia, Canada*

Did you know?
The volume of water expands by 9 per cent when it freezes!

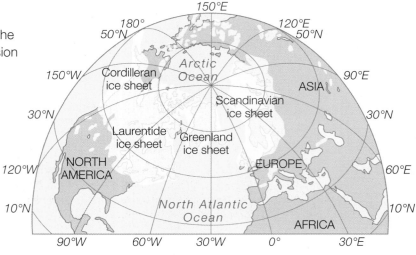

B *Ice coverage of Europe and North America, 18 000 years ago*

How do glaciers move?

In the summer, meltwater lubricates the glacier enabling it to slide downhill. This type of movement, which can be quite sudden, is called *basal slip*. In hollows high up on the valley sides, this movement may be more curved, in which case it is called **rotational slip**.

In the winter, the glacier becomes frozen to the rocky surface. The sheer weight of the ice and the influence of gravity cause individual ice crystals to change shape in a plastic-like way. This process is called *internal deformation* and also causes the glacier to slowly move downhill.

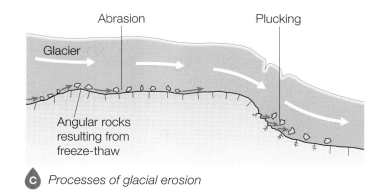

C *Processes of glacial erosion*

How do glaciers transport material?

Rock fragments resulting from freeze-thaw and eroded by the ice are transported by the glacier. This sediment – called **moraine** – can be transported *on* the ice, *in* the ice (buried by snowfall) and *below* the ice.

As the glacier moves forward it pushes loose debris ahead of it effectively transporting it downhill. Not surprisingly, this process is called **bulldozing**.

What causes deposition?

Deposition occurs when the ice melts. As most melting occurs at the front (the *snout*) of a glacier, this is where most deposition takes place.

As a glacier slowly retreats it leaves behind a bed of broken rock fragments called **till** (photo **D**). Due to the lack of water to transport it, till is poorly sorted, with jagged rock fragments of all sizes. Till is also known as *boulder clay*.

Ahead of the glacier, meltwater rivers will carry sediment away. The process of attrition (see page 120) will cause the rock fragments to become smaller and more rounded. Sediment is well sorted, with larger rocks deposited close to the ice and finer material carried many kilometres away. This sandy and gravel material is called **outwash**.

D *Till beneath a glacier in Austria*

E *Outwash stream, Kluane National Park, Canada*

ACTIVITIES

1 **a** Describe the process of freeze-thaw weathering.

   b What is the evidence in photo **A** that this landscape is being affected by freeze-thaw?

2 Copy diagram **C**. Add labels to describe the processes and effects of abrasion and plucking.

3 How do glaciers move?

4 Describe the differences between till and outwash.

Stretch yourself

Find a photo of a *roche moutonnée*. This glacial feature clearly shows the processes of abrasion and plucking. Make a sketch of the photo and add detailed labels to describe its characteristics.

Practice question

Use evidence from photo **D** to show that this is glacial rather than fluvial (river) material.
(4 marks)

On this spread you will find out about the distinctive landforms of glacial erosion

What are the distinctive landforms of glacial erosion?

Ice is a very powerful agent of erosion and can create spectacular landforms in mountainous areas. Photo **A** shows part of the Lake District, a glaciated area in the UK.

Corries

Corries, also known as *cirques* and *cwms*, are large hollowed-out depressions found on the upper slopes of glaciated valleys. They are characterised by a steep back wall and a raised 'lip' at the front. A corrie may contain a lake called a *tarn*.

Diagram **B** describes the formation of a corrie. Snow accumulates in a sheltered hollow on a hillside. *Nivation* (snow-related processes, such as freeze-thaw weathering, meltwater and slumping) enlarges the hollow enabling more snow to collect. Gradually the snow turns to ice and a small corrie glacier is formed. Through rotational slip, the glacier abrades (scoops out) an over-deepened hollow (similar to an ice-cream scoop). Reduced erosion at the front of the corrie, due to the ice being thinner and less erosive, forms a raised lip. Sometimes moraine may be deposited here. A *tarn* (corrie lake) may form in the bottom of the corrie.

A *Red Tarn and Striding Edge in the Lake District, Cumbria*

B *The formation of a corrie*

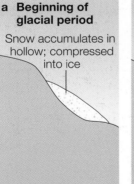

a Beginning of glacial period

Snow accumulates in hollow; compressed into ice

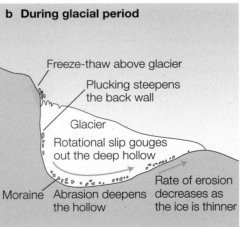

b During glacial period

Freeze-thaw above glacier

Plucking steepens the back wall

Glacier

Rotational slip gouges out the deep hollow

Moraine   Abrasion deepens the hollow

Rate of erosion decreases as the ice is thinner

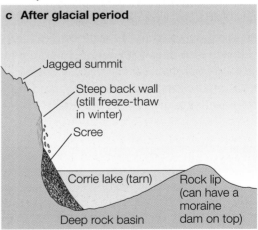

c After glacial period

Jagged summit

Steep back wall (still freeze-thaw in winter)

Scree

Corrie lake (tarn)

Rock lip (can have a moraine dam on top)

Deep rock basin

Arêtes and pyramidal peaks

An arête is a knife-edge ridge often found at the back of a corrie or separating two glaciated valleys (diagram **C**) They are often extremely narrow and popular with hill walkers, although strong winds can make them very dangerous.

A typical arête forms when erosion in two back-to-back corries causes the land in-between to become narrower. If three or more corries have formed on a mountain, erosion may lead to the formation of a single peak rather than a ridge. This feature is called a pyramidal peak.

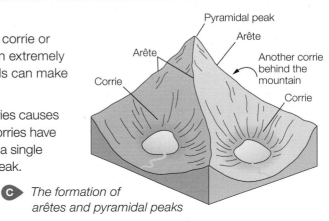

Pyramidal peak

Arête

Arête

Arête

Another corrie behind the mountain

Corrie

Corrie

C *The formation of arêtes and pyramidal peaks*

Glacial valley landforms

Most glaciers flow along pre-existing river valleys. Unable to flow around obstacles, glaciers carve straight courses. Their incredible strength enables them to form dramatic features (diagram **D**):

◆ **Glacial trough** – this is a steep-sided, wide and flat-bottomed valley, formed by abrasion. Most glacial troughs start out as V-shaped river valleys. When the landscape becomes glaciated, individual glaciers occupy the river valleys, eroding them through the process of abrasion to form spectacular U-shaped glacial troughs.

◆ **Truncated spurs** – unable to flow around existing interlocking spurs (see page 118), the glacier cuts straight through them, forming steep-edged truncated spurs.

◆ **Hanging valleys** – these are smaller tributary valleys above the main glacial trough. Smaller glaciers in these valleys were unable to erode down to the same level as the main glacier. Today they are often marked by spectacular waterfalls.

◆ **Ribbon lakes** – these are long, narrow lakes often tens of metres deep. Most ribbon lakes result from severe erosion of the glacial trough. This happens when the ice becomes thicker after a tributary glacier has joined, or where a weaker band of rock has eroded more easily. Sometimes a shallow ribbon lake may form in a glacial trough behind a dam of deposited moraine.

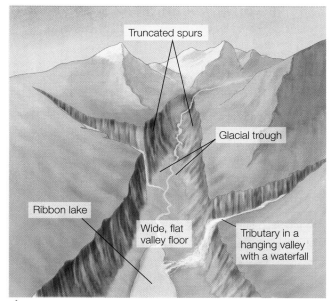

D *Glacial valley landforms*

E *A glacial valley – Nant Ffrancon, Snowdonia*

ACTIVITIES

1 **a** Name the glacial features A–C on photo **A**.
   b What geographical name is given to the lake in the photo? Suggest why a lake has formed here.

2 Draw a series of simple sketches to show how the erosion of two back-to-back corries can form an arête. Add detailed labels to describe the process.

3 Describe the characteristics of a hanging valley (diagram **D**) and explain how it is formed.

Practice question

Use *one or more* examples to show that features formed by glacial erosion dominate areas of highland glaciation. *(6 marks)*

Stretch yourself

Select a photo that shows several glacial landforms in the UK. Label each landform and describe how they were formed.

On this spread you will find out about the distinctive landforms of glacial transportation and deposition

What are the distinctive glacial transportation and deposition landforms?

Glaciers behave like giant conveyor belts carrying weathered and eroded rocks from the mountains to the lowlands. These rocks can be carried *on top of* the ice, *within* the ice (buried by layers of ice and snow) or *beneath* the ice where they carry out the process of abrasion. The term *moraine* is applied to these angular and poorly sorted deposits (see page 133).

When deposited by the melting ice this sediment can form a blanket of material up to several metres thick. Geologists refer to this deposited material as till or *boulder clay* (see page 133), due to the range of sizes of the sediment. Till is found along much of the east coast of England. It forms fertile soils but is weak and easily eroded when exposed on the coast.

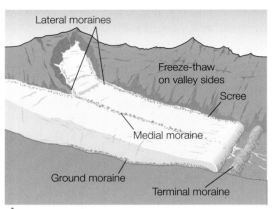

A *Types of moraine*

Moraine

It is possible to identify several types of moraine (diagram **A**):

◆ *Ground moraine* – The material dragged underneath the glacier and left behind when the ice melts. It often forms uneven hilly ground.

◆ *Lateral moraine* – This moraine forms at the edges of the glacier. It is mostly scree material that has fallen off the valley sides due to freeze-thaw weathering. When the ice melts, the moraine forms a low ridge on the valley side.

◆ *Medial moraine* – When a tributary glacier joins the main glacier two lateral moraines will merge to produce a single line of sediment that runs down the centre of the main glacier. On melting, the medial moraine forms a ridge down the centre of the valley. Photo **B** shows medial and lateral moraines on the Mer de Glace glacier in France. Imagine how these will look on the valley floor when the ice finally melts.

◆ *Terminal moraine* – Huge amounts of material pile up at the snout of a glacier to form a high ridge often tens of metres in height across the valley. This is a terminal moraine. It represents the furthest extent of the glacier's advance, hence the name 'terminal'.

As the ice melts and retreats up the valley many of these features are eroded away by meltwater. This explains why there are few such landforms evident in the UK today.

B *Medial and lateral moraines on the Mer de Glace, France*

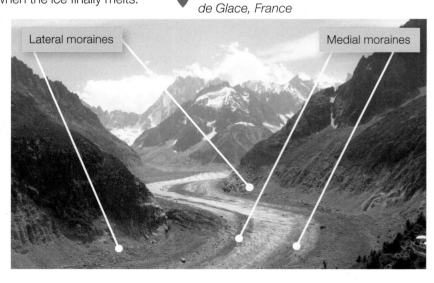

Drumlins

Drumlins are smooth egg-shaped hills about 10 m high and several hundred metres long that are found in clusters on the floor of a glacial trough (figure **C**). They are made of moraine that has been streamlined and shaped by the moving ice. Drumlins usually have a blunt end, which faces up-valley, and a more pointed end facing down-valley. This indicates the direction of movement of a glacier.

Erratics

An erratic is a large boulder that is out of place, resting on a different type of rock. In photo **D** the boulder is completely different to the rock it sits on. So how did it get there?

Scientists have used erratics to trace the history of glaciation in areas like the UK where the ice has long since disappeared. By studying the geology of an area, it is possible to work out where the erratic rocks came from and what route they followed as they were carried by glaciers. Most large erratics were probably transported on or in glaciers.

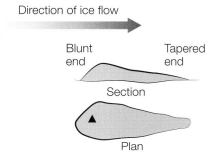

Direction of ice flow

Blunt end — Tapered end

Section

Plan

'Basket of eggs' topography

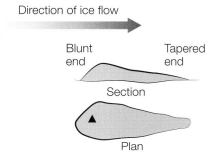

C Drumlins in Swindale, Lake District

ACTIVITIES

1   Make a copy of diagram **A**. Use the text to add detailed labels to describe the characteristics and formation of the different types of moraine.

2   **a**   Why do you think the type of landscape in photo **C** is sometimes called 'basket of eggs topography'?

   b   Can you suggest from the photo which way the ice was moving when it formed these drumlins?

   c   Why do you think it is often hard to interpret the movement of ice when studying drumlins?

3   **a**   What is an erratic?

   b   How do you know that the boulders in photo **D** are erratics?

   c   How can erratics help scientists to piece together the history of a past glacial period?

Stretch yourself

Attempt to draw a sketch based on photo **B** but after the ice has melted.

* Show the moraines as they would appear on the ground surface. (Imagine that meltwater might have eroded part of the features leaving them incomplete.)
* Indicate where you might expect to find drumlins and erratics.

Practice question

Explain how features of glacial deposition can be used as evidence to show the direction of ice movement. *(4 marks)*

D Erratic boulders near Torridon, Scottish Highlands

137

On this spread you will learn how to identify glacial landforms on an OS map

Example

Identifying glacial landforms on an OS map

Cadair Idris has many glacial features such as jagged mountain peaks, narrow ridges, scooped-out corries and bare rocky surfaces. You can identify these features on a 1:50000 OS map. Orientate yourself on photo **A** using the map to help.

◆ The corrie lake (tarn) in the photo is Llyn Cau (grid square 7112).

◆ The ridge in the centre of the photo has the writing 'IDRIS' on the map.

◆ The photo is looking west.

Corries, arêtes and pyramidal peaks

A corrie is shown on the map by a series of semi-circular, horseshoe-shaped contours (brown lines). The very steep sides are indicated by the pattern of bold black lines that you can see clearly in 7112. These black symbols indicate cliffs. Another corrie can be seen south west of Llyn Cau.

The edges of a corrie are marked by arêtes, shown by bold black (cliff) symbols. The arête at the back of the Llyn Cau corrie is called Craig Cau. Look at photo **B** and the map to identify other arêtes.

At the 791 spot height close to 710120 there are three corries back-to-back. This is the intersection of three arêtes – a pyramidal peak. Can you find it on the photo?

Glacial trough

Locate Tal-y-llyn Lake on the map. The lake lies in a wide, flat-bottomed and steep-sided glacial trough (photo **B**).

Truncated spurs

Find the truncated spur on the map at 733106. Note the shape of the contours as they change from semi-circular at the top of the slope, to straighter at the bottom (near the road). This shows where the old interlocking spur has been cut off.

Depositional landforms

Mountain landscapes such as Cadair Idris have few large-scale features of glacial deposition. Features such as drumlins and outwash plains are usually found in lowland areas. But note the black dots on the map in several of the corries. These are rock debris that could be glacial deposits. To the west of Llyn-y-Gadair (7013) you will find some morainic ridges.

A *View of Cadair Idris and Llyn Cau looking west*

B *Tal-y-llyn Lake*

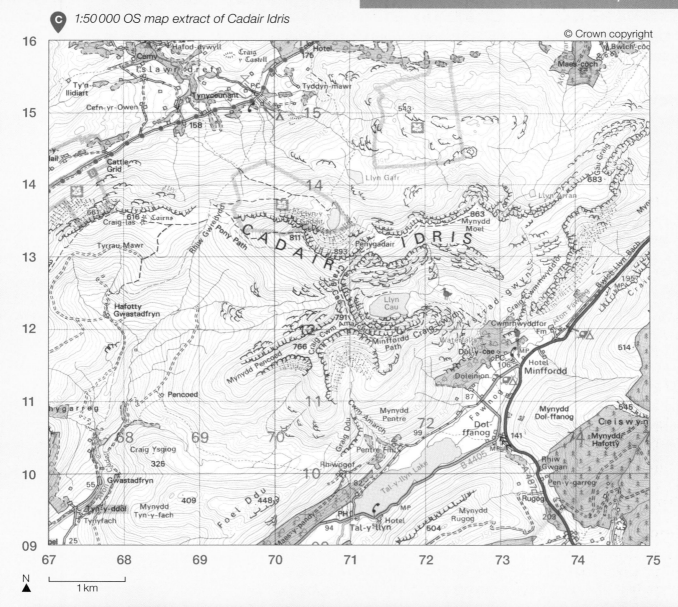

C *1:50 000 OS map extract of Cadair Idris*

© Crown copyright

ACTIVITIES

1 Locate Llyn-y-Gadair in grid square 7013 on map **C**.

a What is the evidence from the map that this lake is in a corrie?

b What do the black dots indicate?

c Draw a simple sketch of this corrie and label its main characteristics. (Hint: don't forget to include a scale and north arrow.)

2 Identify and give a grid reference for another corrie on the map extract.

3 What glacial landform is called Craig Cwm Amarch (south west of Llyn Cau)?

4 Match the following glacial landforms with the grid references: corrie, ribbon lake, corrie lake (tarn), truncated spur, pyramidal peak, glacial trough, arête

a 720100   b 727107   c 710118   d 715124

e 733110   f 709121   g 710125

5 Look at photo **A**.

a Name *three* glacial landforms in the photo.

b Try to suggest the grid reference where the photo was taken.

c Why do you think there are no trees and hardly any vegetation here?

d Write a tweet (maximum 140 characters) to describe this landscape.

6 Look at photo **B**.

a In what direction is the photo looking?

b Describe the characteristics of the glacial trough.

On this spread you will find out about opportunities for economic activities in glaciated areas in the UK

How do glaciated areas provide opportunities for economic activities?

Glaciated areas – both uplands and lowlands – provide opportunities for economic activities.

Farming

In upland areas, glaciers scoured the land stripping away soil and vegetation. As a result, soils in these areas are thin and acidic. Upland areas are not ideal for farming, but are mainly used for grazing. Sheep can tolerate the cold, wet and windy conditions, and the poor vegetation (photo **A**).

Soils in valleys are thicker due to deposition. Flat-bottomed glacial troughs are ideal for using machinery and farming can be more productive. Typical crops include cereals and potatoes. A lot of land is used to grow grass for winter feed (silage and hay) for livestock.

Lowland glaciated areas may be covered by a thick layer of till, which is very fertile. Much of central and eastern Britain has productive farmland growing wheat, barley, potatoes and other crops. With a flat or gently rolling landscape and a warmer, sunnier climate, these areas are well suited to intensive arable farming.

A *Sheep farming near Grasmere, Lake District*

Forestry

Many upland glaciated areas of the UK are well suited to forestry. Large plantations of mostly conifer (cone-bearing) trees have been planted throughout Scotland and across parts of northern England.

Conifer trees are well adapted to cope with the acidic soils. They are one of the few economic ways of utilising steep slopes. Once planted they can be left to grow for 20–30 years before being felled (photo **B**). Conifers produce 'soft' wood used for timber in the construction industry or for making paper.

B *Felled coniferous woodland in Wales*

Quarrying

Upland glaciated areas are made of hard, resistant rock. This can be quarried and crushed to provide stone used in the construction industry and for building roads (photo **C**).

Limestone makes up much of the Pennine Hills in central England. It is a valuable resource used in the chemical industry, for conditioning soils or for making cement.

In lowland areas, glacial deposits of sand and gravel, deposited by meltwater streams, are also valuable in the construction industry. Sand is used for making cement and gravel is used to make concrete.

C Limestone quarry, Yorkshire

D Mountain railway at Aviemore

Tourism

The UK's glaciated uplands provide opportunities for tourism. Tourism can be the most important economic activity and provide employment for thousands of people. Spectacular glacial scenery attracts tourists who enjoy outdoor activities and the cultural heritage.

Aviemore, near the Cairngorm Mountains in Scotland, is one of the UK's main mountain activity centres (photo **D**). Here people can mountain bike, ski and climb. The area is criss-crossed with footpaths and there is lots of wildlife to watch, photograph or draw. There is a wildlife park, a folk museum, an adventure park and a steam railway.

ACTIVITIES

1. Why is the mountain area in photo **A** used for sheep farming?

2. **a** How have economic activities affected the natural mountain landscapes in photos **C** and **D**?

   b Should these areas be left completely natural without any form of economic development? Give reasons for your answer.

3. Complete the table below listing the economic opportunities in glaciated areas. Use the information here, along with your own knowledge and research.

Upland glaciated areas (mountains)	Lowland glaciated areas (covered by ice or meltwater)
• Tourism – opportunities for adventurous activities, for example …	

Stretch yourself

Find out more about the economic activities in a glaciated upland area. For your chosen area, consider the economic importance of farming, forestry, quarrying and tourism. You could present your information in the form of an information poster.

Practice question

Consider why it is necessary to develop a range of economic activities in glaciated areas. Use evidence from the photos in your answer. *(4 marks)*

On this spread you will find out about land use conflicts in glaciated areas

Conflicts in glaciated areas

Glaciated areas provide opportunities for development that can lead to conflicts.

◆ *Quarrying* – rocks such as limestone, slate and granite have economic value, but quarrying can lead to pollution of land and rivers and spoil the landscape.

◆ *Tourism* – can cause conflict with local landowners over access to land. Local people may be affected by traffic congestion and rising house prices.

◆ *Water storage* – building reservoirs can conflict with environmental interests and require the flooding of farmland.

Impact on the local economy with fewer tourists staying in hotels and visiting cafes and pubs

House prices may fall if views are spoilt by wind turbines

Some people think they spoil the natural landscape

A *Wind farm near Bothel, Lake District*

Wind farms in the Lake District

Wind turbines produce renewable energy. Photo **A** shows a wind farm near the village of Bothel in the Lake District. This part of the UK is good for wind farms. Its high elevation and western location expose it to prevailing south-westerly winds from the Atlantic.

The photo shows why local people in the Lake District are concerned about wind farms.

Wind turbines at Kirkstone Pass

Kirkstone Pass is one of the Lake District's most remote and beautiful valleys. The National Park Authority granted permission for three 16 m wind turbines in 2011. The project was completed in April 2012 at a cost of £150 000.

The turbines provide power to the Kirkstone Pass Inn, which had relied on diesel generators for heat and light. Despite opposition to the turbines, local groups like the Friends of the Lake District supported the scheme. They argued that turning to 'green power' was good for the environment and helped secure the future of the pub and its employees.

B *Wind turbines being built in the Kirkstone Pass*

Conflicts between development and conservation

The proposal: Glenridding zip-wire

In 2014, Windermere-based company 'Treetop Trek' put forward a proposal to construct four parallel one-mile long zip-wires above Glenridding in Patterdale (photo **C**).

The zip-wires would run from the disused Greenside mine on land owned by the Lake District National Park Authority, and end in fields above Glenridding.

Consider the evidence and decide if this was the correct decision. You must identify the potential advantages and disadvantages and then make your decision. You must justify it. You can find further information online.

C *The site and route of the proposed zip-wire, Patterdale*

Glenridding zip-wire: yes or no?

'This is another example of the conflict of interest within the Park between the aims of preserving the natural beauty and heritage of the Park and becoming more commercial.
There is much local opposition because it is likely to drive away many visitors who come to enjoy the peace and tranquillity. We already have a huge range of exciting outdoor activities for anyone seeking adventure.'

Graeme Conncher (Steamboat captain and leader of the campaign to stop the scheme)

We are pleased that Treetop Trek has listened to the very strong local opposition to the proposal and has decided against applying to put a zip-wire up on the open fell side in this well-loved valley.'

Kate Willshaw (Policy Officer, Friends of the Lake District)

'Our priority is to balance the need to conserve our spectacular landscape, whilst securing vibrant communities, and being open to opportunities to enhance our economy and improve the offer we have for visitors.'

Mike Turner (Treetop Trek)

ACTIVITIES

1 Why are wind farms controversial in mountain areas like the Lake District (photo **A**)?

2 Work in pairs to complete a table listing the arguments for and against the new wind turbines at Kirkstone Pass. Was the National Park Authority right to grant permission for the project?

Stretch yourself

In 2015, planning permission was sought to replace one of Britain's oldest wind farms on Kirby Moor with new turbines three times as high. Search online for the progress of this planning application. Evaluate whether the developments are good for the area.

Practice question

'Development of glaciated areas must balance the needs of the environment with the economic viability of the area.' Discuss the extent to which it is possible to achieve this balance successfully. *(6 marks)*

Example

On this spread you will find out about tourism in the Lake District – attractions, issues and management

Why do people visit the Lake District?

The Lake District in north-west England (map **A**) became a National Park in 1951. It is famous for its mountains, hills (fells) and lakes. The mountains were carved by giant glaciers during the Ice Age, creating jagged peaks and wide valleys. Some people think it is England's best scenery (photo **B**) – what do you think?

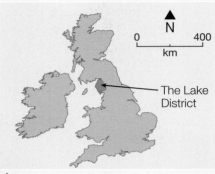

A The location of the Lake District

B Lake Windermere and Ambleside

Physical attractions

◆ Lakes like Windermere and Ullswater offer water sports, cruises and fishing.

◆ Mountain landscapes with peaks like Helvellyn and Scafell Pike are popular for walking and mountain biking.

◆ Adventure activities include abseiling, gorge scrambling and rock climbing.

Cultural/historical attractions

◆ The landscape has inspired writers and poets, such as William Wordsworth.

◆ Beatrix Potter, author of the *Peter Rabbit* stories, lived close to Lake Windermere. Her home, Hill Top, is now a National Trust attraction.

◆ Scenic towns and villages, such as Ambleside and Grasmere, are very popular. Other tourist attractions include monuments like Muncaster Castle at Ravenglass.

Did you know?
Scafell Pike is England's highest mountain (977 m). Windermere is the UK's largest natural lake (17 km long).

Social, economic and environmental impacts of tourism

Social	Economic	Environmental
• In 2014, 14.8 million tourists visited the Lake District. Think about the impact on the 40 000 locals.	• In 2014, tourists spent nearly £1000 million in the Lake District. This supports hotels, shops and restaurants.	• The main tourist ('honeypot') sites and footpaths show signs of overcrowding – footpath erosion, litter, damage to verges by cars.
• Over 89% of visitors arrive by car. Roads are narrow and winding and congestion is a major issue.	• Thousands of local people work in shops, hotels and other services.	• Pollution (oil, fumes) from vehicles and boats can damage ecosystems.
• House prices are high – 20% of property is either holiday rental or second homes.	• New businesses like adventure tourism provide jobs for local people.	• Walkers can damage farmland by trampling crops or leaving litter. Dogs can disturb sheep and cattle.
• Jobs in tourism are mostly seasonal, poorly paid and unreliable.	• Traffic congestion slows down business communications.	

How is tourism managed?

Managing traffic congestion

Traffic congestion causes social, economic and environmental problems. A number of strategies have been adopted to address them.

◆ Several dual-carriageways have been built around the Lake District to improve access.

◆ Transport hubs, like at Ambleside, help create an interchange between parking, buses, ferries, walking and cycling. This helps to relieve congestion elsewhere.

◆ Park-and-ride bus schemes, like the 'Honister Rambler' (photo **D**), have been expanded for tourists.

◆ Traffic calming measures, such as speed bumps, have been introduced in villages.

The use of public transport has been very successful, as the large number of vehicles creates huge problems, particularly in the summer.

Managing footpath erosion

The Upland Path Landscape Restoration Project has successfully repaired paths, created steps, re-surfaced paths with local stone and re-planted native plants.

'Fix the Fells' maintain and repair mountain paths – they're supported by organisations like the National Trust. They use many techniques, including stone pitching, where large stones are dug into a path to create a hard-wearing surface. Local stones and sheep fleece are being used to make a well-drained, solid surface (photo **E**).

However, there are still hundreds of kilometres of footpath in need of constant attention, and their on-going maintenance represents a huge challenge.

C *Tourism around Lake Windermere*

D *The 'Honister Rambler' tourist bus*

E *Sheep fleece used to repair a footpath*

ACTIVITIES

1 **a** Draw a spider diagram to show the tourist attractions of the Lake District.

   b Study photo **C**. What proof is there that tourism can bring economic benefits?

2 List the advantages and disadvantages of tourism to the Lake District.

3 How does public transport help to reduce transport problems?

Stretch yourself

Find out more about the work of 'Fix the Fells'.

• Visit their website to find out about how they repair footpaths.

• What do you think is the best long-term solution for footpath erosion? Produce a presentation to justify your decision.

Practice question

Describe the attitudes to tourism in the Lake District of the following groups:
• local businesses
• farmers
• local inhabitants. *(6 marks)*

Unit 2 Challenges in the human environment

Section A Urban issues and challenges

Ha Dong District, Hanoi, Vietnam

Unit 2 Challenges in the human environment is about human processes and systems, how they change both spatially and temporally. They are studied in a range of places, at a variety of scales and include places in various states of development. It is split into three sections.

Section A Urban issues and challenges includes:

- global patterns of urban change
- urban growth in a city in a newly emerging economy
- urban change in UK cities
- urban sustainability.

You need to study all the topics in Section A – in your final exam you will have to answer questions on all of them.

What if...

1 we all lived to be 100?

2 there was no countryside?

3 no-one migrated?

4 everyone recycled their rubbish?

Your key skills

To be a good geographer, you need to develop important geographical skills – in this section you will learn the following skills:

- Using numerical data.
- Finding evidence from photos.
- Describing population trends from graphs.
- Using a variety of graphic techniques to present data.
- Literacy skills – describing information in photos and preparing a presentation.

Your key words

As you go through the chapters in this section, make sure you know and understand the key words shown in bold. Definitions are provided in the Glossary on pages 346-9. To be a good geographer you need to use good subject terminology.

Your exam

Section A makes up part of Paper 2 – a one and a half-hour written exam worth 35 per cent of your GCSE.

On this spread you will find out how the world's cities are growing

What is urbanisation?

By 1804 global population had doubled from half a billion to one billion in 300 years. By 1999 the total had doubled from 3 billion to 6 billion in just 39 years! The bigger the global population, the faster it grows (graph **A**).

Urbanisation – the proportion of the world's population who live in cities – is also growing. It is the result of the natural increase of a population (births minus deaths) plus **migration**. Urban growth is the increase in the area covered by cities.

Urbanisation has taken place at different times and at different speeds in different parts of the world. The UK was one of the first countries in the world to become urbanised.

Did you know?
More than half the world's population now live in towns and cities. In the UK the figure is currently 82 per cent!

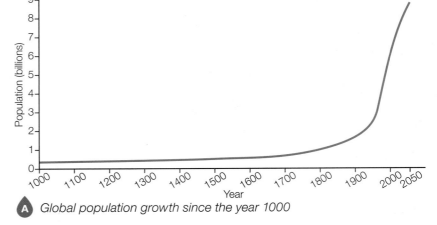

A Global population growth since the year 1000

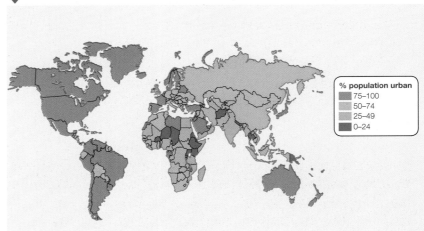

B Global urban population, 2014

% population urban
- 75–100
- 50–74
- 25–49
- 0–24

Maths skills

1 Complete a copy of the table by filling in the missing values.

2 Use bar graph **D** to state which continent will have the biggest change in its share of world urban population by 2050.

Type of country	Country	% urban population, 1950	% urban population, 2050 (estimated)	% change in urban population 1950–2050
HIC	United Kingdom	79		+9
NEE	Nigeria		75	+65
LIC	Botswana	3	81	

How does urbanisation vary around the world?

The proportion of people living in towns and cities varies in different parts of the world (map **B**).

◆ In most of the world's richer countries over 60 per cent of the population live in cities.

◆ In South and South East Asia around half the population live in towns and cities.

◆ All but six countries in Africa have urban populations of more than 20 per cent (Niger, Uganda, Burundi, Ethiopia, South Sudan and Malawi). The average is almost 40 per cent.

In different regions of the world the urban population is growing at different rates (graph **C**).

The distribution of the world's urban population

Different rates of urbanisation around the world have changed the distribution of the world's urban population. The projected changes between 1950 and 2050 are shown in graph **D**.

◆ The largest growth in urban population by 2050 will take place in India, China and Nigeria.

◆ These three countries will account for 37 per cent of the projected growth of the world's urban population between 2014 and 2050.

◆ By 2050, India is projected to add 404 million urban dwellers, China 292 million and Nigeria 212 million.

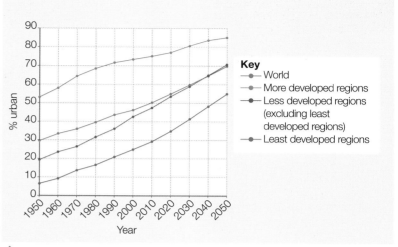

C Urban population, 1950–2050

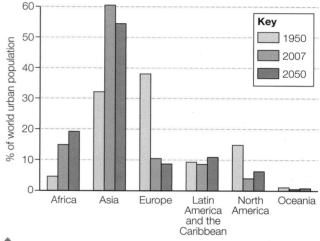

D Distribution of the world's urban population in 1950, 2007 and 2050 (estimated)

ACTIVITIES

1 a Name the continent outside Europe, North America and Oceania which has the highest percentage of its population living in urban areas (map **B**).

  b With the aid of an atlas name one country in Europe with an urban population of less than 39.9 per cent.

2 Describe the trends shown by each of the lines on graph **C**. Support your answer with evidence from the graph.

3 Suggest why Asian countries like India and China are likely to have a higher urban population percentage in 2050 than in 2000.

4 Give examples of how the process of urbanisation has happened at different times and speeds.

Stretch yourself

Produce a presentation (5 slides maximum) to illustrate urban trends in different parts of the world.

Practice question

Suggest why there is such a low rate of urbanisation in rich countries and why some show evidence of counter-urbanisation. (*6 marks*)

On this spread you will find out what factors make cities grow

Why do cities grow?

More than half the world's population now live in urban areas, and cities all over the world are continuing to grow. There are two main reasons why cities are getting bigger.

◆ *Rural–urban migration* – the movement of people from the countryside into towns and cities.

◆ **Natural increase** – where the birth rate is higher than the death rate.

A natural increase in population occurs when there is high proportion of young adults aged 18–35. Therefore, more children will be born. The smaller proportion of older people means the death rate is lower. Improvements to health care, particularly in urban areas of poorer countries, can also result in a lower death rate. Natural increase therefore tends to be higher in LICs (such as Cambodia) and in some NEEs (such as India).

Rural–urban migration is caused by *push* and *pull* factors. These are the real or imagined disadvantages of living in a rural area and advantages of living in a town or city (diagram **A**).

Sunita's story

My name is Sunita. Two years ago my parents, my brother Rakesh and I came to live in Mumbai, in an area called Dharavi. Everyone here is poor. Our house only has two rooms, but we have electricity. My father says at least we have work. One day maybe my brother and I will be rich!

Dharavi is crowded, noisy and very busy. Outside our house people wash laundry, sew clothes and bash dents out of oil cans to recycle them. There are 15 000 small workshops here.

It is very smelly, with open sewers. I like to walk to the biscuit factory because it smells nicer there!

I go to school every morning and learn maths and literacy. In the afternoon I help my mother clean the house. Then I go rag picking with my friends to earn some money.

B *Dharavi, Mumbai*

'Push' factors

People want to leave the countryside because:

◆ farming is hard and poorly paid
◆ desertification and soil erosion make farming difficult
◆ drought and other climate hazards reduce crop yields
◆ farming is often at subsistence level, producing only enough food for the family, leaving nothing to sell
◆ poor harvests may lead to malnutrition or famine
◆ there are few doctors or hospitals
◆ schools provide only a very basic education
◆ rural areas are isolated due to poor roads.

'Pull' factors

People are attracted to the city because:

◆ there are more well-paid jobs
◆ a higher standard of living is possible
◆ they have friends and family already living there
◆ there is a better chance of getting an education
◆ public transport is better
◆ a range of entertainments are available
◆ there are better medical facilities.

A *Push and pull factors*

Higher quality of life

Drought and flooding

Lack of services

Higher-quality services, e.g. education, health and entertainment

More opportunities

Urban pull

Rural poverty

Few opportunities

Higher-paid jobs

Improved housing

Low pay

What are megacities?

These are cities with a population of over 10 million. In 2015 there were 28 of these megacities (map **C**), and the United Nations estimates that by 2050 there may be as many as 50. There are three types of megacities.

Slow-growing

Where?
South East Asia, Europe and North America

Features
Population at 70%+ urban
No squatter settlements

Examples
Osaka-Kobe
Tokyo
Moscow
Los Angeles

Growing

Where?
South America and South East Asia

Features
Population 40–50% urban
Under 20% in squatter settlement

Examples
Beijing
Rio de Janeiro
Shanghai
Mexico City

Rapid-growing

Where?
South/South East Asia and Africa

Features
Population under 50% urban
Over 20% in squatter settlements

Examples
Jakarta
Lagos
Mumbai
Manila

C *The distribution of megacities in 2014 and 2030 (projected)*

Key
• Megacities, 2014
• Additional megacities by 2030

ACTIVITIES

1 a Give three factors which might explain why Sunita's family moved to Mumbai.

  b Explain whether these are push or pull factors.

2 Draw a table with three columns headed Social, Economic and Environmental. List each of the push and pull factors listed in diagram **A** under the correct heading.

3 Describe how the three types of megacities are different from each other.

Stretch yourself

Using population data, calculate the rate of natural increase of Malawi, the Philippines, and Colombia. Is there a pattern between these countries?

Practice question

Use map **C** to describe the changes in the distribution of megacities between 2014 and 2030. *(6 marks)*

On this spread you will find out why the city of Rio de Janeiro is growing so rapidly

What is Rio like?

Rio de Janeiro is situated on Brazil's Atlantic coast at 23°S and 43°W. It has grown up around a large natural bay called Guanabara Bay (photo **A**). Until 1960 Rio was the capital of Brazil — it is now Brasilia. It is the cultural capital of Brazil, with over 50 museums, and its famous annual carnival is one of the world's biggest music and dance celebrations. It is a UNESCO World Heritage Site. The staging of the 2014 soccer World Cup and the 2016 Olympics have increased its global importance.

Brazil's second most important industrial centre, producing 5 per cent of Brazil's Gross Domestic Product (GDP)

A major port – main exports are coffee, sugar and iron ore.

The Statue of Christ the Redeemer is one of the Seven New Wonders of the World.

Sugar Loaf Mountain

Main service industries are banking, finance and insurance.

Stunning natural surroundings and amazing beaches make it one of the most visited cities in the southern hemisphere.

Guanabara Bay

Main manufacturing industries are chemicals, pharmaceuticals, clothing, furniture and processed foods.

Rio hosted matches during the 2014 World Cup and will host the 2016 Olympic Games.

A *Some facts about Rio*

Rio has become a 'global city' because of its importance in the global economy as an industrial and financial centre. It is a major regional, national and international centre for many important companies and industries. It is an important international hub, with five ports and three airports.

How and why has Rio de Janeiro grown?

Rio de Janeiro is the second largest city in Brazil (the largest is São Paulo). In 2014 Rio had a population of 6.5 million people in the city itself and 12.5 million in the surrounding area (the population of Greater London is about 8 million).

B *Ipanema Beach*

Rio has grown rapidly in the last 50 years to become a major industrial, administrative, commercial and tourist centre. These economic activities have attracted many migrants from Brazil and other countries to swell the population of the city. These migrants have contributed to Rio's continuing economic development. As a result Rio has a racially mixed population. Migrants have come to Rio from many different places.

- From other parts of Brazil such as the Amazon Basin.

- From other countries in South America, such as Argentina and Bolivia.

- More recent migrants have come from South Korea and China seeking new business opportunities.

- The common language still attracts migrants from Portugal, Brazil's former colonial power.

- Rio's industry attracts skilled workers from the USA and UK.

Land uses in Rio de Janeiro

Rio has mountains, coast and large **squatter settlements** (page 160). The city is divided into four main zones: Centro (centre), South Zone, West Zone and North Zone (map **C**). These are Rio's main industrial and commercial areas.

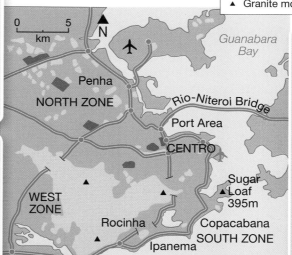

C *Rio's main zones*

Key
- Tijuca National Forest Park
- Squatter settlements (favelas)
- Industrial areas
- Motorways/expressways
- Junctions
- International airport
- Granite mountains

North Zone

- The city's main industrial and port area.
- The city's International Airport and Maracanã soccer stadium are here.
- An area of low-quality housing and favelas.
- The location of the Tijuca National Park.

West Zone

- Barra da Tijuca has changed from a lower-class area into a wealthy coastal suburb with luxury apartments, shopping malls, recreational and tourist facilities.
- The industrial area of Campo Grande has low-quality housing around the steelworks.
- The main Olympic stadiums and competitor village for 2016 are located here (photo **D**).

Centro

- The oldest part of the city, with many historic buildings.
- The city's CBD and main shopping area.
- The financial centre with the headquarters of Petrobras and CVBB, Brazil's largest oil and mining companies.

South Zone

- Developed after tunnels were cut through the mountains.
- Rio's main tourist hotels and beaches such as Copacabana and Ipanema (photo **B**).
- Wealthy area dominated by luxury flats: it has the wealthiest district in the whole of South America.
- Overlooked by Rocinha, the largest favela in South America (page 160).

D *The new Olympic stadium*

ACTIVITIES

1 **a** Describe Rio's natural surroundings (photo **A** and text).

   b What are the advantages of this site for Rio's development?

2 Why do you think Rio is such a popular tourist destination?

3 **a** Suggest why many historic buildings are found in the Centro zone (figure **C**).

   b Use the map to give the direction the camera was facing in photo **A**.

   c How does the South Zone show Rio's inequalities of wealth?

   d Suggest why the West Zone was chosen as the site for the 2016 Olympic Park.

Stretch yourself

Investigate what functions Rio de Janeiro has kept since losing its status as the capital of Brazil.

Practice question

Explain how migration has been responsible for the growth and racial make-up of Rio's population. *(6 marks)*

On this spread you will find about the social challenges facing Rio

Rio faces many challenges in providing important services for its rapidly-growing population:

◆ health care ◆ education ◆ water supply ◆ energy.

These are made more difficult because of the contrasts between areas, which are often very close to one another (photo **A**). This causes great **inequalities**.

Now you'll consider the problems faced in providing each of these services in Rio, and how the authorities have tried to create social opportunities.

 Copacabana Beach with a squatter settlement (favela) on the hillside above

Health care

Challenges

In 2013 only 55 per cent of the city had a local family health clinic. Services for pregnant women and the elderly were very poor, especially in the West Zone.

District	Zone	Infant mortality rate	Pregnant females getting medical care	Average life expectancy
Cidada de Deus	West	21 per 1000	60%	45
Barra de Tijuca	South	6 per 1000	100%	80
Rio de Janeiro (as a whole)		19 per 1000	74%	63

 Comparing health in two contrasting districts with Rio as a whole

Solutions

One example of how the authorities have tried to improve health care is the favela of Santa Marta. Set on a steep hillside, with a population of 8000, it has few roads and the main means of access is an overcrowded cable car. It is 13 km to the nearest hospital. Medical staff took a health kit into people's homes, and were able to detect twenty different diseases and treat them. As a result, infant mortality has fallen and life expectancy increased.

Education

Challenges

Education in Brazil is compulsory for children aged 6–14. In Rio only half of all children continue their education beyond the age of 14. Many drop out of school and some get involved in drug trafficking.

The level of school enrolment in Rio is low. The main reasons for this are:

◆ a shortage of nearby schools

◆ a lack of money and a need to work

◆ a shortage of teachers

◆ low pay for teachers

◆ poor training for teachers.

Solutions

The authorities have tried to improve access to education by:

◆ encouraging local people to volunteer to help in school

◆ giving school grants to poor families to help meet the cost of keeping their children in school

◆ making money available to pay for free lessons in volleyball, football, swimming and squash in Rocinha favela

◆ opening a private university in Rocinha favela.

Water supply

Challenges

Around 12 per cent of Rio's population did not have access to running water. It is estimated that 37 per cent of water is lost through leaky pipes, fraud and illegal access. The situation has become worse in recent years.

Drought-hit Rio braces for Carnival water shortages

S E Brazil is experiencing its worst drought for 80 years

Paraibuna and Santa Branca reservoirs are declared empty

Water to take priority over energy: less water to be taken from the River Paraiba do Sol for electricity generation

C *Newspaper headlines from 2015*

D *Improved water supply to Olympic Park in West Zone*

Solutions

Most of the work has been on improving the quantity or quality of the water in the favelas and in the Olympic Park (photo **D**). Seven new treatment plants were built between 1998 and 2014, and over 300 km of pipes were laid. By 2014, 95 per cent of the population had a mains water supply.

Energy

Challenges

The whole city suffers frequent blackouts due to a shortage of electricity. The growing population and the demands of the forthcoming Olympics will make the situation worse.

Many people living in the poorer parts of Rio de Janeiro get their electricity by illegally tapping into the main supply, which is risky and unsafe (photo **E**).

E *Illegal electricity connections in a favela*

Solutions

The electricity supply to Rio has been improved by:

- installing 60 km of new power lines
- building a new nuclear generator
- developing the new Simplício hydro-electric complex which will increase Rio's supply of electricity by 30 per cent. It took 6 years to build and cost over US$ 2 billion.

ACTIVITIES

1 **a** Name two differences between the housing areas of Rio shown in photo **A**.

   b Suggest problems with providing services to the favela in photo **A**.

2 Read the newspaper headlines (**C**). Why were 2014 and 2015 difficult years for Rio?

3 Why do you think the authorities were keen to improve the water supply to the West Zone?

Stretch yourself

Suggest reasons for the situation referred to in the newspaper headlines in figure **C**.

Practice question

Explain why the authorities in Rio have to cope with such a range of social challenges. *(4 marks)*

On this spread you will find out about the economic opportunities and challenges facing Rio

The growth of Rio's urban industrial areas has boosted the city's economy. Economic development has brought improvements to Rio's roads, transport, services and environment. The policy to improve the city's favelas has improved the quality of life for many people. Growing economic prosperity has attracted large companies to Rio from other parts of Brazil and South America, as well as from abroad. These developments have created a range of new **economic opportunities** in the **formal economy**.

The effects of economic growth in Rio?

Rio is Brazil's second most important industrial centre after São Paulo. Its large population, financial sector, port facilities and industrial areas (photo **A**) have contributed to Rio's rapid economic development. The city now provides more than 6 per cent of all employment in Brazil.

Rio has one of the highest incomes per head in the country, and the city's retail and consumer sector is a major source of employment. A growing number of jobs are provided by service industries, such as finance (pie chart **B**). Oil has been discovered just off the coast and this has encouraged the growth of oil-related industries.

A One of Rio's largest steelworks, at Volta Redonda

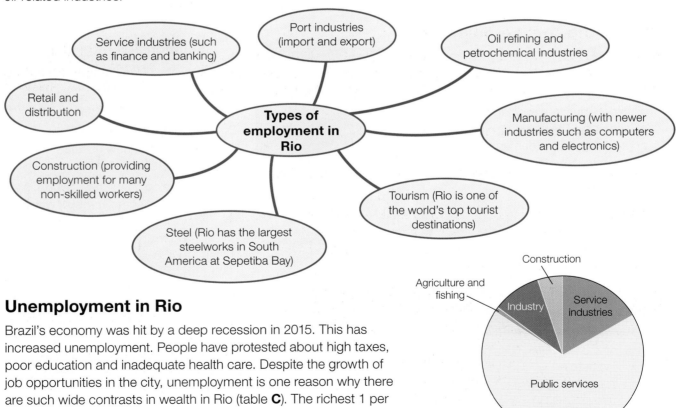

Service industries (such as finance and banking)

Port industries (import and export)

Oil refining and petrochemical industries

Retail and distribution

Types of employment in Rio

Manufacturing (with newer industries such as computers and electronics)

Construction (providing employment for many non-skilled workers)

Tourism (Rio is one of the world's top tourist destinations)

Steel (Rio has the largest steelworks in South America at Sepetiba Bay)

Unemployment in Rio

Brazil's economy was hit by a deep recession in 2015. This has increased unemployment. People have protested about high taxes, poor education and inadequate health care. Despite the growth of job opportunities in the city, unemployment is one reason why there are such wide contrasts in wealth in Rio (table **C**). The richest 1 per cent of the population earns 12 per cent of the total income. But the income of the poorest 50 per cent is only 13 per cent of the total.

Construction

Agriculture and fishing

Industry

Service industries

Public services

B Types of employment in Rio

Unemployment rates in the favelas are over 20 per cent. Most work in the **informal economy**, making a living however they can. People work as street vendors (photo **D**), drivers, labourers, maids or in the production of sewing and handicraft work for the local street market. Work in the informal sector is poorly paid (less than £60 a month) and irregular. About one-third of Rio's 3.5 million workers don't have a formal employment contract, and many are without any insurance cover or unemployment benefit. They do not pay any taxes and the government receives no income from them.

What is being done about unemployment?

The local government is using education to try to reduce youth unemployment. The Schools of Tomorrow programme aims to improve education for young people in the poor and violent areas of the city. There are also practical skills-based courses.

Courses are available for adults who have temporarily left education but want to continue their studies. Free child care is provided for teenage parents to enable them to return to education.

C Rio's unemployment rates

District	Zone	Unemployment rate
Barra da Tijuca	South	2%
Complexo do Alemão	North	Estimated 37%

D A Rio street vendor

What is the crime problem in Rio?

Robbery and violent crime present great challenges in Rio. Murder, kidnapping, carjacking and armed assault occur regularly. Street crime is a problem, especially at night. Powerful gangs control drug trafficking in many of the favelas. The police have taken steps to control crime.

◆ In 2013 Pacifying Police Units (UPPs) were established to reclaim favelas from drug dealers.

◆ Police have taken control of crime-dominated Complexo do Alemão and 30 smaller favelas.

There has been criticism that the police are targeting favelas near the Olympic sites. People living in these areas think this is an attack on their freedom. But the police argue that a lower crime rate, increased property values and growing tourism are positive results of their fight against crime in the favelas.

Think about it

How does crime affect peoples' everyday lives?

ACTIVITIES

1 a Approximately what percentage of the industries in Rio are services (pie chart **B**)?

  b What are the advantages and disadvantages of street vendors working in the informal economy?

  c The main manufacturing areas are around the port and on the outskirts of the city. Suggest what the advantages might be for manufacturing in these two locations.

2 a What is the vendor selling in photo **D**?

  b Why are these goods likely to be typical of the types sold by many street vendors?

Stretch yourself

Imagine you are living in one of Rio's favelas. Make a case for or against the police moving into your favela to deal with the drug gangs.

Practice question

'A city of great contrasts.' Explain why this fact makes it difficult for Rio to overcome its economic challenges. *(6 marks)*

On this spread you will learn how Rio is responding to its environmental challenges

What are Rio's environmental challenges?

The environmental challenges which affect the quality of life for people in Rio are caused by the physical geography of the city as well as by human activities (diagram **A**). The city authorities have developed solutions to many of these problems.

 *Environmental challenges in Rio*

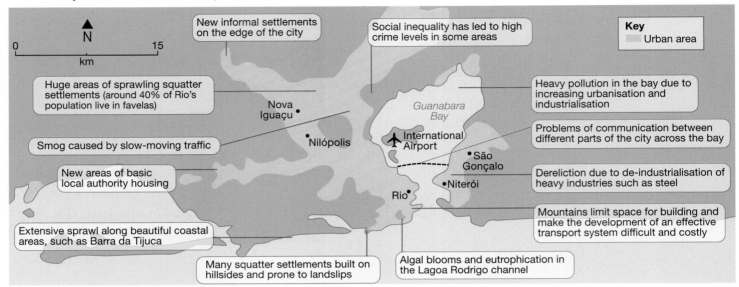

New informal settlements on the edge of the city

Social inequality has led to high crime levels in some areas

Key
Urban area

Huge areas of sprawling squatter settlements (around 40% of Rio's population live in favelas)

Nova Iguaçu

Nilópolis

Guanabara Bay

International Airport

Heavy pollution in the bay due to increasing urbanisation and industrialisation

Problems of communication between different parts of the city across the bay

São Gonçalo

Niterói

Rio

Smog caused by slow-moving traffic

New areas of basic local authority housing

Dereliction due to de-industrialisation of heavy industries such as steel

Mountains limit space for building and make the development of an effective transport system difficult and costly

Extensive sprawl along beautiful coastal areas, such as Barra da Tijuca

Many squatter settlements built on hillsides and prone to landslips

Algal blooms and eutrophication in the Lagoa Rodrigo channel

Air pollution

Air pollution is estimated to cause 5000 deaths per year in Rio. The city is often covered with brown *smog*. This happens because:

◆ heavy traffic and congestion on roads causes build-up of exhaust fumes.

◆ mist from the Atlantic mixes with vehicle exhaust fumes and pollutants from factory chimneys

Traffic congestion

Rio is the most congested city in South America (photo **C**). Traffic congestion increases stress and pollution levels and wastes time for commuters and businesses.

◆ Steep mountains – roads can only be built on coastal lowland. Main transport routes become very congested.

◆ Tunnels through the mountains are needed to connect different areas of the city.

◆ The number of cars in Rio has grown by over 40 per cent in the last decade.

◆ High crime levels mean that many people prefer to travel by car.

Solutions

Improvements have been aimed at reducing traffic congestion (map **B**) and improving air quality:

◆ expansion of the metro system under Guanabara Bay, to South Zone and Barra da Tijuca

◆ new toll roads into city centre to reduce congestion

◆ making coast roads one-way during rush hours, to improve traffic flow.

 *Improvements to Rio's transport system*

Key
Major roads
Metro lines

Via Outra

Avenida Brasil

Line 2

Linha Vermella

Linha Amarela

Ilha do Governador

International Airport

Guanabara Bay

Ilha do Fundão

Rio-Niterói, Bridge

Centro

Domestic airport

Line 1

Rebouças Tunnel

Lagoa Rodrigo de Freitas

Urca

Botafogo

Leme

Lagoa-Barra Tunnel

Rocinha

Copacabana

Elevado do Joá

Ipanema

Barra da Tijuca

Leblon

São Conrado

C *Traffic congestion in Rio*

Water pollution

Guanabara Bay is highly polluted, causing a major threat to wildlife. Commercial fishing has declined by 90 per cent in the last 20 years. There is a danger that pollution could affect Ipamena and Copacabana Beaches which would damage tourism and the local economy. The authorities have promised to clean up the bay in time for the Olympics but there will still be problems.

There are several sources of water pollution:

◆ many of the 55 rivers flowing into the bay are heavily polluted

◆ rivers are polluted by run off from open sewers in the favelas

◆ over 200 tonnes of raw sewage pours into the bay each day

◆ over 50 tonnes of industrial waste enters the bay each day

◆ there have been oil spills from the Petrobras oil refinery

◆ ships empty their fuel tanks in the bay because there are no facilities to dispose of the fuel properly.

Solutions

Overseas aid has been used to reduce the amount of sewage being released into the bay.

◆ 12 new sewage works have been built since 2004 at a cost US$ 68 million.

◆ Ships are fined for discharging fuel into the bay illegally.

◆ 5 km of new sewage pipes have been installed around badly polluted areas.

Waste pollution

The worst waste problems are in the favelas. Many are built on steep slopes and have few proper roads, making access difficult for waste collection lorries. Most waste is therefore dumped and pollutes the water system. This causes diseases like cholera and encourages rats.

Solutions

A power plant has been set up near the University of Rio using methane gas (biogas) from rotting rubbish. It consumes 30 tonnes of rubbish a day and produces enough electricity for 1000 homes.

ACTIVITIES

1 List the problems shown on map **A** under the headings 'Physical' and 'Human' (some may be under both headings).

2 What are the main causes of water pollution in Rio?

3 What impact could coastal pollution have on Rio's tourism?

4 Why is traffic congestion such a problem in Rio?

Stretch yourself

Write a speech agreeing or disagreeing with this statement: 'Rio's hosting of the Olympic Games in 2016 proved to be beneficial to the city's environment'.

Maths skills

Use an appropriate method to present this data about daily journeys to work in Rio.
30% by bus
60% by car
5% by metro
3% by rail
2% by cycle or on foot

Practice question

Outline how the quality of life for Rio's population can be improved. *(6 marks)*

159

On this spread you will find out about housing the poor in Rio

Why have favelas grown?

Squatter settlements in Brazil are called *favelas*. They are illegal settlements where people have built homes on land that they did not own. The favelas are areas of great social deprivation.

People leave Amazonia and the drought-hit areas of north east Brazil countryside in the hope of finding a better life in the city. Many are young adults so the birth rates are higher than in the more prosperous parts of the city.

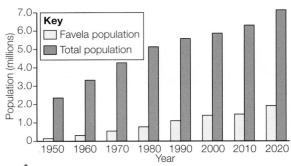

A The growth of the total and favela populations of Rio, 1950–2020

Where are the favelas located?

There are up to a 1000 favelas in the greater Rio area:

◆ 60 per cent are in the suburbs

◆ 25 per cent are in the outer parts of the city

◆ some are being built up to 40 km from the city centre.

The authorities have cleared many of those near the city centre to make Rio more attractive to businesses and tourists.

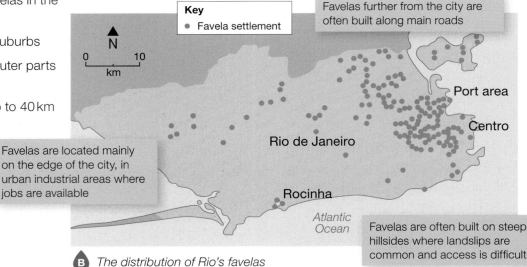

Favelas further from the city are often built along main roads

Favelas are located mainly on the edge of the city, in urban industrial areas where jobs are available

Favelas are often built on steep hillsides where landslips are common and access is difficult

B The distribution of Rio's favelas

Rocinha

Rocinha is the largest favela in Rio. It had a population of 75 000 in the 2010 census but that is now likely to be three times higher. The favela is built on a very steep hillside overlooking the wealthy areas of Copacabana and Ipanema where many of its inhabitants work. More regular work allows improvements to be carried out by the people themselves as well as those done by the local authorities.

As a result of improvements, the favela now has:

◆ 90 per cent of houses built with brick and with electricity, running water and sewage systems

◆ many houses with TVs and fridges

◆ its own newspapers and radio station

◆ retail facilities including food, clothes and video rental shops, bars, travel agent and MacDonald's

◆ schools, health facilities and a private university.

C Rocinha favela overlooking the South Zone beach area

The challenges of squatter settlements

Construction

- Houses are poorly constructed, as they were built illegally with basic materials such as iron, broken bricks and plastic sheets.
- Many favelas are built on steep slopes and heavy rain from storms can cause landslides. In 2010, 224 people were killed and 13 000 lost their homes when houses were swept away.
- There is limited road access due to the steepness of the slopes.

Services

- In the non-improved favelas, around 12% of homes do not have running water, over 30% have no electricity and around 50% have no sewage connections.
- Many homes use illegal connections to electricity pylons.
- Sewers are often open drains.
- Drinking water is often obtained by tapping into a city water main. Taps are often at the bottom of steep slopes and require several trips each day to fetch water.

Unemployment

- Unemployment rates are as high as 20%.
- Much employment is poorly paid with irregular jobs in the informal sector.
- Average incomes may be less than £75 a month.

Crime

- There is a high murder rate of 20 per 1000 people in many favelas.
- Drug gangs dominate many favelas.
- Many inhabitants distrust the police because of violence and corruption.

Health

- There are population densities of 37 000 per km^2.
- Infant mortality rates are as high as 50 per 1000.
- Waste cannot be disposed of and builds up in the street, increasing the danger of disease.
- Burning rubbish often sets fire to the wooden houses. Smoke is harmful to health.

D *A favela on a Rio hillside*

Favelas face many problems, and are often portrayed in a negative way. But inhabitants such as Maria Tanos sometimes have a different view to that of outsiders.

'I like living in Jacarezinho. I have a house and there is a tap down the street. There is a real community spirit here, and my neighbours are helping us improve our house. I work part time as a cleaner and my husband has a stall downtown. My five children go to school here. Impossible if we still lived in northern Brazil. The rains failed for several years and all our animals died.'

Stretch yourself

Research everyday life for the residents of Rocinha. Focus on the challenges facing people living in Rocinha and on the improvements that have been made. Write an article describing your findings.

ACTIVITIES

1 Explain the main locations of favelas in Rio.
2 What are the main challenges facing people living in squatter settlements?

Practice question

Evaluate why housing the urban poor will prove to be a great challenge for the authorities in Rio. *(6 marks)*

Planning for Rio's urban poor

On this spread you will find out about improvements to squatter settlements

Example

How are favelas being improved?

Until 1980, the authorities in Rio did not acknowledge the existence of favelas. They were not shown on any maps. In the mid-1980s city planners felt that something needed to be done for the city's poorest citizens. Rather than destroy the favelas and squeeze their large populations into public housing, the city decided to upgrade them and provide essential services. Since Rio was awarded the 2016 Olympics there has been a move to destroy favelas, especially in areas where Olympic facilities are being built.

Figure **A** shows different plans and approaches that have been used to improve conditions in the favelas.

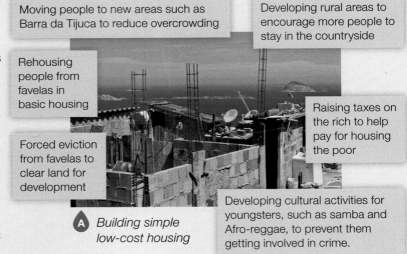

Moving people to new areas such as Barra da Tijuca to reduce overcrowding

Developing rural areas to encourage more people to stay in the countryside

Rehousing people from favelas in basic housing

Raising taxes on the rich to help pay for housing the poor

Forced eviction from favelas to clear land for development

Developing cultural activities for youngsters, such as samba and Afro-reggae, to prevent them getting involved in crime.

A *Building simple low-cost housing*

Favela Bairro Project – improving life in the favelas

This is a *site and service scheme*, where the local authority provides land and services for residents to build homes. For example, Complexo do Alemão is a group of favelas in Rio's North Zone with more than 60,000 people. Here, the local authority have been responsible for many new improvements (figure **B**).

B *Improvements in Complexo do Alemão*

- Paved and formally named roads
- Access to a water supply and drainage system for improved **sanitation**.
- Hillsides secured to prevent landslides, or people relocated where necessary
- Building of new health, leisure and education facilities

- Installation of a cable car system across the Complexo do Alemão hillsides – inhabitants are given one free return ticket a day
- Access to credit to allow inhabitants to buy materials to improve their homes
- 100 per cent mortgages available for people to buy their homes
- A Pacifying Police Unit (UPP) set up, with police patrolling the community to help reduce crime

Has the Favela Bairro Project been a success or failure?

The quality of life, mobility and employment prospects of the inhabitants of the favelas have improved because of the developments made possible by the project. It has been recognised as a model by the UN and been used in other Brazilian cities.

However, it has not been a complete success, and there are still problems:

- the budget of US$1 billion may not cover every favela
- the newly-built infrastructure is not being maintained
- residents lack the skills and resources to make repairs

- more training is needed to improve literacy and employment
- rents rise in the improved favelas and the poorest inhabitants are even worse off.

The effect of the Olympics on the favelas

Some favelas have been demolished to make way for the developments for the Olympic Games. About 1000 people have lost their homes to make way for a new road. There are plans to demolish about 3000 houses ahead of the Games.

The small town of Campo Grande in the West Zone is a 90 minute drive from the city centre. Eight hundred new houses have been built there for people whose homes were demolished. For some residents, the houses are better than in the favelas. But Campo Grande lacks a sense of community, has no shops, nowhere for children to play, and is a long way from the city.

C *Favelas demolished to build new roads*

Another view

Many favela residents have benefited from the Olympic Games. The favelas near the Olympic Park are being redeveloped, and many people have found employment building Olympic facilities.

D *The settlement of Campo Grande*

ACTIVITIES

1 How and why did the Rio city authorities change their attitudes towards the favelas?

2 What are the advantages of the housing being built in photo **A**?

3 List and classify the developments in the Complexo do Alemão under the headings: 'Economic', 'Environmental' and 'Social'.

4 Use a table to list the advantages and disadvantages of the Favela Bairro Project.

Practice question

Discuss whether the inhabitants of the favelas or the city authorities have gained the most from the attempts to improve the conditions of the poor of Rio. *(6 marks)*

Stretch yourself

Evaluate whether the Favela Bairro Project will have more long-term benefits for favela inhabitants or for the city authorities.

On this spread you will find out about the distribution of population in the UK and the location of the major cities

How many people live in the UK?

Graph **A** shows how the population of the UK has changed since 1900 and how it is likely to change the future. The total population in 2015 was 64.6 million (map **B**).

How is the population distributed?

The UK population is unevenly distributed, with 82 per cent living in urban areas. One in four of those live in London and the south east. In contrast, many highland regions of Scotland and Wales are very sparsely populated. These are upland areas that are remote and can experience harsh climatic conditions.

The UK's urban areas

The distribution of the UK's major cities and most densely populated areas reflect its industrial past (map **C**). This was shaped by the Industrial Revolution in the eighteenth century. There was a development of heavy industries and concentration of population near supplies of coal and raw materials such as iron ore. For example:

◆ the Central Lowlands of Scotland (Glasgow)

◆ north east England (Newcastle and Sunderland)

◆ Lancashire (Manchester)

◆ West Yorkshire (Leeds and Sheffield)

◆ South Wales (Cardiff and Swansea).

London developed because of its position as the capital of the UK with associated political and administrative functions. From being the capital of a large empire it became a global city and a financial centre. Belfast, Cardiff and Edinburgh grew because of their function as capital cities of Northern Ireland, Wales and Scotland. The UK's second city, Birmingham, grew mainly as a centre of industrial innovation due to its key position in the centre of the country. The UK's position as an important trading nation explains the growth of ports such as London, Liverpool and Bristol.

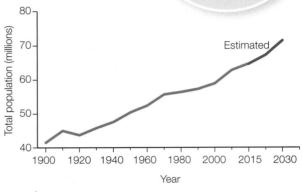

A *The population of the UK since 1900*

B *The population of the UK and its major cities, 2015*

How might this distribution change?

There are key factors affecting the distribution of the UK population.

◆ There has been a general drift towards south east England and London – one of the world's financial, business and cultural centres.

◆ Since 1997 annual inward migration to the UK has been greater than outward migration. Between 2009 and 2014 this increased the population by an average of 243 000 each year. Immigrants generally settle in larger cities where there are more job opportunities.

◆ There has recently been a movement from urban to rural areas. The UK has an increasing proportion of older people, many choosing to retire to live on the coast or in the country.

C *The distribution of the population in the UK*

UK's lowest density – Scottish Highlands/Eileen Siar: 9 people per km²

Lowest density in Northern Ireland – Moyle: 35 people per km²

Lowest density in England – Eden, Cumbria: 25 people per km²

Lowest density in Wales – Powys: 26 people per km²

UK's highest density outside London – Portsmouth: 5141 people per km²

Maths skills

Use map **B** to draw a bar graph to show the population of the UK's major cities. What trends can you infer from the data?

Think about it

London is a World City. What defines a World City, and why is London a good example?

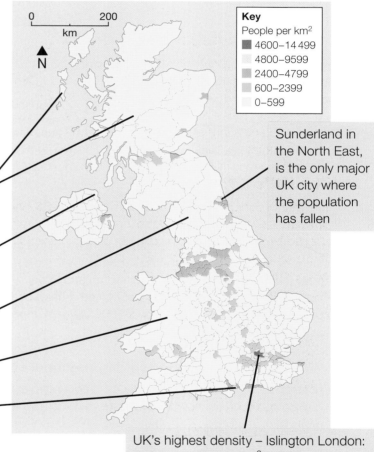

Key
People per km²
■ 4600–14 499
■ 4800–9599
■ 2400–4799
■ 600–2399
□ 0–599

Sunderland in the North East, is the only major UK city where the population has fallen

UK's highest density – Islington London: 14 517 people per km²

ACTIVITIES

1 Describe how the size of the UK population has changed over the last 100 years (graph **A**).

2 How did the development of industry influence the distribution of population in the UK?

3 Suggest reasons for the uneven distribution of the UK's population.

4 What factors may affect the distribution of the UK population in the future?

Stretch yourself

Choose one of the cities on map **B**. Do some research to find out when, how and why the city developed. Why is the city important today?

Practice question

Explain how the distribution of population in the UK reflects both physical and human geographical factors.
(6 marks)

On this spread you will find out about the importance of Bristol as a major UK city

What makes Bristol a major UK city?

Bristol is the largest city in the south west of England. It has a population of 440 500. The population is expected to reach half a million by 2029.

Bristol is a city of regional and national importance, and is one of the UK's ten 'core cities'.

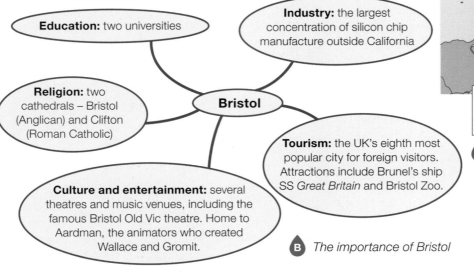

Education: two universities

Industry: the largest concentration of silicon chip manufacture outside California

Religion: two cathedrals – Bristol (Anglican) and Clifton (Roman Catholic)

Bristol

Tourism: the UK's eighth most popular city for foreign visitors. Attractions include Brunel's ship SS *Great Britain* and Bristol Zoo.

Culture and entertainment: several theatres and music venues, including the famous Bristol Old Vic theatre. Home to Aardman, the animators who created Wallace and Gromit.

B *The importance of Bristol*

A *The location of Bristol*

Bristol developed in the eighteenth century as part of the triangular trade linking West Africa and the West Indies. Today it has two major docks, Avonmouth and Royal Portbury, and the UK's most centrally-located deep-sea container port (photo **C**). Around 700 000 cars are imported each year from Japan, Germany and Korea.

Why is Bristol an important international city?

Bristol has recently experienced a lot of economic and social change. The recent growth and development as an important international city are due to a number of factors.

◆ It holds a strategic position on the M4 corridor, with good road and rail links, and easy access to London and rail and ferry services to Europe.

◆ Bristol airport links the city to major European centres and the USA.

◆ There has been a change from dependence on traditional industry like tobacco and paper, to the development of global industries such as financial and business services, defence, aerospace, technology, culture and media (see page 171).

◆ There has been a high level of inward investment, including FDI (Foreign Direct Investment), in manufacturing (companies such as Airbus, BMW and Siemens), finance and high-tech businesses.

◆ Bristol University attracts students from all over the world, providing graduates for professional, managerial and knowledge-based jobs.

C *The port of Bristol*

The impact of migration in Bristol

Between 1851 and 1891 Bristol's population doubled as people arrived looking for work. In recent years migration from abroad has accounted for about half of Bristol's population growth (graph **D**). This has included large numbers from EU countries, in particular Poland and Spain. Migrant workers are employed in a wide range of sectors:

- hospitality
- manufacturing
- construction
- retail
- health
- transport.

Compared to elsewhere in the UK, a higher proportion of migrants coming to Bristol intend to stay permanently. Inward migration has had a significant impact on Bristol (diagram **E**).

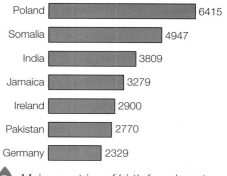

Poland	6415
Somalia	4947
India	3809
Jamaica	3279
Ireland	2900
Pakistan	2770
Germany	2329

D *Main countries of birth for migrants to Bristol, 2011 census*

E *The impact of migration on Bristol*

- A hard-working and motivated workforce
- Enriching the city's cultural life
- Challenge of integration into the wider community
- The mainly young migrants help to balance the ageing population
- Improving the level of skills, where there are shortages
- **What impact have migrants had in Bristol?**
- Contributing to both the local and national economy
- Pressures on housing and employment
- The need to provide education for children whose first language is not English

Fifty countries are represented in Bristol's population. As well as their economic impact, migrants contribute to the cultural life of the city in music, art, literature and food. Bristol's large African and Afro-Caribbean population has had a significant impact, and has created a strong community spirit. The St Paul's Carnival (photo **F**) attracts around 40 000 people each year. Its aim is to help improve relations between the European, African, Caribbean and Asian communities.

F *St Paul's Carnival*

ACTIVITIES

1. **a** Describe Bristol's location (map **A**).
   b Describe Bristol's network of communications.
   c Explain the regional importance of Bristol.
2. Suggest the locational advantages of the Port of Bristol.
3. Assess the importance of Bristol as an international city.
4. Suggest how Bristol's multicultural population may affect the city.

Stretch yourself

Carry out research to find out more about some of the new industries in Bristol and their role in the city's development.

Practice question

Explain how Bristol's growth has been affected by migration. *(4 marks)*

On this spread you will find out how changes in Bristol have created social opportunities

What changes are affecting Bristol?

Bristol is changing. Some of the factors that are bringing about these changes are shown in diagram **A**.

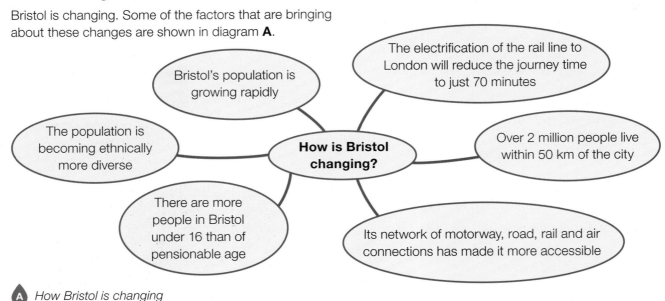

Bristol's population is growing rapidly

The electrification of the rail line to London will reduce the journey time to just 70 minutes

The population is becoming ethnically more diverse

How is Bristol changing?

Over 2 million people live within 50 km of the city

There are more people in Bristol under 16 than of pensionable age

Its network of motorway, road, rail and air connections has made it more accessible

A *How Bristol is changing*

Cultural opportunities

Bristol's youthful population means there is a vibrant underground music scene in addition to the usual range of nightclubs and bars. The Colston Hall has concerts and entertainment by major names in rock, pop, jazz, folk, world and classical music. The Bristol Old Vic, the Bristol Hippodrome and the Tobacco Factory provide a wide choice of entertainment in the form of plays, dance, opera and musical theatre.

Leisure and recreation in the city

Sport

Bristol has two professional soccer teams, City and Rovers, and a rugby union team. It is also the headquarters of Gloucestershire County Cricket. All these teams are developing their stadiums to provide a range of leisure and conference facilities and accommodation. Rovers plan to move to a proposed site on the outskirts of the city (photo **B**). They aim to develop facilities attractive to a wider range of people than just sports enthusiasts.

B *An impression of the proposed new UWE Stadium for Bristol Rovers Football Club*

Shopping

Shopping is a growing leisure activity. Bristol has seen major changes in its shopping provision. The out-of-town retail park at Cribbs Causeway affected the Broadmead shopping development in the city centre which had become outdated.

Improved shopping facilities were needed to:

◆ reduce crime ◆ improve the environment

◆ compete with other cities ◆ attract employment.

Developments to encourage people to come back to shopping in the CBD include:

◆ pedestrianising the area and installing CCTV to improve safety

◆ providing a more attractive shopping environment with new street furniture, floral displays and landscaping

◆ the development of open street markets

◆ improving public transport into the centre, e.g. park and ride

◆ promoting tourism to encourage greater spending, by making the nearby Old Market area of the city into a conservation area.

Cabot Circus

This development opened in September 2008 at a cost of £500 million. Shops and leisure facilities take up two-thirds of its floor space. As well as shops there are offices, a cinema, a hotel and 250 apartments.

C *The interior of Cabot Circus Shopping Centre*

Bristol's Harbourside

This is part of the project to regenerate the central part of the city. Former workshops and warehouses have been converted into bars and nightclubs and cultural venues. These include an art gallery, a media and arts centre, a museum and the At-Bristol science exhibition centre. The free three-day annual Harbourside Festival attracts around 300 000 spectators.

D *Bristol's annual Harbourside Festival*

ACTIVITIES

1 a Suggest *three* advantages of the proposed location for the new Rovers soccer stadium.

   b Suggest what the old soccer ground could be used for.

   c Why do new sports developments provide more than just a new pitch and seats for the spectators?

2 How has the out-of-town Cribbs Causeway affected shops in the city centre?

3 How might the developments in central Bristol make it more likely to attract people?

Stretch yourself

Give a short presentation on the social opportunities offered by Bristol to one of the following:
• a teenager
• a parent of a young family
• a pensioner.

Practice question

Explain how the changes in Bristol can prove positive for the people of the city. *(4 marks)*

On this spread you will find out how urban changes in Bristol can create economic opportunities

How has Bristol's industry changed?

Bristol's traditional industries were based on its function as a port. Cigarettes were made using tobacco from the West Indies and sherry from wine imported from Bordeaux. The closure of the city centre port left empty warehouses. Some have been turned into flats (photo **A**) with some re-used for new industry.

Bristol's employment structure is shown in chart **B**. Major developments have been in the *tertiary* (services) and *quaternary* (high-tech) sectors. In 2015 Bristol's unemployment rate was below the UK average. Employment growth in Bristol is projected to be higher than for the UK as a whole.

Why have high-tech industries developed in Bristol?

The major change in Bristol's industry has been the growing number of people employed by high-tech companies. There are 50 micro-electronic and silicon design businesses in the Bristol area – the largest concentration outside California's Silicon Valley.

Bristol is home to global companies such as Aardman Animations, Hewlett-Packard and Toshiba as well as smaller firms working in robotics, 3D printing and other advanced technologies. Chinese telecommunications giant Huawei has invested in the city. The following factors attract high-tech businesses to Bristol:

- a government grant of £100 million to become a Super-Connected City with broadband download speeds of at least 80 Mbps
- close links between the city council and the university
- an educated and skilled workforce
- advanced research at the university
- different industries working collaboratively in research and development
- a clean and non-polluted environment.

> **Did you know?**
> 30 per cent of jobs in Bristol are in the financial services sector

A Warehouses converted into flats along Bristol's harbour front

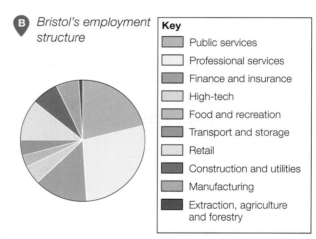

B Bristol's employment structure

Key
- Public services
- Professional services
- Finance and insurance
- High-tech
- Food and recreation
- Transport and storage
- Retail
- Construction and utilities
- Manufacturing
- Extraction, agriculture and forestry

Maths skills

Look at chart **B**.

1 What is the approximate total percentage of people employed in public and professional services in Bristol?

2 Which of the following is the approximate percentage of people employed in high-tech industries?

   5   10   15   20

Aardman Animations

Aardman Animations are based in Bristol. The company was set up in 1972. The studio has become well-known for its films using stop-motion clay animation techniques, particularly those featuring the characters Wallace and Gromit. After making some experimental short films in the late 1990s, it entered the computer animation market. Its films have won an Oscar and many other awards.

C ▶ *The Aardman characters Wallace and Gromit*

Defence Procurement Agency

For some time it has been government policy to decentralise agencies away from London. The Ministry of Defence Procurement Agency (DPA), employing over 10 000 people, was established on a **greenfield site** in 1996 (photo **D**). It supplies the army, air force and navy with everything they need from boots to aircraft carriers.

As the number employed by the DPA increased, there was a need for more housing. This has contributed to the city's **urban sprawl**. Bradley Stoke, with a population of over 21 000, was Europe's largest private housing development when building commenced in the late 1980s. The development is an example of the link between economic opportunities and urban change.

D ▶ *The purpose-built DPA headquarters at Filton*

The aerospace industry

Fourteen of the fifteen main global aircraft companies are found in the Bristol region. These include Rolls-Royce, Airbus and GKN Aerospace. Supply chains have grown up in the region to supply these high-tech companies.

Developments like the Filton Enterprise Area have become established hubs for cutting-edge aviation technology. The area produces parts for aircraft, as well as electronic systems such as those for communications and navigation. There is a 100-year tradition for the aircraft industry in Bristol and this is supported by world-class aerospace courses at local universities.

ACTIVITIES

1 **a** Describe the warehouses shown in photo **A**.

  b How can warehouses like these be redeveloped?

2 What is the evidence that the DPA building has been built on a greenfield site (photo **D**)?

3 **a** Why are universities in Bristol important for the development of high-tech industry in the city?

  b Why is a non-polluted atmosphere important for high-tech industry?

4 Explain why manufacturing is likely to remain less important than service industries in Bristol.

Stretch yourself

How are developments in the aerospace industry complementary to the growth of the high-tech industry in Bristol?

Practice question

How is Bristol making use of changes in the city to promote economic growth? *(6 marks)*

On this spread you will find out how changes in Bristol's economy have created opportunities to improve the environment

How are changes affecting Bristol's environment?

In 2015 Bristol became the first UK city to be awarded the status of European Green Capital (diagram **A**). There is a plan to achieve the following by 2020:

◆ transport improvements

◆ improved energy efficiency

◆ development of renewable energy.

Bristol plans to increase the number of jobs in low-carbon industries from 9000 to 17 000 by 2030. Recent annual growth in the city's green economy was as high as 4.7 per cent. In 2015, Bristol's first year as European Green Capital:

◆ 175 businesses created a 'Green' action plan

◆ major events included an international festival on leadership in green technology and an international competition to develop mobile apps and environmental awareness games

◆ the first 100 electric car charging points were installed in the city

◆ every primary pupil in Bristol planted a tree to improve the city's green coverage.

Improve energy efficiency – reduce energy use by 30% and CO$_2$ emissions by 40% by 2020

Increase the use of renewable energy from 2% (2012)

Reduce water pollution by improved monitoring and maintenance

Establish an Air Quality Management plan to monitor air pollution

Increase the use of **brownfield sites** for new businesses and housing

BRISTOL 2015 EUROPEAN GREEN CAPITAL

A *What is Bristol doing to improve the environment?*

B *The integrated transport plan for Bristol*

An integrated transport system for Bristol

In 2012 Bristol was the second most congested city in the UK. A journey during rush hours takes 31 per cent longer than at other times of the day. Bristol has a higher percentage of people walking and cycling than any other UK city (57 per cent). It aims to double the number of cyclists by 2020.

The key to the city's plans is the development of an integrated transport system (ITS), linking different forms of public transport within the city and the surrounding areas (map **B**).

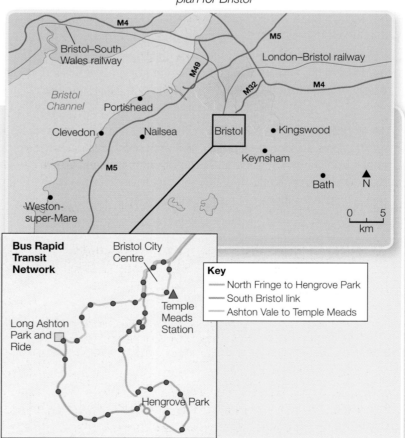

Bristol–South Wales railway
London–Bristol railway
Bristol Channel
Portishead
Clevedon
Nailsea
Bristol
Kingswood
Keynsham
Bath
Weston-super-Mare
M4, M5, M49, M32

0 — 5 km

Bus Rapid Transit Network
Bristol City Centre
Temple Meads Station
Long Ashton Park and Ride
Hengrove Park

Key
— North Fringe to Hengrove Park
— South Bristol link
— Ashton Vale to Temple Meads

An ITS connects different methods of transport, making journeys smoother. The aim is to encourage people to switch from using cars to public transport. This makes transport more sustainable, as well as reducing traffic congestion. The Rapid Transit Network (map **B**) consists of three bus routes linking the main Temple Meads railway station with the city's Park and Ride sites. Construction on the network started in early 2015. The first services will start operating in late 2016.

As part of its transport developments, Bristol is planning many new rail improvements. These include the electrification of the line to London. Electrification will mean greener transport, more reliable journeys and improved connections across southern England and South Wales.

Urban greening

More than a third of Bristol is open space and over 90 per cent of the population live within 350 m of parkland or waterways. There are eight nature reserves and three hundred parks in the city. Queen Square was once a dual carriageway, but has now been transformed into an open space with cycle routes (photo **C**).

There have been a number of green initiatives for the city:

◆ Sites of Nature Conservation Interest (SNCI) to be raised to top conservation condition by 2026

◆ 27 per cent of the city to be part of a wildlife network

◆ objectives set for wildlife in non-natural habitats, e.g. cemeteries

◆ 30 per cent of the city to be covered with trees.

A new housing development at Portbury Wharf was allowed by the local council on condition that the neighbouring area was made into a nature reserve (photo **D**).

The areas of open water and meadow provide an invaluable habitat for wildlife, birds and plants.

C *Queen Square, Bristol*

D *The nature reserve, Portbury Wharf*

ACTIVITIES

1 a Why was it important for the ITS (map **B**) to cover both the city and surrounding areas?

b What does the plan suggest about the link between transport improvements and economic activities?

2 Describe how different forms of transport are linked in Bristols' ITS.

3 How does photo **C** show Bristol's attempts to link transport improvements with urban greening?

4 Why should money be spent on nature reserves in an urban environment?

Stretch yourself

Find out more about the Bristol ITS. How are road and rail linked? What advantages does it bring to Bristol?

Practice question

Describe how environmental changes can improve the quality of life for the people of Bristol.
(6 marks)

On this spread you will find out how change has created environmental challenges in Bristol

Bristol's environmental challenges

Changes in the economy and industry of Bristol have created problems and challenges for the city's environment.

◆ Many industrial buildings that are no longer used have become derelict.

◆ Demand for new homes has led to urban sprawl – new housing developments in rural areas on the edge of the city.

Where are the areas of dereliction in Bristol?

The areas that have become run-down are mainly in the inner city and where there is a concentration of redundant industrial buildings (photo **A**). When the port function moved downstream from the city, many warehouses were abandoned and fell into decay.

A *Disused industrial buildings in Bristol*

Stokes Croft

This inner-city area consisted of high-density housing built in the nineteenth century for industrial workers. The area became notorious for its derelict housing and abandoned properties, including Perry's Carriage Works, which is now a *listed building*. Many empty houses have been taken over by squatters, and the area has suffered from riots and antisocial behaviour.

B *Graffiti on an abandoned building in Stokes Croft*

What is being done to improve the area?

Bristol City Council obtained lottery grants to help improve the poor economic activity and environmental decay in the area. Activists and artists wanted to revitalise the area through community action and public art. It is now well known for its music, independent shops, nightclubs and numerous pieces of graffiti art (photo **B**). There have been protests about the possible *gentrification* of the area, which would mean many people could no longer afford to live there.

How has urban growth led to urban sprawl?

Bristol's growing population towards the end of the twentieth century, and the demolition of older areas of slum dwellings, have led to an increased demand for new housing. Bristol was heavily bombed during the Second World War. Over 3200 houses were lost and 1800 badly damaged. In 1955, 43 families per week were moving into brand new homes on new estates like Hartcliffe on the edge of the built-up area. Many new homes were owned by the council. Private houses were also built and the city's boundaries were extended outwards.

Urban sprawl has extended particularly to the north west of the city. The new town of Bradley Stoke has extended the city to the north.

C *Bristol's Harbourside – a brownfield development*

D *Finzels Reach – redeveloping an old industrial site*

What is being done to reduce urban sprawl?

Bristol has done well in developing brownfield sites.

◆ Between 2006 and 2013 only 6 per cent of new housing developments were on greenfield land.

◆ By 2026 over 30 000 new homes are planned on brownfield sites.

◆ Planned brownfield developments will be high-density with an average of 210 houses per hectare compared with 60 on greenfield sites.

Bristol has successfully developed many smaller-scale brownfield sites, such as Temple Meads, Templegate, Harbourside (photo **C**) and Finzels Reach (photo **D**).

Finzels Reach

This is a 2-hectare brownfield site near the CBD with a redundant sugar refinery and old brewery buildings. The facades of the old industrial buildings have been retained. It is a high-density development with a variety of uses, including:

◆ office space

◆ shops

◆ 400 apartments.

ACTIVITIES

1 a Describe the features of the buildings in photos **A** and **B** which suggest they once had an industrial use.

b Suggest one possible way to re-use these buildings.

c What would be the advantages of re-using the buildings this way?

2 a Explain the advantages of developing brownfield sites (photos **C** and **D**), apart from reducing urban sprawl.

b What are the possible disadvantages of developing a brownfield site?

Stretch yourself

Who are the winners and losers when an area is regenerated?

Practice question

How successful has Bristol been in overcoming environmental challenges? *(4 marks)*

On this spread you will find out how Bristol is responding to its problems of waste disposal and atmospheric pollution

What is Bristol's waste disposal problem?

The amount of waste produced per head in Bristol is 23 per cent lower than the UK average. However, the city still produces over half a million tonnes of waste per year. It is among the worst cities in the country in terms of the amount of food waste it produces.

How is Bristol reducing the environmental impact of waste disposal?

A range of strategies have been adopted to cope with the problem of waste disposal in Bristol and reduce pollution. These include:

◆ reducing the amount of waste that has to be sent to landfill sites

◆ reducing the amount of waste generated per household by 15 per cent

◆ increasing the amount of **waste recycling** to 50 per cent.

Bristol's population has grown by 9 per cent since 2000. The amount of household waste has been reduced by 18 per cent in the same period. A major factor has been the increase in the recycling rate. This has been achieved by:

◆ agreeing higher targets with contractors who handle household waste

◆ doing more to teach pupils in schools about the importance of recycling and how to recycle at home

◆ introducing specialised kerbside collections and facilities for recycling different kinds of household waste

◆ making technological improvements in recycling.

These strategies generate income when recycled materials are sent to reprocessing plants in England and Wales (map **C**). A recycling plant will create around 4.2 million litres of diesel each year by treating 6000 tonnes of waste plastics.

The Avonmouth waste treatment plant treats 200 000 tonnes of waste per year. Any non-recyclable waste is used to generate enough electricity to meet the needs of nearly 25 000 homes in the Bristol area.

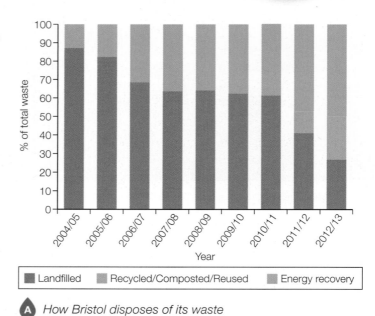

A How Bristol disposes of its waste

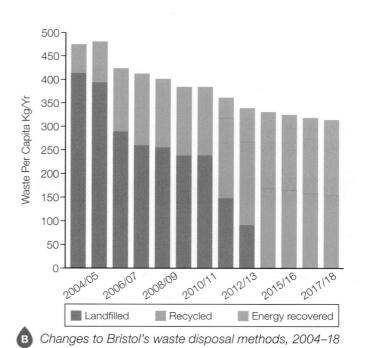

B Changes to Bristol's waste disposal methods, 2004–18

Bristol's atmospheric pollution

Vehicle emissions are the main cause of air pollution in the city (graph **D**). Bristol is the most congested city in England and the main bus routes are often the most polluted.

An estimated two hundred people die in the city as a result of air pollution each year. The prevailing winds are from the south west and at times pollutants are blown over the city from the industrial area around the port at Avonmouth.

Steps are being taken to improve the air quality in Bristol. The whole of the city has been made a smoke control area. Other plans to reduce air pollution include:

◆ reducing speeds limits on motorways and in residential areas

◆ the Frome Gateway, a walking and cycling route to the city centre

◆ an electric vehicle programme with charging points in 40 public car parks

◆ a smartphone app with information about public transport.

Bristol's eco-friendly 'poo bus'

Britain's first bus to be powered by human and food waste will transport people between Bath and Bristol Airport. The bus will run on bio-methane gas produced at a sewage treatment works. The eco-friendly vehicle can travel up to 300 km (186 miles) on one tank of gas, which takes the annual waste of about five people to produce!

All of Bristol's recycling is sent to reprocessors within England and Wales

N
0   150
km

Aluminium cans and foil to Warrington, Cheshire

Plastic to Corby, Northants

Batteries to Chester

Glass to Harlow, Essex

Cardboard and cartons to Sittingbourne, Kent

Textiles and shoes to Bilston, West Midlands

Garden waste and food waste to Sharpness, Gloucestershire

● **Bristol**

Steel cans and aerosols to Port Talbot

Spectacles to Crawley, Sussex

Paper to Aylesford, Kent

Engine oil to Newport

C *What happens to Bristol's recyclable waste?*

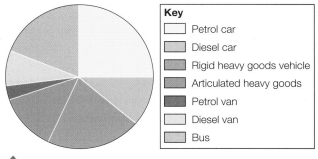

Key
- Petrol car
- Diesel car
- Rigid heavy goods vehicle
- Articulated heavy goods
- Petrol van
- Diesel van
- Bus

D *Nitrogen oxide (NO_2) emissions in Bristol from different types of vehicle*

ACTIVITIES

1 **a** What was the percentage of Bristol's waste sent to landfill in 2004/05 and 2012/13 (graph **A**)?

   b What percentage of waste was used for energy recovery in 2011/12?

2 Describe changes in Bristol's waste disposal methods between 2004 and 2018 (graph **B**).

3 Describe what happens to Bristol's recycleable waste (map **C**).

4 What steps are being taken to reduce air pollution in Bristol?

Stretch yourself

Investigate whether regeneration and environmental improvements always have positive outcomes.

Practice question

Assess the success of Bristol's attempts to reduce the environmental effect of waste disposal. *(6 marks)*

On this spread you will find out how changes in Bristol have created social challenges

Inequality in Bristol

Bristol's population, like that of most UK cities, shows great social variations between different areas. These can be measured by looking at a range of factors that affect people's lives, including housing, education and health.

Lack of investment in the city has led to social inequalities between different areas. In some areas there are high levels of **social deprivation**. On this spread you will find out about two contrasting areas of Bristol – Filwood and Stoke Bishop (map **A**) – to show the inequalities that can be found within the city.

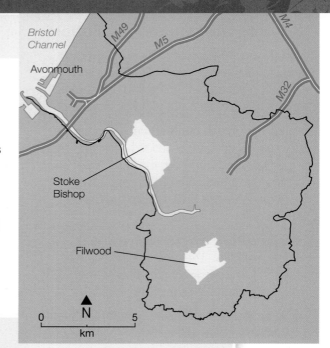

A *The location of Filwood and Stoke Bishop in Bristol*

Filwood

In 2010 a survey by Bristol City Council revealed that more than a third of people living in Filwood and over half the children were in very low-income households. It is in the top 10 per cent of the most socially deprived areas in the country. Bullying, crime, drug use, poor environment, lack of transport and dumped cars are identified as problems facing local residents.

Homes in Filwood are split equally between owner-occupied properties and those rented from the city council. Most of the council houses in the area were built in the 1930s and 1940s. They replaced the slums that had been cleared, and homes bombed during the war. Many are poorly insulated. The designs of the new council areas were not successful and there were plans to replace 1000 homes. But these plans were abandoned after local opposition.

Here are some more facts about Filwood.

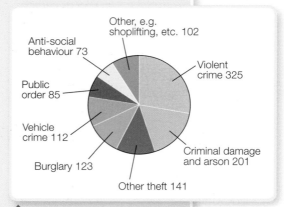

Anti-social behaviour 73
Other, e.g. shoplifting, etc. 102
Violent crime 325
Public order 85
Vehicle crime 112
Burglary 123
Other theft 141
Criminal damage and arson 201

C *Breakdown of reported crime in Filwood, 2014–15*

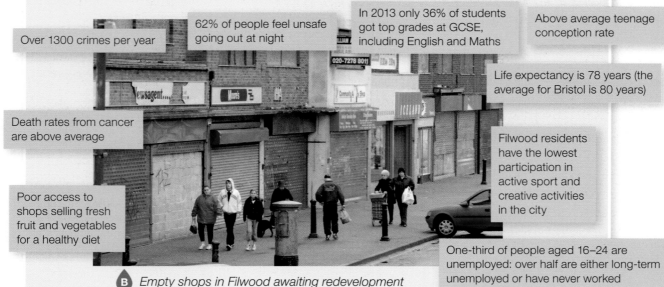

Over 1300 crimes per year

62% of people feel unsafe going out at night

In 2013 only 36% of students got top grades at GCSE, including English and Maths

Above average teenage conception rate

Death rates from cancer are above average

Life expectancy is 78 years (the average for Bristol is 80 years)

Poor access to shops selling fresh fruit and vegetables for a healthy diet

Filwood residents have the lowest participation in active sport and creative activities in the city

One-third of people aged 16–24 are unemployed: over half are either long-term unemployed or have never worked

B *Empty shops in Filwood awaiting redevelopment*

Stoke Bishop

Stoke Bishop is a very affluent suburb to the north west of the city. It includes Sneyd Park, an area that is home to many millionaires who live in large Victorian and Edwardian villas. The area overlooks the open space of Clifton Downs and the gorge of the River Avon. Despite the proportion of older people being 8 per cent higher than the average for Bristol, the death rate is comparatively low.

Here are some more facts about Stoke Bishop.

D *The centre of Stoke Bishop*

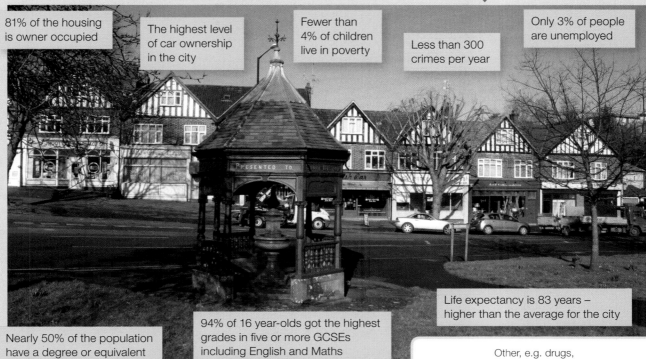

81% of the housing is owner occupied

The highest level of car ownership in the city

Fewer than 4% of children live in poverty

Less than 300 crimes per year

Only 3% of people are unemployed

Life expectancy is 83 years – higher than the average for the city

94% of 16 year-olds got the highest grades in five or more GCSEs including English and Maths

Nearly 50% of the population have a degree or equivalent

ACTIVITIES

1 Compare the two urban environments shown in photos **B** and **D**. Comment on the likely social inequalities.

2 Look at pie charts **C** and **E**.

   a Name the biggest category of crime in *each* area.

   b Suggest why these are not the same in both areas.

   c What do the different levels of vehicle crime suggest about the two areas?

3 Why may statistics not give the complete picture of life in these two parts of Bristol?

Stretch yourself

Use a range of census data to analyse the development of these two areas. How have the two areas changed over time?

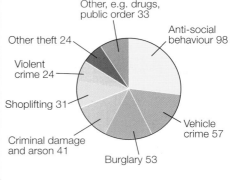

Other, e.g. drugs, public order 33

Anti-social behaviour 98

Other theft 24

Violent crime 24

Shoplifting 31

Vehicle crime 57

Criminal damage and arson 41

Burglary 53

E *Breakdown of reported crime in Stoke Bishop, 2014–15*

Practice question

Assess the causes and consequences of social inequality in Bristol. *(6 marks)*

On this spread you will find out about plans for new housing in Bristol

The Bristol and Bath green belt

The *green belt* was set up to prevent urban sprawl on the **rural-urban fringe** and the merging of the cities of Bath and Bristol.

Only 5 per cent of the green belt around Bristol is controlled by the city authorities. Three neighbouring local authorities are in charge of planning on the majority of the protected land. There is a lot of opposition from local people to building houses on green belt land.

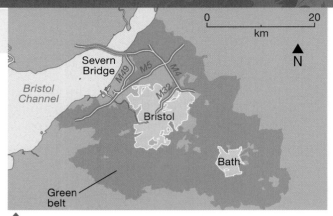

A *The green belt around Bristol*

Bristol's importance as a regional centre means that many people travel from the surrounding areas to work in the city. Towns to the north and south of the city, such as Wotton-under-Edge and Clevedon, have expanded to become commuter settlements.

Housing development on greenfield sites

The national shortage of new houses has meant that recent government policy has encouraged the use of greenfield sites (photo **B**).

B *New housing on a greenfield site*

C *OS map of the Harry Stoke greenfield site*

© Crown copyright

South Gloucestershire

This authority controls the area north of the city. Housing developments have taken place for a number of years, including the building of the new town of Bradley Stoke, built in the late 1980s. A new development of 1200 homes has been built on land at Harry Stoke (map **C**) with a further 2000 homes planned for 2016–17.

Local people objected to the Harry Stoke development. They were concerned about:

- increased congestion, road traffic noise and poor air quality
- the impacts on ecology and loss of habitats (especially the Great Crested Newt population)
- the loss of open space and informal recreational areas
- the impact on existing community services and facilities
- the effect of development on the local flood risk.

Housing development on brownfield land

Bristol has a good record for re-using **brownfield sites**. Between 2006 and 2013, 94 per cent of new housing was built on brownfield sites. Nearly 8000 new homes could be built on 89 identified brownfield sites. They include former office buildings, public houses, coach depots, factories, dockyards and listed buildings.

Demand for brownfield land within Bristol comes from the growing need for student accommodation. The city needs 30 000 new homes by 2026 and the city council is confident that this can be achieved without using any greenfield sites.

Bristol Harbourside

Bristol's dockland declined when cargo ships became too large to come up the River Avon from the Bristol Channel. The closure of several industries around Bristol docks – such as tobacco factories, sand dredging and lead-shot works – left several listed buildings empty and unused. The regeneration of the area has taken 40 years. The re-use of industrial buildings for residential purposes was only part of the scheme, which also included facilities for culture and leisure. Developing the area required cooperation between the council, the landowners – who included British Gas and British Rail – private developers and the South West Regional Development Agency.

D *New apartments on Bristol Harbourside*

Not everybody is happy about the architecture of the waterfront properties. The cost of the flats for sale is too high for most of the people on the city's housing waiting list.

However, the scheme has been successful because:

- a very run-down area of the city has been redeveloped
- several listed buildings have been preserved
- people still live in the centre, so the city does not have a 'dead heart' in the evenings.

ACTIVITIES

1 Describe Bristol's green belt (map **A**).
2 Look at OS map **C**.
   a What is situated at grid reference 625796? (Go to Unit 3 to check on how to use 6-figure grid references.)
   b Using evidence from the map, suggest the advantages of building a new housing development at Harry Stoke.
3 Describe the differences between the new homes shown in photos **B** and **D**. Which social groups are they targeted at?
4 Assess the advantages and disadvantages of housing developments on greenfield and brownfield sites.

Stretch yourself

You are a local planning officer. Explain why the continuing UK shortage of homes justifies building on greenfield sites.

Practice question

Describe how the Bristol region is providing homes for its growing population. *(4 marks)*

On this spread you will find out why the Temple Quarter is in need of urban regeneration

Example

Why regenerate run-down urban areas?

Run-down areas are known as brownfield sites. They are more expensive to build on than greenfield sites because the land and buildings often need to be cleared first. They may also be contaminated from previous industrial use. But there are advantages in developing sites like these.

◆ Existing buildings can be put to a range of uses on any one site.

◆ The land is often disused or in a state of **dereliction**.

◆ The site has already been developed and so reduces urban sprawl.

◆ Using unsightly areas for building developments improves the urban environment.

◆ Sites are often in urban areas, so building there may reduce car use.

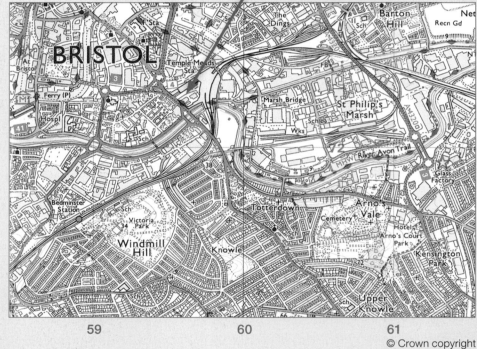

Temple Meads Station

© Crown copyright

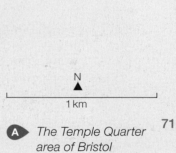

N

1 km

A The Temple Quarter area of Bristol

Why did the Temple Quarter need regeneration?

The Temple Quarter was very run down. It gave a bad impression to visitors, as it was the first part of the city seen by anyone driving from Wells to the south or from Bath to the south east. It is also the area that many visitors see when they first arrive at Temple Meads, the city's main railway station.

What was the area like before regeneration?

The Temple Quarter developed as an industrial area in the eighteenth century. The area was often flooded until the construction of the 'Floating Harbour' and the Feeder Canal in the nineteenth century. The water level in the harbour was no longer affected by the tide, but remained constant. This made more industrial development possible. In 1841, Brunel built the first railway station. More railway sidings were added, until eventually they covered 40 per cent of the area. In the twentieth century the remaining terraced housing was removed in the process of slum clearance.

There are four separate areas within the Temple Quarter (photo **B**).

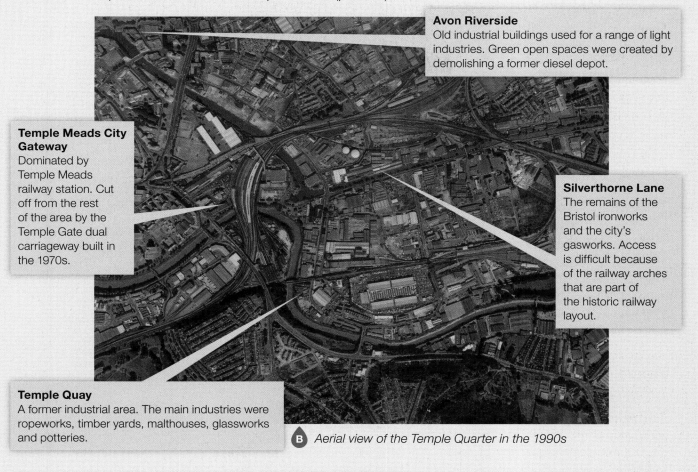

Avon Riverside
Old industrial buildings used for a range of light industries. Green open spaces were created by demolishing a former diesel depot.

Temple Meads City Gateway
Dominated by Temple Meads railway station. Cut off from the rest of the area by the Temple Gate dual carriageway built in the 1970s.

Silverthorne Lane
The remains of the Bristol ironworks and the city's gasworks. Access is difficult because of the railway arches that are part of the historic railway layout.

Temple Quay
A former industrial area. The main industries were ropeworks, timber yards, malthouses, glassworks and potteries.

 Aerial view of the Temple Quarter in the 1990s

Regeneration in the Temple Quarter

The area includes the remains of the ironworks and Brunel's original railway tracks and several listed buildings. The surviving cobbled streets are historically important and give character to the area. The former gasworks is now a car showroom with the former industrial yards used as car parks. A wood company now operates from a former warehouse.

C *Temple Quarter after redevelopment*

ACTIVITIES

1 Describe the location and extent of the Temple Quarter area (figure **A**). Use evidence from the OS map extract and the photo.

2 Why does its location make it an important area to regenerate?

3 In what ways does the Temple Quarter have the characteristics of a brownfield site?

Stretch yourself

Find out why Isambard Kingdom Brunel is so important for Bristol.

Practice question

Explain why the Temple Quarter of Bristol was in need of regeneration. *(4 marks)*

On this spread you will find out how regeneration is improving Bristol's Temple Quarter

Example

How successful has the Temple Quarter regeneration been?

Successful **urban regeneration** must improve an area economically, environmentally and socially.

Regeneration of brownfield sites is an expensive option so there must be evidence of 'value for money'. Bristol's Temple Quarter covers 70 hectares and is one of the largest urban regeneration projects in the UK. The three key aspects of the project are shown in diagram **A**.

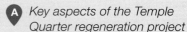
A *Key aspects of the Temple Quarter regeneration project*

Enterprise Zone status
Enterprise Zones encourage economic growth and create jobs. They offer a range of incentives to businesses to move to the area, including business rate relief, low rents and easier planning procedures.

Improved access from in and around Bristol
- Electrification will shorten the rail journey time to London.
- Improvements to Temple Meads station to encourage more people to travel by train.
- Improved road layout with links to the rapid transit network and the Bristol–Bath cycle path.

How has the area been regenerated?

The target is to create 4000 new jobs by 2020 and 17 000 by 2037. There will 240 000 m^2 of either new or refurbished buildings, creating offices, homes, shops and the redeveloped railway station. In addition to the general regeneration of the area, there is a focus on several key projects (map **B**).

Temple Quarter Regeneration

New bridge across the River Avon to the site of the former diesel depot
This gives access to the new Bristol Arena.

B *Key regeneration projects in Bristol's Temple Quarter*

Engine Shed – for high-tech and creative businesses

Glass Wharf – new office development

Electrification of London to Bristol railway line

Temple Studios – new technical and digital enterprises

'Arena Island' – with bridge to new Bristol Arena

Temple Meads Station – major redevelopment to turn the station into a modern transport hub for the city

Paintworks – new mixed use development

Bristol Arena

The team behind London's Millennium Dome and Olympic Stadium designed the new Bristol Arena (image **C**). This is due to open in 2018. Bristol is the largest city in the UK without such a venue. Access will be by the new bridge over the river as well as a pedestrian and cycling bridge to 'Arena Island'. This route is to be redeveloped with cafés, offices and flats.

The arena will allow for smaller capacity theatre-style events with seating for 4000 people. But it can also be used for major conventions, exhibitions and sporting events with up to 12 000 spectators.

It is planned that the area around the arena also becomes somewhere people visit, whether they have a ticket to a show or not. This area will host outdoor events such as an ice rink in the winter or outdoor theatre productions.

C An artist's impression of how the Bristol Arena could look

Brunel's Engine Shed

This is an example of the re-use of a listed historic building.
The new £1.7 million Innovation Centre is being developed in Isambard Kingdom Brunel's historic engine shed at Temple Meads station.

The new centre is home to high-tech, creative and low-carbon sector companies. This will add to Bristol's importance as a major UK high-tech centre. The centre includes:

- 18 micro-electronics, media and digital production companies

- a further 44 companies who use the facilities

- a company developing the next generation of wi-fi

- the use of superfast broadband as part of the Bristol Gigabit project.

ACTIVITIES

1 Give examples of a new building and of the re-use of an old building in the Temple Quarter of Bristol.

2 Why is the Bristol Arena so important for the success of the regeneration project?

3 How important is Temple Meads railway station to the regeneration of the area?

4 In what ways can a regeneration scheme be considered a success?

Stretch yourself

How can listed and other historic buildings in an area affect regeneration?

Practice question

'The regeneration of the Temple Quarter of Bristol is a success.' Discuss. *(6 marks)*

15.1 Planning for urban sustainability

On this spread you will find out that the sustainable development of urban areas requires social, economic and environmental planning

Urban sustainability

You have seen In Chapters 13 and 14 that cities have many problems. There are lots of people and buildings competing for space, consuming huge quantities of energy, water and other resources. There are problems with waste disposal and traffic congestion. But there are ways that towns and cities can tackle these problems and become more sustainable (diagram **A**). This requires social, economic and environmental planning.

Freiburg – a sustainable city

In 1970 the German city of Freiburg set a goal of urban **sustainability**. While environmental concerns were important, the new approach had to consider also how the inhabitants were affected socially and economically.

In this chapter you will learn about Freiburg as an example of a sustainable city.

Social planning in Freiburg

Social planning takes into account people's needs. It is important that people take part in decision making on the things that will affect their lives. There is also a need to provide enough affordable homes.

In Freiburg, local people are involved in urban planning at both local and city level. Possible sites for building are discussed and recommendations made to the council. Special groups put forward the views of children.

◆ Local people can invest in renewable energy resources, e.g. in one district they have invested over £5 million in 9 windmills, 8 solar energy systems (one at the football stadium), a hydro-electric plant, and an energy conservation scheme at the local school.

◆ In addition to financial returns investors received free football season tickets.

◆ Financial rewards are given to people who compost their green waste and use textile nappies.

Economic planning in Freiburg

Economic planning involves providing people with employment. Freiburg is a city where people come to attend conferences on sustainability, and this provides jobs for local people. Many jobs have also been created in the research and manufacture of solar technology.

Recycling water to conserve supplies

Providing green spaces

Reducing the reliance on fossil fuels – and rethinking transport options

Keeping city wastes within the capacity of local rivers and oceans to absorb them, and making 'sinks' for the disposal of toxic chemicals

Sustainable urban strategies

Involving local communities and providing a range of employment

Conserving cultural, historical and environmental sites and buildings

Minimising the use of greenfield sites by using brownfield sites instead

A *Strategies for a sustainable city*

B *The location of Freiburg*

C *Sustainable urban living in Freiburg*

Freiburg's 'Solar Valley'

More than 10000 people are employed in 1500 environmental businesses in the city. More than 1000 people are employed in the solar technology industry producing advanced solar cells and the machinery to make them.

D ▸ *Freiburg's Solar Factory*

The Solar Factory employs 250 people making solar panels

A Solar Training Centre provides training in the skills needed for the new solar technology

Many solar institutions have their HQs in Freiburg, and the city hosts major European solar energy conferences

The Institute for Solar Energy Systems conducts research and has developed new systems for solar cooling and air conditioning

Environmental planning in Freiburg

Environmental planning ensures that resources are not wasted and the environment is protected for future generations. One of the key strategies for making cities more sustainable is to reduce the amount of waste being produced by re-using and recycling as much as possible.

Environmental planning also involves the use of brownfield sites.

Freiburg has...

- a biogas digester for organic food and garden waste which is collected weekly
- more than 1 million corks recycled each year
- 90 kg per head of non-recyclable waste (Germany's average is 122 kg)
- provided energy for 28000 homes from burning waste
- 350 community collection points for recycling
- more than 88% of packing waste recycled
- reduced annual waste disposal from 140000 tons 50000 tonnes tonnes in 12 years

Vauban

The inner city district of Vauban in Freiburg was built on the site of a former army barracks. It now houses 5500 people in low-energy buildings. All existing trees have been retained, with green spaces between the houses providing play areas for children. Green roofs covered in vegetation store water, which is collected and then reused.

Stretch yourself

Investigate why a sustainable urban city needs to conserve its historic environment. What has Freiburg done to conserve its history?

ACTIVITIES

1 Define the term 'urban sustainability' in your own words.
2 List the features of a sustainable city (diagram **A**) under the following headings: 'Economic', 'Environmental', 'Social'.
3 Explain why involving local people in decision-making can make sustainable living more successful.
4 Explain how Vauban shows examples of social and environmental planning.

Practice question

Explain why planners must consider more than just the environment to achieve urban sustainability *(6 marks)*

On this spread you will find out how urban water supply, energy and green spaces can be made sustainable

Sustainable water supply in urban areas

A sustainable water supply depends on individuals using as little water as possible. This involves collecting and recycling water rather than relying solely on water pumped from reservoirs. Homes need to have roof gardens, with facilities for rainwater harvesting and wastewater recycling. Groundwater is the most important source of drinking water and has to be protected from pollution. As water soaks through green open spaces they filter out pollutants.

Freiburg's waste water system allows rainwater to be retained, reused or to seep back into the ground. There are financial incentives for inhabitants to use water sparingly. In the Vauban district (see page 191), in addition to the many open spaces, water conservation involves:

- collecting rainwater for use indoors
- green roofs (photo **A**)
- pervious pavements that allow rainwater to soak through
- unpaved tramways
- drainage wetlands.

Water in the River Dreisam which flows through Freiberg is managed using flood retention basins. These reduce the danger of flooding by storing excess water, which can then be used in the city. These have been designed to fit naturally into the scenery of the Black Forest.

Providing sustainable energy in urban areas

Cities make great demands on energy supply. Up to now, burning fossil fuels has provided this energy. Pollution and climate change have meant that the production of a **sustainable energy supply** is becoming more important.

Freiburg has a strict energy policy based on:

- energy saving
- efficient technology
- use of **renewable energy sources**.

The city plans to be 100 per cent powered by renewable energy by 2050. This will require halving energy consumption by increasing energy efficiency in homes, offices and factories (photo **C**).

Freiburg is one of the sunniest cities in Germany so solar power is an important form of renewable energy. There are about 400 solar panel installations in the city. These include the main railway station and the football stadium.

A *Green roofs look attractive and are used to harvest rainwater*

B *Freiburg's Solar Settlement and Solar Business Park*

C *The 'Heliotrope' rotates to follow the sun*

Freiburg produces 10 million kilowatts of electricity per year from **solar energy**. Homes often produce more than they need, and can sell any excess. The largest proportion of Freiburg's renewable electricity comes from biomass using waste wood and rapeseed oil. Biogas is produced from organic waste. This produces enough energy to heat Freiburg's three swimming pools!

Green spaces in urban areas

The provision of open spaces contributes to sustainability in both economic and environmental terms. These areas serve as the city's 'green lungs' and help keep the air clean. The soil is protected and prevents runoff of water during heavy rainfall. Green spaces provide a natural and free recreational resource as well as providing a habitat for wildlife.

D *Freiburg – the 'green city'*

44% of wood from the city's forests is used for timber but 75% grows back within a year

40% of the city is forested

56% of forests are nature conservation areas of which 50% is managed and the remaining 6% left wild

In the Riselfeld District only 78 hectares are built on, leaving 240 hectares of open space

44 000 trees have been planted in parks and streets

Only native trees and shrubs are planted in the 600 hectares of parks

River Dreisam is allowed to flow unmanaged to provide natural habitats for flora and fauna

ACTIVITIES

1 Give *two* advantages of green roofs (photo **A**).
2 **a** In which direction are the roofs with solar panels likely to be facing (photo **B**)? Explain why.
   b Describe the design of the buildings in the photo.
   c How would this style of building design contribute to sustainability?
3 How do you think local people have benefited from the greening of Freiburg?

Stretch yourself

Only 3.7 per cent of Freiburg's electricity comes from locally generated, renewable resources. This is well short of the target of the 10 per cent set in 2004 and 100 per cent by 2050. Suggest reasons why increasing the proportion of energy from renewable resources is proving difficult.

Practice question

Explain why Freiburg must reduce its use of resources in order to be sustainable. *(6 marks)*

On this spread you will find out how urban transport strategies can reduce traffic congestion

Why is there a need to reduce traffic congestion?

Traffic congestion can lead to air pollution. There are also the n egative economic effects of increased journey times, higher fuel consumption and greater risk of accidents. Bristol's plans to tackle traffic congestion are outlined on page 172. On this spread you will find out how Singapore, Beijing and Freiburg are dealing with the congestion problem.

Freiburg

The city has an integrated traffic plan which is updated every ten years (see page 172 for Bristol's ITS). The most important part of Freiburg's integrated transport system is the tram network. This provides efficient, cheap and accessible public transport.

Compared with other German cities Freiburg has a low car density with less than 500 cars per 1000 residents.

As well as the integrated transport system, there are:

◆ 400 km of cycle paths with 9000 parking spaces for bikes including 'bike and ride' facilities at railway and bus stations

◆ restrictions on car parking spaces; in Vauban district each one costs £20 000!

As a result of Freiburg's transport plan, tram journeys have increased by over 25 000 in one year, while car journeys reduced by nearly 30 000.

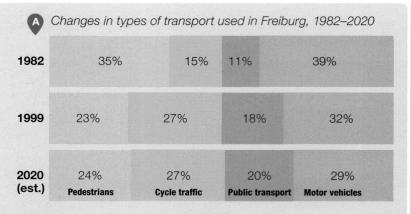

A Changes in types of transport used in Freiburg, 1982–2020

	Pedestrians	Cycle traffic	Public transport	Motor vehicles
1982	35%	15%	11%	39%
1999	23%	27%	18%	32%
2020 (est.)	24%	27%	20%	29%

Low fares allow unlimited travel in the city and surrounding district

Any ticket for a concert, sports or other event is also valid for use on public transport

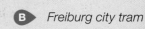
B Freiburg city tram

The tram network covers 30 km and is connected to the 168 km of city bus routes

70% of the population live within 500 m of a tram stop with a tram every 8 minutes

ACTIVITIES

1 Describe how transport use has changed in Freiburg since 1982 (chart **A**).

2 Suggest why an integrated transport plan may encourage more people to use public transport.

3 For each of the three examples, list the strategies described in two columns, headed 'Carrot' and 'Stick'.

Singapore

Singapore, in south east Asia, is a small island state with limited space, so traffic congestion is a major problem. A range of measures have been introduced to reduce the volume of traffic and the number of cars on the roads (**C**).

As a result of the transport policies in Singapore:

- there is 45 per cent less traffic and 25 per cent fewer accidents in the city centre
- traffic on the roads into the city centre has reduced by 40 per cent
- two-thirds of all daily journeys are now by public transport
- car ownership has declined by nearly 1 per cent since 2000.

Restricted entry to the city centre during rush hours

Electronic road pricing on major roads

Quota system to reduce the number of car owners

High petrol prices

Development of an overhead railway system and efficient bus network

High vehicle registration fees and strict requirements for obtaining a driving licence

Advanced electronic monitoring and control of traffic signals to keep traffic flowing

Financial incentives for using cars only at weekends

Government car-sharing schemes

C *Electronic road (ERP) pricing in Singapore*

Beijing

Beijing, the capital city of China, has an estimated 5 million cars. The city centre is often gridlocked. Congestion is predicted to get worse as the number of cars on the roads continues to grow. Since the 2008 Olympic Games, a wide range of strategies (**D**) have been introduced to reduce the high level of traffic congestion.

Limiting car sales. Only 20% of people who apply to own a vehicle are allowed to do so.

Restrictions on vehicle use. Cars are banned from the city one day a week, based on a number plate system. Non-residents cannot bring a car into the city.

Increased parking fees. Congestion charge and pollution tax introduced to help improve air quality.

Expansion of the public transport system. Thirty new metro lines and a rapid bus transit system to be built by 2020. The metro currently serves only half of Beijing's population.

D *The world's worst traffic jam in Beijing, 2010*

Did you know?

The world's worst traffic jam happened in Beijing in 2010. More than 100 km long and lasting 11 days!

The restrictions in Beijing have led to a 20 per cent drop in car use. There has been a 12 per cent drop in the use of car parks in the city centre. But building and widening roads has also resulted in increased car use at the expense of cycling

Stretch yourself

Suggest why the traffic plans in Singapore and Beijing may have a better chance of being effective than those in Bristol and Freiburg.

Practice question

For any named city you have studied evaluate the strategies employed to manage traffic congestion.
(6 marks)

16.1 Our unequal world

On this spread you will find out about global variations in economic development and quality of life

What is development?

Development means positive change that makes things better. As a country develops it usually means that people's standard of living and quality of life will improve. Different factors affect a country's level and speed of development.

◆ *Environmental* factors such as natural hazards, e.g. earthquakes

◆ *Economic* factors such as **trade** and debt

◆ *Social* factors such as access to safe water and education

◆ *Political* factors such as stable government or civil war.

The **development gap** is the difference in standards of living between the world's richest and poorest countries.

A *Does this family enjoy a good quality of life?*

Measuring development

Gross National Income (GNI)

Wealth and income can be used to describe a country's level of economic development. A common measures used by the World Bank is **Gross National Income (GNI)**.

GNI is the total value of goods and services produced by a country, plus money earned from, and paid to, other countries. It is expressed as per head (per capita) of the population.

The World Bank uses four different levels of income to divide the countries of the world into: high, upper-middle, lower-middle, and low.

The UK, most of Europe, North America and Australia, Argentina and Japan are all HICs (map **B**). Most, but not all, LICs are in Africa.

This *economic* indicator is one way of showing development. Some countries may seem to have a high GNI as they are relatively wealthy and have a small population. But this does not always mean that their citizens enjoy a good quality of life. Equally, some people in LICs are well off and enjoy a high standard of living.

Some countries have begun to experience higher rates of economic development, with a rapid growth of industry. These are known as **Newly-Emerging Economies (NEEs)**, for example, Brazil, Russia, India, China and South Africa (the BRICS countries) and the MINT countries (Malaysia, Indonesia, Nigeria and Turkey).

Key

■ High
■ Higher middle
■ Lower middle
■ Low
■ No data

PPP (purchasing power parity) – adjustment made to income to equate what can be purchased for the same amount in different countries

B *Gross National Income per capita in PPP terms, 2013*

Human Development Index (HDI)

Devised by the United Nations, HDI links wealth to health and education. It aims to show how far people are benefiting from a country's economic growth. It is a *social* measure. Measures used to produce the HDI are:

◆ **life expectancy** at birth

◆ number of years of education

◆ GNI per head.

The HDI is expressed in values 0–1, where 1 is the highest. This enables countries to be ranked. The highest-ranked country in 2014 was Norway (0.944), followed by Australia (0.935) and Switzerland (0.930). In 2014 the UK was 14th (0.930). The lowest-ranked country in 2014 was Niger (188th) with an HDI of 0.348. The lowest ten countries were all in Africa.

Map **C** shows the pattern of development according to the HDI, using four categories of development.

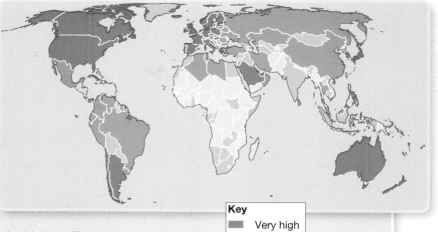

Key
- Very high
- High
- Medium
- Low

C *World HDI scores, 2014*

How can we measure quality of life?

Economic and social measures use broad statistics to measure standard of living for whole countries. But they cannot give an accurate measure of an individual's quality of life.

A good quality of life will mean different things in different countries. Consider, for example, safety and security, freedom and the right to vote, women's rights and … happiness!

The strange-looking map (map **D**) is called a *topological* map. Instead of using true scale, it has been drawn to show the size of each country in proportion to the number of people living on US$10 a day or less.

D *Topological map showing the number of people living on US$10 a day or less*

ACTIVITIES

1 What factors affect the quality of life of the family in photo **A**?

2 Describe the pattern of high and low income countries (map **B**).

3 **a** Describe the global pattern of HDI (map **C**).

  b What are the advantages of HDI as a measure of development?

4 Comment on the usefulness of map **D** in classifying people's quality of life.

Stretch yourself

Investigate other ways of subdividing the world, for example the PQLI (page 196), world peace or gender equality.

Practice question

To what extent is the HDI the most effective measure of development?
(6 marks)

On this spread you will find out about the economic and social indicators of development

What are the measures of development?

There are many economic and social measures of development. For example:

◆ Gross National Income (GNI), used by the World Bank to measure economic development.

◆ The United Nation's Human Development Index (HDI), involving both economic and social factors.

Table **A** identifies several economic and social measures of development and lists examples from selected countries. You will study Nigeria in Chapter 17.

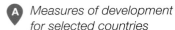 **A** *Measures of development for selected countries*

Country	GNI per head (US$)	HDI	Birth rate (per 1000 per year)	Death rate (per 1000 per year)	Infant mortality (per 1000 live births per year)	Number of doctors (per 1000 people)	Literacy rate (%)	% of population with access to safe water
USA	55200	.915	12.49	9.35	5.87	2.5	99.0	99
Japan	42000	.891	7.93	9.51	2.08	2.3	99.0	100
UK	43430	.907	12.17	9.35	4.38	2.8	99.0	100
Brazil	11530	.755	14.46	6.58	18.60	1.9	92.6	98
Turkey	10830	.761	16.33	5.88	18.87	1.70	95.0	100
China	7400	.727	12.49	7.53	12.44	1.90	96.4	95
Nigeria	**2970**	**.514**	**37.64**	**12.90**	**72.70**	**0.40**	**59.6**	**69**
Ivory Coast	1450	.462	28.67	9.55	58.70	.01	43.1	82
Bangladesh	1080	.570	21.14	5.61	44.09	.40	61.5	87
Zimbabwe	840	.509	32.26	10.13	26.11	.10	86.5	77

How useful are the measures of development?

Some of the measures shown in table **A** are more useful than others.

◆ **Birth rate** is a reliable measure. As a country develops, women are likely to become educated and want a career rather than staying at home. They marry later and have fewer children.

◆ **Death rate** is a less reliable measure. Developed countries such as the UK, Germany and Japan tend to have older populations and death rates will be high. In less developed countries, such as the Ivory Coast or Bangladesh, death rate may be lower because there are proportionally more young people.

◆ **Infant mortality** rate is a useful measure of a county's health care system.

◆ The number of doctors per 1000 people indicates how much money a country has for medical services.

◆ A high **literacy** *rate* shows a country has a good education system.

◆ A high percentage of access to clean water shows a country has modern infrastructure. such as dams, reservoirs and water treatment plants

An indicator with several variables, such as the Human Development index (HDI) or the Physical Quality of Life Index (PQLI), combines social and economic factors. These indicators are generally more useful.

What are the limitations of economic and social measures?

A single measure of development can give a false picture, as it gives the *average* for the whole country. Both photos in **B** were taken in an Arab country with a high GNI. But these two people clearly have a very different quality of life.

Other factors may limit the usefulness of economic and social measures of development:

◆ Data could be out of date or hard to collect.

◆ Data may be unreliable (the level of infant mortality is well above the figures given by some countries).

◆ They focus on certain aspects of development, and may not take into account subsistence or informal economies, which are important in many countries.

◆ Government corruption may mean that data are unreliable.

In many countries the top 10 per cent of the population may own 80 per cent of the wealth. It may also be concentrated in cities while rural areas remain very poor.

How can we compare people's quality of life?

People in different countries have very different ideas of what affects their quality of life. Consider refugees fleeing war-torn Syria in 2016 to seek sanctuary in Europe. They have virtually nothing but they are at least relatively safe. This is why it is very difficult to use social indicators to compare different countries' level of development.

B *Variations in development within a rich country.*

Maths skills

Studying correlation (relationship) between data sets
Graph **C** is called a scattergraph (see page 338). It shows the relationship between HDI and birth rate. You might expect that the higher the level of development (HDI) the lower the birth rate. This is shown by the negative relationship indicated by the 'best fit' line.

1 Draw a scattergraph to show the correlation between HDI and *one* of the other measures in table **A**. Describe the relationship (if there is one) and draw a best fit line.

2 Use the data in the table to do the same to show the correlation between a *social* index and an *economic* one.

C *Correlation between HDI and birth rate*

ACTIVITIES

1 a List the measures where a higher figure indicates a higher level of development (table **A**).

   b List the measures where a lower figure indicates a higher level of development.

   c Which figures for Turkey appear to be anomalies (exceptions)?

   d Why is death rate a poor measure of development?

2 Suggest why there are such clear differences in the quality of life of the two people in the photos in **B**.

Stretch yourself

Why is it so difficult to give a true picture of the level of development in many Arab countries?

Practice question

Use a range of development indicators to explain the difference between standard of living and quality of life.
(6 marks)

On this spread you will find out how levels of development can be linked to the Demographic Transition Model

What is the Demographic Transition Model?

The Demographic Transition Model (DTM) (diagram **A**) shows changes over time in the population of a country. It is based on the changes that took place in western countries such as the UK.

The gap between birth rate and death rate is called *natural change*. This usually shows a natural increase in population but in Stages 1 and 5 a *natural decrease* happens.

The total population of a country responds to variations in birth and death rates (natural change). It will also be affected by migration, both *immigration* (people moving in) and *emigration* (people moving out). This is not shown in the DTM.

What links the DTM with development?

As a country becomes more developed its *population characteristics* change. Graph **A** shows the general increase in level of development from Stage 1 to Stage 5.

A *The Demographic Transition Model*

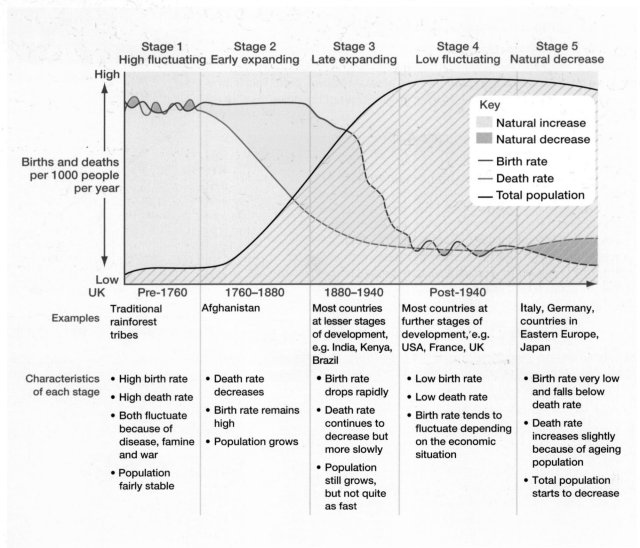

	Stage 1 High fluctuating	Stage 2 Early expanding	Stage 3 Late expanding	Stage 4 Low fluctuating	Stage 5 Natural decrease
Examples	Traditional rainforest tribes	Afghanistan	Most countries at lesser stages of development, e.g. India, Kenya, Brazil	Most countries at further stages of development, e.g. USA, France, UK	Italy, Germany, countries in Eastern Europe, Japan
Characteristics of each stage	• High birth rate • High death rate • Both fluctuate because of disease, famine and war • Population fairly stable	• Death rate decreases • Birth rate remains high • Population grows	• Birth rate drops rapidly • Death rate continues to decrease but more slowly • Population still grows, but not quite as fast	• Low birth rate • Low death rate • Birth rate tends to fluctuate depending on the economic situation	• Birth rate very low and falls below death rate • Death rate increases slightly because of ageing population • Total population starts to decrease

Countries at different stages of development

Stage 1: Traditional rainforest tribes
In parts of Indonesia, Brazil and Malaysia, small groups of people live separately with little contact with the outside world. They have high birth and death rates (photo **B**).

Stage 2: Afghanistan
Afghanistan is one of the poorest and least developed countries in the world. Its birth rate is 39 per 1000 and its death rate is 14 per 1000. About 80 per cent of the population are farmers who need children to support them in the fields and tending livestock.

B *Rainforest tribe*

Stage 3: Nigeria
Nigeria is a newly-emerging economy (NEE) experiencing economic growth. The death rate is much lower than birth rate (see page 220). The country's population is growing rapidly.

C *Germany's ageing population*

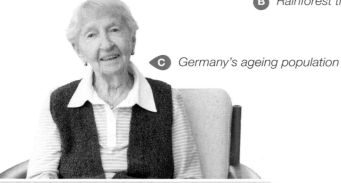

Stage 4: USA
The USA is one of the most developed countries in the world. Good-quality health care means death rates are low (8 per 1000). Women tend to have small families, choosing to study and follow careers. Therefore, birth rate is low (13 per 1000) Population growth is due mainly to immigration.

Stage 5: Germany
Germany is a well-developed country experiencing population decline as death rate exceeds birth rate. The birth rate is 8.2 per 1000 – the lowest in the world. Women have careers and have few children. With an ageing population, Germany's death rate (11.2 per 1000) will continue to rise (photo **C**).

ACTIVITIES

1 Draw a sketch to show the DTM. Indicate on your sketch the links with economic development.

2 How does a falling birth rate reflect increased economic development?

3 How can an increasing death rate reflect high levels of development?

4 Consider the possible impacts for Germany of being in Stage 5 of the DTM. How might it have a negative effect on development?

Stretch yourself

Investigate the population and developmental characteristics of Afghanistan. Do you agree that it is in Stage 2 with a low level of development?

Practice question

Evaluate how far economic development can be linked to the DTM.
(6 marks)

On this spread you will find out how the population structures of two contrasting countries are changing

Population pyramids

Geographers don't only look at total population numbers – they also look at the *structure* of a population. That means thinking about how many babies are being born and how many people are dying – and how the number of people in different age groups is changing. This is done using graphs called *population pyramids*.

A population pyramid is a type of graph which shows the percentage, or number, of males and females in each age group – how many aged 0–4 years, 5–9 years, and so on.

Understanding population pyramids

It is important to know how to 'read' a population pyramid.

◆ *Understanding the overall* shape. For example, if the pyramid is wide at the bottom – like those for Mexico in graph **A** – it means that there in a high proportion of young people in the population.

◆ *Interpreting details* – for example, bars that are longer or shorter than those above and below them. Shorter bars could indicate high death rates in those age groups – perhaps through war or famine.

The dependency ratio

The *dependency ratio* is the proportion of people below (aged 0–14) and above (over 65) normal working age. This is calculated by adding together the numbers for both groups, then dividing by the number aged 15–64 (the 'working population'), and multiplying by 100.

The lower the number, the greater the number of people who work and are less dependent. The higher the number, the greater the number who are dependent on the working population. Low dependency ratios are more common in HICs than NEEs and LICs.

Dependency ratios change as a country develops.

Why is Mexico's population structure changing?

Mexico has a large proportion of young people. Under-15s currently make up 28 per cent of the population, and just over 7 per cent are over 65. The average age is 27. But Mexico's population structure is slowly changing:

◆ Death rate is falling – just 5 deaths per 1000. More babies are being born, and people are living longer, due to an increase in childhood vaccination and improved health care.

◆ Birth rate is 19 per 1000, and falling rapidly. Even if people have fewer children than their parents, the population of Mexico will continue to rise for some time to come.

It is expected to be at least 50 years before Mexico's population levels out. Today's young people will then be moving into old age.

Key

Age 65 and over

Age 15–64

Age 0–14

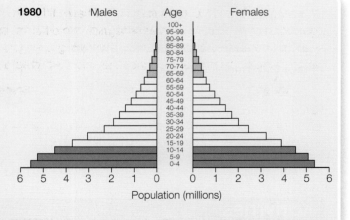

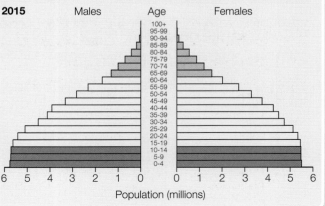

A *Mexico's changing population structure*

How is Japan's population structure changing?

By contrast with Mexico, Japan has an ageing population which is getting smaller (figure **B**). Japan has the oldest population in the world – 27 per cent of the population are over 65 (with under-15s just 13.1 per cent). The average age is 46. Japan's population structure is also changing:

◆ People are living longer. Death rate is 10 per 1000. Average life expectancy in Japan is 80 for men and 87 for women, due to a healthy diet (low in fat and salt) and a good quality of life. Japan is one of the richest countries in the world and has good health care and welfare systems. There are 230 doctors for every 100 000 people (compared with 281 in the UK).

◆ Birth rate in Japan is 8 per 1000, and has been falling since 1975. The average age when women have their first child rose from 26 in 1970 to 30 in 2012. The number of couples getting married has fallen, and the age at which they get married has increased.

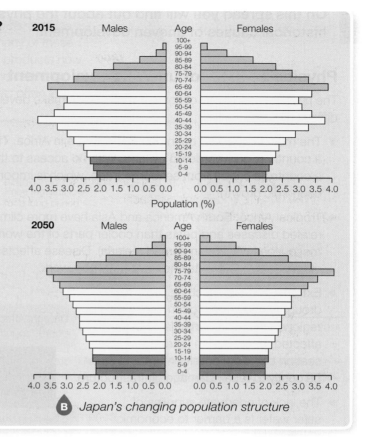

B *Japan's changing population structure*

Population pyramids and the DTM

Countries at different stages of the DTM have different shaped population pyramids. If you can recognise the different basic shapes, and understand what they're showing, then you can tell which stage of the model a country has reached (diagram **C**).

C *Population pyramid shapes for the stages of the DTM*

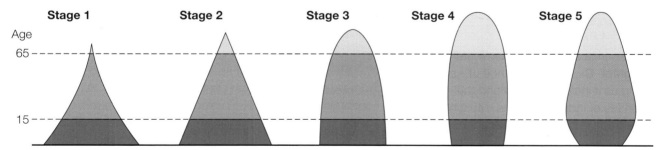

ACTIVITIES

1 Explain what population pyramids show.

2 Look at the two population pyramids for Mexico (diagram **A**).

  a Describe the shape of the pyramid in 1980.

  b What changes took place between 1980 and 2015?

  c Suggest reasons for the changes you have identified.

Practice question

Compare the population structure of an LIC or NEE with one for an HIC. *(6 marks)*

Stretch yourself

Use the population pyramids for Japan (graph **B**) to explain why the country's population is getting older, and is declining.

Uneven development – wealth and health

On this spread you will find out how uneven development leads to inequalities of wealth and health

What is the imbalance between rich and poor?

There is a global imbalance between rich and poor. Some countries, particularly in Africa and parts of the Middle East, have lower levels of development and a poorer quality of life than richer western countries.

Imbalances also exist within countries. Areas of considerable poverty can be found in parts of the UK and USA, and great wealth in some of the world's poorest countries. Inequalities exist at all scales and in all countries.

How does uneven development lead to disparities in wealth?

There is a clear link between a country's development and the wealth of its people. The most developed countries enjoy the greatest wealth. Wealth, in the form of **Gross National Income (GNI)** is often used as a measure of levels of development (figure **B** page 194).

There are significant differences between the wealth of different global regions (graph **B**).

◆ In 2014 the fastest growth in wealth was in North America, which now holds 35 per cent of total global wealth. This wealth is held by just over 5 per cent of the world's adult population!

◆ The USA is not the world's wealthiest country (that is Qatar), but it is the world's most important economic 'engine of growth'.

A *A cartoon highlighting global inequalities*

◆ Of the newly emerging economies, China has recorded the highest growth since 2000. Personal wealth in India and China has quadrupled since 2000, yet its global share of wealth is still well below that of its population.

◆ Africa's share of global wealth is very small (about 1 per cent).

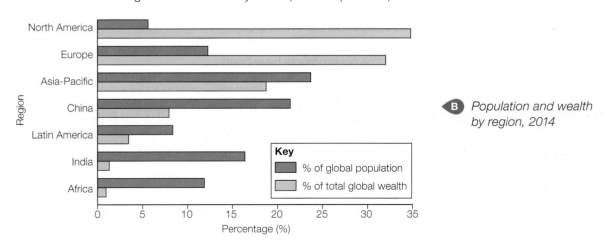

B *Population and wealth by region, 2014*

Disparities in health

Levels of development are closely linked to health. LICs are unable to invest in good-quality health care. In the world's poorest countries health care is often very patchy. There is a wide disparity between causes of death in HICs and LICs.

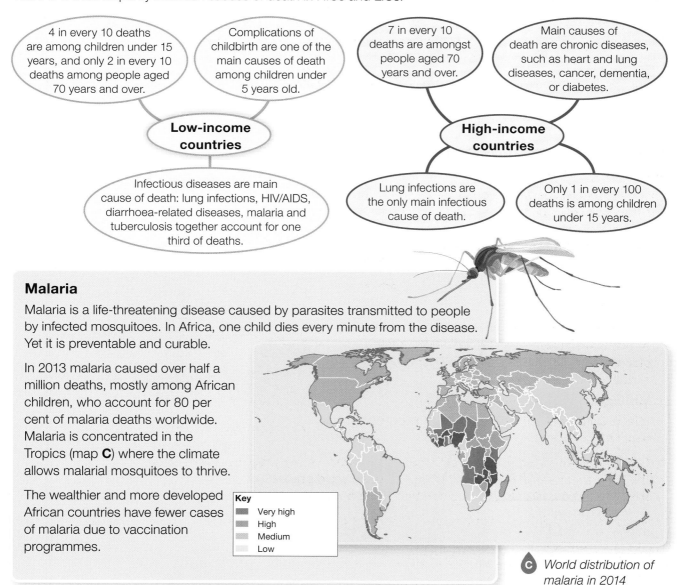

4 in every 10 deaths are among children under 15 years, and only 2 in every 10 deaths among people aged 70 years and over.

Complications of childbirth are one of the main causes of death among children under 5 years old.

7 in every 10 deaths are amongst people aged 70 years and over.

Main causes of death are chronic diseases, such as heart and lung diseases, cancer, dementia, or diabetes.

Low-income countries

High-income countries

Infectious diseases are main cause of death: lung infections, HIV/AIDS, diarrhoea-related diseases, malaria and tuberculosis together account for one third of deaths.

Lung infections are the only main infectious cause of death.

Only 1 in every 100 deaths is among children under 15 years.

Malaria

Malaria is a life-threatening disease caused by parasites transmitted to people by infected mosquitoes. In Africa, one child dies every minute from the disease. Yet it is preventable and curable.

In 2013 malaria caused over half a million deaths, mostly among African children, who account for 80 per cent of malaria deaths worldwide. Malaria is concentrated in the Tropics (map **C**) where the climate allows malarial mosquitoes to thrive.

The wealthier and more developed African countries have fewer cases of malaria due to vaccination programmes.

Key
- Very high
- High
- Medium
- Low

C *World distribution of malaria in 2014*

ACTIVITIES

1. How does cartoon **A** show the causes and effects of global disparities of wealth?
2. **a** Which two regions in graph **B** have a higher share of global wealth than their share of the world's population?

   b Which region has the greatest disparity between wealth and population?
3. Describe and suggest reasons for the pattern of malaria cases (map **C**).
4. To what extent is malaria a disease of poverty?

Stretch yourself

Investigate why malaria is such a devastating disease in Africa. What factors will influence its future eradication from the continent?

Practice question

How does uneven development lead to disparities of global wealth? *(4 marks)*

On this spread you will find out how uneven development leads to international migration

What are the different types of migration?

Photo **A** shows a group of migrants fleeing from poverty, war and persecution in Afghanistan and Syria. They are seeking safety and the chance of a better life in Europe. International migration is one of the main consequences of uneven development, as people seek to improve the quality of their lives.

Migration is the movement of people from place to place. It can be voluntary, where people consider the advantages and disadvantages of moving. Or it can be forced, where people have little or no choice to escape natural disasters, wars or persecution. It is important to make sure you understand the following terms.

◆ *Immigrant* – a person who moves into a country.

◆ *Emigrant* – a person who moves out of a country.

◆ *Economic migrant* – a person who moves voluntarily to seek a better life, such as a better-paid job or benefits like education and health care.

A *Afghan and Syrian migrants in a temporary shelter in Greece, 2015*

◆ *Refugee* – a person forced to move from their country of origin often as a result of civil war or a natural disaster such as an earthquake.

◆ *Displaced person* – a person forced to move from their home but who stays in their country of origin.

Middle East refugee crisis, 2015

In the last few years hundreds of thousands of desperate refugees have fled their homes in Syria, Afghanistan and Iraq in search of a better life in Europe. They are responding to uneven levels of development.

In Syria a civil war has raged since 2011. In five years the war has claimed 470 000 lives, and 11.5 per cent of Syria's population has been killed or injured. Four million have fled the country to temporary camps in Turkey, Jordan and Lebanon. Here there are no jobs and few prospects of a better life.

Thousands have made the dangerous journey across the Mediterranean in overcrowded and unsafe boats. Some of these have capsized and many lives have been lost. Some people made the long journey by land through Turkey and into Eastern Europe (map **B**).

B *Main migration routes from Syria into Europe*

In August 2015 Germany announced that it would process asylum claims for anyone who reached Germany. This sparked a mass exodus across Europe (photo **C**). Many more people left Syria to escape the war.

It's estimated that 1.1 million migrants entered Germany in 2015. German Chancellor Angela Merkel came under pressure to slow the number of arrivals. In March 2016, the EU and Turkey signed a deal to give Turkey political and financial benefits in return for taking back refugees and migrants.

Apart from Germany many refugees have travelled to Sweden and through France towards the UK. The UK government has pledged to accept 20 000 refugees.

In January 2016, Sweden announced it was going to deport 80 000 migrants.

 Syrian refugees walking through Europe to Germany

Economic migration to the UK

The UK has a long history of accepting migrants from all over the world. The country is known for its tolerant approach and many parts of the UK benefit from being multicultural.

Since 2004 over 1.5 million economic migrants have moved to the UK, two-thirds of whom are Polish. The unemployment rate in Poland is over 10 per cent, and they can earn up to five times as much in the UK. Money is often sent home to friends and relatives.

Most migrants pay tax, which is good for the UK economy. They are prepared to work hard, often doing manual jobs such as working on farms (photo **D**). However, they do put pressure on services such as health and education.

 Migrant workers on a Lincolnshire farm

ACTIVITIES

1 What is the difference between an economic migrant and a refugee?

2 **a** Describe and suggest reasons for the routes taken by Syrians fleeing their country (map **B**).

  b Why was there a surge in migration to Europe in August 2015?

3 What are the arguments for and against the UK accepting economic migrants?

Stretch yourself

Investigate the migration crisis involving people from Syria, Afghanistan and Iraq. What are the latest trends? Try to find some statistics to support your research.

Practice question

How does uneven development cause international migration? *(4 marks)*

On this spread you will find out how investment, industrial development and tourism can reduce the development gap

What strategies can reduce the development gap?

Reducing the development gap can involve a range of strategies that aim to improve a country's economy and the quality of life of its people.

Investment

Many countries and TNCs choose to invest money and expertise in LICs to increase their profits. Investment can involve:

◆ the development of infrastructure such as water, roads and electricity

◆ the construction of dams to provide electricity

◆ improvements to harbours and ports

◆ the development of new industries.

Investment can support a country's development by providing employment and income from abroad. As economies grow, poverty decreases and education improves. People become more politically involved, leading to better government. Investment is not the same as a **loan**, which is simply the provision of money with agreed terms of repayment.

Industrial development

Industrial development brings employment, higher incomes and opportunities to invest in housing, education and **infrastructure**. This is called the *multiplier effect* (diagram **B**). Countries such as Malaysia, Brazil, Mexico and China have all followed programmes of industrialisation to achieve their current levels of development.

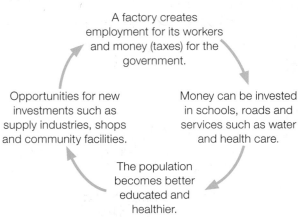

A factory creates employment for its workers and money (taxes) for the government.

Money can be invested in schools, roads and services such as water and health care.

The population becomes better educated and healthier.

Opportunities for new investments such as supply industries, shops and community facilities.

B *The multiplier effect*

Foreign investment in Africa

China has now become Africa's most important trading partner, overtaking the USA. However, in recent years a number of US companies have invested in the continent (map **A**).

More than 2000 Chinese companies have invested billions of dollars in Africa, mainly in energy, mining, construction and manufacturing. They have invested in a power plant in Zimbabwe, hydro-electricity in Madagascar and railway construction in Sudan.

Chinese investment has led to new roads, bridges, stadiums and other projects being built all over Africa. The building of the new headquarters of the African Union was funded entirely by China, at a cost of $US200 million.

There are many benefits to Chinese investment in Africa, but some people think it is exploiting the continent's resources to benefit China's own economy.

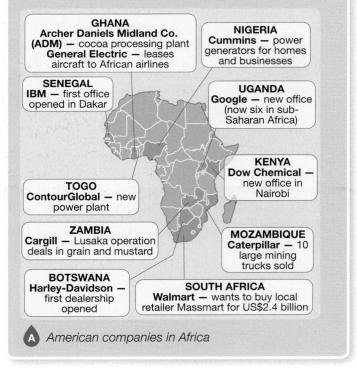

GHANA
Archer Daniels Midland Co. (ADM) — cocoa processing plant
General Electric — leases aircraft to African airlines

NIGERIA
Cummins — power generators for homes and businesses

SENEGAL
IBM — first office opened in Dakar

UGANDA
Google — new office (now six in sub-Saharan Africa)

KENYA
Dow Chemical — new office in Nairobi

TOGO
ContourGlobal — new power plant

ZAMBIA
Cargill — Lusaka operation deals in grain and mustard

MOZAMBIQUE
Caterpillar — 10 large mining trucks sold

BOTSWANA
Harley-Davidson — first dealership opened

SOUTH AFRICA
Walmart — wants to buy local retailer Massmart for US$2.4 billion

A *American companies in Africa*

Industrial development in Malaysia

Malaysia is one of the richest countries in south-east Asia. Since the 1970s it has seen a dramatic growth in its wealth and the quality of life of its population. This is due to the development of its natural resources such as oil and gas, palm oil and rubber. It has made use of foreign investment to exploit these resources and develop a thriving manufacturing sector. One of Malaysia's leading products is the Proton car (photo **C**).

Today Malaysia has a highly-developed mixed economy with growing financial and service sectors (table **D**) and flourishing trade links with the rest of the world.

C *Proton automated car assembly line*

D *Malaysia's economic profile*

Sector	Average % annual growth rate 2011–15	% share of GDP in 2015
Services	7.2	58.3
Manufacturing	5.7	26.3
Construction	3.7	2.9
Agriculture	3.3	6.6
Mining	1.1	5.9

Tourism

For some countries, tourism has helped to reduce the development gap. Countries with tropical beaches, spectacular landscapes or abundant wildlife have become tourist destinations. This has led to investment and increased income from abroad, which can be used for improving education, infrastructure and housing.

Several countries in the Caribbean, such as the Bahamas and the British Virgin Islands, and Indian Ocean islands such as the Seychelles (photo **E**) and the Maldives have become highly dependent on tourism. This can be an advantage and a disadvantage. Tourism can generate a lot of income but is vulnerable in times of economic recession.

E *A tourist village in the Seychelles*

ACTIVITIES

1 What is the difference between an investment and a loan?

2 Why do countries like China and the USA choose to invest in Africa? What are the advantages and disadvantages of foreign investment for African countries?

3 What is the multiplier effect and how can it help to reduce the development gap (diagram **B**)?

4 How can the development of tourism help to close the development gap (photo **E**)?

Stretch yourself

Use the internet to research Chinese and American investment in Africa. Use labels to locate examples of these investments on an outline map of Africa.

Maths skills

Draw two appropriate graphs to display the data in table **D**.

Practice question

How can industrial development reduce the development gap? *(4 marks)*

On this spread you will find out how aid and intermediate technology can reduce the development gap

What is aid?

Aid is when a country or non-governmental organisation (NGO) such as Oxfam donates resources to another country to help it develop or improve people's lives. Aid can take the form of:

◆ money (grants or loans)

◆ emergency supplies (tents, medicines, water, etc.)

◆ food such as rice or wheat, technology (tools or machinery)

◆ skills (people with special skills such as doctors or engineers).

There are different types of aid, in different circumstances and sometimes with specific conditions attached (diagram **A**).

Short-term – emergency help usually in response to a natural disaster, such as a flood or earthquake

Long-term – sustainable aid that seeks to improve resilience, e.g. wells to reduce the effects of drought, or improvements in agriculture

Bilateral – aid from one country to another (which is often tied)

Types of aid

Tied – aid may be given with certain conditions, e.g. that the recipient has to spend the aid money on the donor country's products

Multilateral – richer governments give money to an international organisation such as the World Bank, which then redistributes the money as aid to poorer countries

Voluntary – money donated by the general public in richer countries and distributed by NGOs such as Oxfam

A *Different types of aid*

How can aid reduce the development gap?

Only aid that is long-term and freely given can really address the development gap. Aid can enable countries to invest in development projects such as roads, electricity and water management that can bring long-term benefits. On a local scale aid can help improve people's quality of life if it focuses on health care, education and services.

UK aid

The UK currently spends 0.7 per cent of its Gross Domestic Product (the measure of the wealth of a country) on overseas aid – the target set by the United Nations. In 2013 the top three recipients of UK Official Development Assistance (ODA) were Pakistan (£338 million), Ethiopia (£329 million), and Bangladesh (£272 million).

B *A UK charity providing health and hygiene education in Pakistan*

UK aid to Pakistan

Pakistan receives more aid from the UK than any other country (photo **B**). There are currently 66 million people in Pakistan living in poverty, equivalent to the entire population of the UK. The population is set to rise by 50 per cent in less than 40 years. In 2013 aid was spent mainly in the education sector and to reduce hunger and poverty.

Goat Aid from Oxfam

Goat Aid Oxfam is a project set up to help families in African countries like Malawi. The money donated is used to buy a family a goat, which produces milk, butter and meat. This has many advantages for the family and the local community, because:

◆ goats are an excellent food source, providing both milk and meat

◆ manure can be used as a crop fertiliser

◆ milk can be sold as a source of income to pay for food and education

◆ goats can be bred easily and kids sold at market or given to other families

◆ care of the goats builds community spirit.

This helps to improve people's quality of life and to raise the level of development.

C *Goat aid provides money to buy school uniforms*

What is intermediate technology?

Intermediate technology is sustainable technology that is appropriate to the needs, skills, knowledge and wealth of local people. It must be suitable for the local environment and must not put people out of work.

How can intermediate technology reduce the development gap?

Intermediate technology takes the form of small-scale projects often associated with agriculture, water or health. These involve local communities, and can make a real difference to the quality of people's lives.

ACTIVITIES

1 Outline the different types of aid and suggest which are most appropriate in reducing the development gap.

2 **a** To what extent is Oxfam's Goat Aid project sustainable?

   b Can you suggest any problems with the scheme?

3 What evidence is there that the Adis Nifas project is sustainable?

4 Write a paragraph arguing the case for the use of aid to fund intermediate technology projects.

Stretch yourself

Find out more about the Tekeze Dam in Ethiopia. Will the dam help Ethiopia's development needs?

Practice question

Explain why the use of aid must be sustainable if it is to be effective in raising a poor country's level of development.
(6 marks)

Irrigation at Adis Nifas, Ethiopia

The village of Adis Nifas is in northern Ethiopia, north Africa. Here a small dam (about 15 metres high and 300 metres long) was built to create a reservoir close to the village's fields. Appropriate machinery and money were given and the village provided the labour.

Each family has been given an area of irrigated land with fruit trees. Elephant grass is grown to divide the fields and help prevent soil erosion. The irrigated land is now providing a permanent food supply for the villagers.

The project made use of intermediate technology to build and run the dam scheme (photo **D**).

Employment for local people

Local tools and knowledge

Local materials such as stones and sand

Treadle pumps used to lift water to the fields

Reduction in soil erosion stops the reservoir silting up

D *Taking water from the Adis Nifas dam*

On this spread you will find out how fair trade can reduce the development gap

Is trade fair?

Richer countries benefit more from world trade than poorer countries. This explains why in some cases the development gap is widening. Rich countries are powerful enough to protect their trade using two main systems.

◆ *Tariffs* are taxes paid on imports. They make imported goods more expensive and less attractive than home-produced goods

◆ *Quotas* are limits on the quantity of goods that can be imported. They are usually applied to primary products so they affect mainly poorer countries.

What is free trade?

Free trade is when countries do not charge tariffs and quotas to restrict trade with each other. This has the potential to benefit the world's poorest countries and help reduce the development gap.

The World Trade Organisation (WTO) aims to make trade easier and remove barriers. One of the main barriers to trade is *agricultural subsidy*. This is financial support from governments to help their farmers. Rich countries can afford to pay subsidies and so their products are cheaper than those produced by poorer countries. This goes against free trade.

Trading groups are countries which have grouped together to increase the level of trade between them by cutting tariffs and discouraging trade with non-members. The European Union (EU) is an example.

There are advantages for poor countries in joining a trading group.

◆ It encourages trade between member countries.

◆ Richer countries cannot shop around for cheaper prices.

◆ Members can command a greater share of the market.

◆ Members are able to get higher prices for their goods.

Cocoa from Ghana

Ghana in West Africa is the world's largest producer of cocoa beans (photo **A**). Most of the processing and packaging of the cocoa is done in Europe. The EU charges 7.7 per cent import tariff on cocoa powder and 15 per cent on chocolate. But no tariff is charged on raw cocoa beans. So Ghana is forced to export the beans rather than develop its own industry making chocolate, which would be more valuable.

A *Cocoa farmer in Ghana*

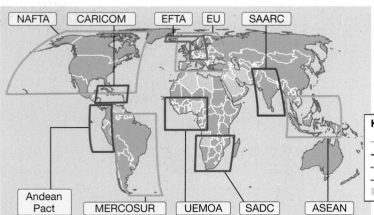

NAFTA	CARICOM	EFTA	EU	SAARC

Andean Pact	MERCOSUR	UEMOA	SADC	ASEAN

Key
— Major trade group
— Loose-knit trade group
— Smaller trade group of HICs
— Smaller trade group of LICs
▨ OPEC countries

B *Global trading groups*

Poorer countries have formed trading groups (map **B**) like CARICOM (Caribbean Community), UEMOA (West African Economic and Monetary Union) and ECOWAS (Economic Community of West African States – see page 222).

Fairtrade

You may have seen the Fairtrade logo (image **C**). Fairtrade sets standards for trade with poorer countries. It seeks to reduce the development gap by improving the quality of life for ordinary farmers. Fairtrade is an international movement. It helps to ensure that producers in poor countries get a fair deal.

◆ The farmer gets all the money from the sale of his crop.

◆ It guarantees the farmer a fair price.

◆ Part of the price is invested in local community development projects.

◆ In return the farmer must agree to farm in an environmentally-friendly way.

◆ The product gains a stronger position in the global market.

C The Fairtrade logo

D A Gumutindo farmer from Uganda

Ugandan coffee farmers

Over 90 per cent of small coffee farmers in eastern Uganda have joined the Gumutindo Coffee Cooperative to gain economies of scale. This means making savings by buying and selling larger amounts of coffee. The farmers also earn extra income from the Fairtrade Premium. This would not be possible if individual farmers tried to sell their coffee.

The first stage of processing the coffee beans is done on the farm. The semi-processed beans are worth more to the farmer than unprocessed beans. They are then sent to a nearby warehouse for milling, before being packed for export abroad, where the final roasting takes place. The processing of the coffee beans adds value to the product and increases the farmer's income.

Last year the money from the Fairtrade Premium helped pay my son's school fees. Other farmers have tried to copy what we are doing and the quality of the coffee is getting better!

ACTIVITIES

1 Describe how the cocoa farmer in photo **A** harvests his crop.

2 **a** What is the difference between tariffs and quotas?

  b What effect do EU tariffs have on Ghanaian farmers?

3 Which countries are members of the following trading groups: EU; NAFTA; OPEC (map **B**)?

4 **a** Write a paragraph explaining how Fairtrade benefits poorer countries.

  b How does it help reduce the development gap?

Stretch yourself

Investigate whether there are any disadvantages of the Fairtrade scheme.

Practice question

Discuss whether trade or aid is the best way for poorer countries to develop. *(9 marks)*

On this spread you will find out how debt relief and microfinance loans can help reduce the development gap

Loans and debt

One country may borrow money from another country, or from an international organisation such as the World Bank, in order to invest in development projects. This loan has to be repaid with interest.

How have poor countries built up debt?

Many of the world's poorest countries built up debt in the 1970s and 1980s. This led to a **debt crisis**. Many poor countries borrowed money to develop their economies by investing in industry, manufacturing and infrastructure. Low commodity prices reduced the value of their exports and high oil prices increased the price of imports. Both these factors increased the debt of poor countries.

The highly indebted poor countries (HIPCs) are the 39 countries with the highest level of poverty and debt (map **A**). They are unable to repay their debt and the high level of interest.

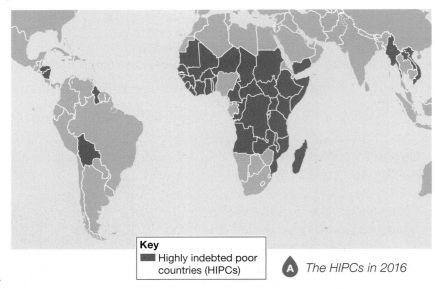

Key
▇ Highly indebted poor countries (HIPCs)

A *The HIPCs in 2016*

Debt relief

At their meeting in 2005 the world's richest countries (known as the 'G8') agreed to cancel the debts of many of the HIPCs (text **B**). To qualify for debt relief countries had to:

◆ demonstrate they could manage their own finances

◆ show there was no corruption in their government

◆ agree to spend the saved debt money on education, health care and reducing poverty.

'On 6 January 2006, the IMF cancelled the debts owed to it by 19 of the world's poorest countries. This will change the lives of millions of people. In Ghana the money saved is being used for basic infrastructure, including rural feeder roads, as well as increased expenditure on education and health care. In Tanzania, the government is using the money saved to import vital food supplies for those affected by drought. Across Africa, lifting the burden of debt is allowing millions of dollars to be directed to fighting poverty instead of repaying rich countries.'

B *Announcement by the IMF in 2006*

By 2015, 36 of the HIPCs had met these conditions and were receiving full debt relief from the International Monetary Fund (IMF). The total amount of debt relief for all HIPCs is around US$75 billion. The three countries yet to satisfy the terms for debt relief are Eritrea, Somalia and Sudan.

Despite the debt relief, African countries still have debt of over US$300 billion and are unlikely to ever be able to repay it.

How can debt relief reduce the development gap?

Debt relief can help poor countries invest money in development projects, such as industry, resources or infrastructure. By cancelling their debts, some countries have used the money saved to improve the quality of life for their people. For example, in Tanzania free education is now available, resulting in a 66 per cent increase in attendance. In Uganda the government has spent money to provide safe water to over 2 million people.

However, debt relief can also lead to problems.

◆ Countries may get into further debt expecting that this will also be written off in the future.

◆ Corrupt governments may keep the money rather than use it to help the poor.

What is microfinance?

Microfinance is small-scale financial support available directly from banks set up especially to help the poor. Small **microfinance loans** enable individuals or families to start up small businesses, and helps them to become self-sufficient. Many borrowers are women. As small businesses thrive, employment opportunities increase and incomes rise.

Did you know?
The world's poorest countries pay more than $US1.5 billion a day in interest repayments.

Grameen Bank, Bangladesh

This bank was set up in Bangladesh in 1976. The name comes from the Sanskrit word for 'village'. The bank was founded to help local people, especially women, use their skills to develop small businesses. Borrowers have a share in the ownership of the bank, so there is a good rate of repayment. Loans are often less than $100 with low interest. The bank has so far lent over $11 billion to 7 million members.

The bank lends US$200 to village women to buy a mobile phone. Other villagers then pay the women to use the phones. The loan can then be repaid and the borrower makes a small profit. The phones help people to check prices before they go to market, keep in touch with relatives who have moved to the city, and receive health advice. Halima Khatun owns 15 hens and sells their eggs for a living. She uses the village phone to try to get a better price. (photo **C**).

C *Using the village phone*

'Last week, they wanted to pay me 12 taka per hali [four eggs]. I checked the prices using the village phone. The price was 14 taka in nearby markets. We agreed to buy and sell at 13 taka per hali.'

ACTIVITIES

1 **a** Which continent has the most HIPCs (map **A**)?

   b Use an atlas to name two HIPCs from this continent and one from each of two other continents.

2 Do you think debt relief for HIPCs benefits richer countries? Give your reasons.

3 Explain how the use of a mobile phone has helped Halima Khatun and her egg business. Can you suggest other ways that she might use the phone to support her business?

Stretch yourself

Investigate the 'Make Poverty History' movement. What are its links with debt relief?

Practice question

How can debt relief help to improve the status of women? *(4 marks)*

On this spread you will find out how tourism in Jamaica can help reduce the development gap

What is the state of Jamaica's economy?

Jamaica is one of the largest islands in the West Indies. Its population is 2.7 million, just over a third of the size of London. Its economy is based upon a range of minerals (such as bauxite and oil), agricultural products (sugar and rum) and some manufacturing. It is classed as an 'upper middle-income country', but has suffered from slow growth, debt and high unemployment over a long period.

How has tourism contributed to Jamaica's development?

Tourism, along with bauxite and energy, is one of the few growth sectors of Jamaica's economy. The country has become a popular tourist destination (map **B**), offering beautiful beaches (photo **A**), a warm sunny climate and rich cultural heritage. Jamaica enjoys good international air communications and is a hub for cruise ships.

Tourism is important to the Jamaican economy, generating taxes, employment and income. Over the last few decades it has helped raise the level of development in Jamaica and reduce the development gap.

A Turtle Beach, Ocho Rios, Jamaica

B Tourist attractions in Jamaica

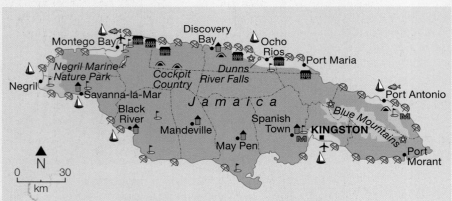

Key
- Main tourist areas
- Hotels outside main tourist areas
- Beach/bathing areas
- Watersports
- Deep-sea fishing
- Caves
- Botanic garden
- Golf course
- Museum
- Bird sanctuary/wildlife reserve
- National Park
- Parish boundary
- Airport
- Plantation house

Economy

In 2014 tourism contributed 24 per cent of Jamaica's GDP – one of the highest proportions of any country in the world. This is expected to rise to 32 per cent by 2024. Income from tourism is US$2 billion each year and taxes paid to the government contribute further to the development of the country (graph **C**). This in turn helps to reduce the development gap.

The increase in tourism from cruises has brought many benefits. However, the annual 1.1 million cruise passengers only spend an average US$70 per day. This compares with an average US$120 per day spent by the 2.5 million other visitors.

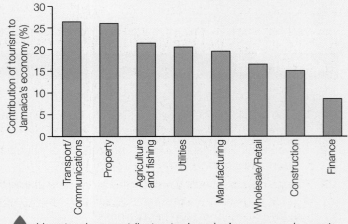

C How tourism contributes to Jamaica's economy by sector

Employment

Tourism is the main source of employment in Jamaica. It provides jobs for 200 000 people either directly in hotels, transport and tourist attractions or indirectly in shops, manufacturing and banking. These are mainly in or around the main tourist towns.

Employment in tourism provides income which helps to further boost the local economy as people spend money in shops and on services and recreation (photo **D**). Those in employment learn new skills which can improve their prospects of better-paid jobs in the future. The quality of life for many people has improved.

Infrastructure

Tourism has led to a high level of investment on the north coast where much of the country's tourism is centred. New port and cruise-liner facilities have been built at Trelawney together with new hotel accommodation. However, improvements in roads and airports have been slower and some parts of the island remain isolated.

Quality of life

In the northern tourist areas of Montego Bay and Ocho Rios, wealthy Jamaicans live in high-quality housing with a high standard of living. These areas have benefited from the tourist industry. However, large numbers of people live nearby in poor housing with limited food supply and inadequate access to fresh water, health care and education.

The environment

Mass tourism can create environmental problems such as footpath erosion, excessive waste and harmful emissions. It can also bring environmental benefits. Conservation and landscaping projects provide job opportunities and encourage people to visit the island.

Montego Bay on the north coast has been improved by landscaping, and a new water treatment plant at Logwood has reduced pollution from hotels. The Negril Marine Nature Park attracts many tourists and brings direct and indirect income. Community tourism and sustainable **ecotourism** is expanding in more isolated regions, with people running small-scale guesthouses or acting as guides.

D Tourism boosts the local economy

> ### Think about it
> *What disadvantages might tourism bring to countries like Jamaica?*

ACTIVITIES

1 **a** Describe the distribution of the main tourist areas of Jamaica (map **B**).

   b What are Jamaica's main tourist attractions?

2 **a** Name the two sectors which benefit most from tourism (graph **C**). Give the percentages.

   b Suggest why these sectors benefit so much.

3 How can tourism boost the economy of local communities (photo **D**)?

4 How will improvements to infrastructure and the environment help to increase tourism and boost the economy?

Stretch yourself

Investigate how increased tourism at resorts such as Ocho Rios and Montego Bay can have a multiplier effect on Jamaica's development.

Practice question

Explain why the Jamaican government sees tourism as a way to reduce the development gap. *(4 marks)*

17.1 | Exploring Nigeria (1)

On this spread you will find out about Nigeria's location and its global and regional importance

Where is Nigeria?

Nigeria is a country in West Africa. Nigeria borders Benin, Niger, Chad and Cameroon (map **A**). It is almost due south of the UK, just one hour ahead of Greenwich Mean Time.

At latitude 4°14', Nigeria extends from the Gulf of Guinea in the south to the Sahel in the north. It has a tropical climate with variable rainy and dry seasons in different parts of the country. It is hot and wet most of the year in the south, but inland there is a long, dry season.

What is the global importance of Nigeria?

Nigeria is a Newly-Emerging Economy (NEE) (see page 194). This means that it is one of a number of countries experiencing a period of rapid economic development. In 2014, Nigeria became the world's 21st largest economy – by 2050 it should be in the top twenty. Nigeria is predicted to have the world's highest average GDP growth for 2010–15.

Nigeria supplies 2.7 per cent of the world's oil – the 12th largest producer. Much of the country's economic growth has been based on oil revenues. But it has also developed a very diverse economy which now includes financial services, telecommunications and the media. In common with cities around the world, the centre of Lagos is a thriving global economic hub (photo **B**).

Politically, Nigeria has a significant global role. It currently ranks as the fifth largest contributor to UN peacekeeping missions around the world (photo **C**).

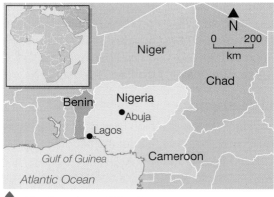

A *The location of Nigeria*

Did you know?
Lagos is the fourth most densely populated city in the world, with a population of 13.5 million and 18 150 people per km² – London is 43rd!

B *The city business skyline in Lagos*

C *A Nigerian peacekeeper in Liberia*

Year	Population	Annual change %	Fertility rate	Urban population %	Urban population	% of world pop
2015	182 201 962	2.71	5.74	48.10	87 680 500	2.63
2010	159 424 742	2.69	5.91	43.60	69 440 943	2.45
2005	139 611 303	2.59	6.05	39.10	54 541 496	2.28
2000	122 876 723	2.53	6.17	34.80	42 810 252	2.14
1995	108 424 822	2.55	6.37	32.20	34 918 670	2.04
1990	95 617 345	2.65	6.6	29.70	28 379 229	1.97

D *How has Nigeria developed in 25 years?*

Nigeria's importance in Africa

Nigeria has one of the fastest-growing economies in Africa. In 2014 it had the highest GDP in the continent and the third largest manufacturing sector. With a population of more than 182 million people, it has the largest population of any African country.

Nigeria has low levels of productivity and there are widespread issues over land ownership. But it still has the highest farm output in Africa. About 70 per cent of the population are employed in agriculture. Most are subsistence farmers growing food crops like yams, cassava, sorghum and millet or keeping livestock. Nigeria has over 19 million cattle, the largest number in Africa (photo **E**).

Nigeria could lead the way in Africa's future development. Despite its problems with internal corruption and lack of infrastructure – with poor roads and frequent power cuts – the country has huge potential. US President Barack Obama said Nigeria is 'critical to the rest of the continent and if Nigeria does not get it right, Africa will really not make more progress'.

E *Cattle herding in Nigeria*

ACTIVITIES

1 Describe the location of Nigeria (map **A**).
2 Give reasons for the global importance of Nigeria.
3 a Work in pairs to find out more about Nigeria. Start by writing out some questions, such as 'How large is Nigeria compared to the UK?' or 'Who are Nigeria's famous sportspeople?'
  b Do some online research to answer your questions.
  c Present your findings as a poster display, and add some data and photos you find interesting.
4 Do you agree with US President Obama that 'if Nigeria does not get it right, Africa will really not make more progress'?

Stretch yourself

Find out more about the international role of Nigeria. How are the country and its people having a global impact?

Practice question

Discuss how Nigeria has a growing influence in Africa. *(6 marks)*

On this spread you will find out about political, social, cultural and environmental aspects of Nigeria

Political context

The political map of Africa was drawn by a small group of powerful European countries at the Berlin Conference in 1883. These countries literally carved up control of Africa between them. This explains why many country borders are straight lines. Europeans exploited Africa's resources, including its people, who were traded as slaves.

In the 1960s many African countries gained their independence. Nigeria became fully independent from the UK in 1960. However, bitter power struggles resulted in a series of dictatorships and a civil war between 1967 and 1970. Lack of political stability affected Nigeria's development and led to widespread corruption. It is only since 1999 that the country has had a stable government. Recent elections in 2011 and 2015 were seen as free and fair.

Several countries are now starting to invest in Nigeria.

◆ China is making major investments in construction in the capital, Abuja.

◆ South Africa is investing in businesses and banking.

◆ American companies such as General Electric are investing in new power plants.

◆ American corporations such as Wal-Mart, and IT giants IBM, Microsoft and Oracle are operating in Nigeria (photo **A**).

Social context

Nigeria is a multiethnic, multifaith country. Ethnic groups in Nigeria include the Yoruba (21 per cent of the population), Hausa and the Fulani (29 per cent), and Igbo (18 per cent) as well as many smaller groups. Christianity, Islam, and traditional African religions are practised widely. This social diversity is one of Nigeria's great strengths, but has also been a source of conflict.

In 1967 the Igbo-dominated south-east tried to separate from Nigeria to become the Republic of Biafra. As a result, the country was torn by civil war until the Biafrans were defeated in 1970.

More recently, economic inequality between the north and south of Nigeria has created new religious and ethnic tensions, with the rise of the Islamic fundamentalist group Boko Haram. This has created an unstable situation in the country, and has had a negative impact on the economy, with a reduction in investment from abroad and a rise in unemployment.

Table **B** compares some key social indicators with those for the UK.

A Technological development in Nigeria

 B Key facts: Nigeria and the UK compared

Fact	Nigeria	UK
Land area	924 000 km^2	244 000 km^2
Population (millions)	182 (largest in Africa)	65
Population growth rate (% per year)	2.4	0.6
Birth rate (per 1000)	38	12
Death rate (per 1000)	13	9
Infant mortality rate (per 1000 live births)	73	4
Life expectancy (years)	52	81
Literacy rate (%)	61	99
GNI per head (US$)	2970	43 430
Capital	Abuja (2 million)	London (8.6 million)
Largest city	Lagos (11 million)	London
Internet users (%)	38	90
Percentage in poverty	70	15

Regional variations

There are huge variations in levels of wealth and development within Nigeria. Urban areas have a greater share of public services and facilities. For example, 60 per cent of children in urban areas attend secondary school, but only 36 per cent in rural areas. This encourages widespread rural–urban migration.

GDP per person varies greatly across the country. It is highest in the south (US$3617), US$8343 in the north-east, but only US$292 in the south-east.

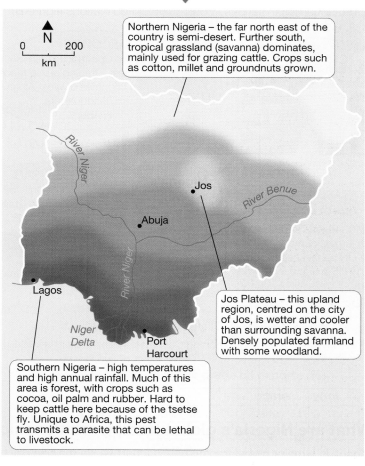

C *'Nollywood' – Nigeria's thriving film industry*

Cultural context

With such a diverse population, Nigeria enjoys a rich and varied culture.

D *Nigeria's natural environments*

◆ Nigerian music is enjoyed across the continent and beyond. Have you heard of Fela Kuti?

◆ Nigerian cinema – known as 'Nollywood' – is the second largest film industry in the world, ahead of the United States and behind India (photo **C**).

◆ In literature, well-known Nigerian writers include Wole Soyinka, Chinua Achebe, Chimamanda Ngozi Adichie and Nnedi Okorafor.

◆ In sport, the Nigerian football team has won the African Cup of Nations three times, most recently in 2013. Several Nigerian football players have played for Premier League sides including Victor Moses, John Obi Mikel, Jay-Jay Okocha and Kanu.

Northern Nigeria – the far north east of the country is semi-desert. Further south, tropical grassland (savanna) dominates, mainly used for grazing cattle. Crops such as cotton, millet and groundnuts grown.

Jos Plateau – this upland region, centred on the city of Jos, is wetter and cooler than surrounding savanna. Densely populated farmland with some woodland.

Southern Nigeria – high temperatures and high annual rainfall. Much of this area is forest, with crops such as cocoa, oil palm and rubber. Hard to keep cattle here because of the tsetse fly. Unique to Africa, this pest transmits a parasite that can be lethal to livestock.

Environmental context

Nigeria's natural environments form a series of bands across the country (map **D**). This reflects the decreasing rainfall towards the north in West Africa. These environmental regions extend to the east and west of Nigeria. To the north is the Sahel and the Sahara Desert.

ACTIVITIES

1 How important is political stability to the development of Nigeria's economy?

2 Working in pairs, carry out some research into Nigeria's diverse culture (music, film, books and sport). What role has Nigeria's culture played in its recent economic development?

3 Describe the challenges and opportunities of Nigeria's natural environment for promoting economic growth.

4 Suggest reasons for the variations in wealth (GDP) across Nigeria.

Stretch yourself

How have social and political conflict affected development in Nigeria in recent years?

Practice question

Describe briefly how politics has shaped Nigeria's economic development.

(4 marks)

The impacts of international aid

On this spread you will find out about the impact of international aid on Nigeria

What is aid?

Aid (see page 210) can be defined as 'assisting people'. The providers of aid can be individuals, charities, non-governmental organisations (NGOs), governments and international (multi-lateral) organisations, like the EU or the UN.

There are two main types of aid.

◆ *Emergency aid* – this usually follows a natural disaster, war or conflict. Aid may take the form of food, water, medical supplies and shelter.

◆ *Developmental aid* – this is long-term support given by charities, governments and multi-lateral organisations. It aims to improve quality of life by providing safe water, education or improvements to infrastructure such as roads and electricity supplies.

A *Poverty in Nigeria*

Why does Nigeria receive international aid?

Despite rapid economic growth and wealth from oil reserves, many people in Nigeria are poor. They have limited access to services such as safe water, sanitation and a reliable electricity supply (photo **A**).

Almost 100 million people (over 60 per cent of the population) live on less than a US$1 (£0.63) a day. Birth rates and infant mortality rates are high and life expectancy is low, particularly in the north-east of the country.

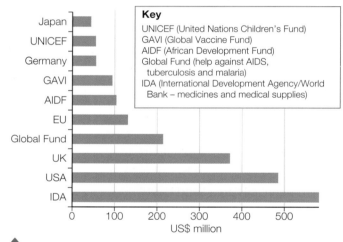

Key
UNICEF (United Nations Children's Fund)
GAVI (Global Vaccine Fund)
AIDF (African Development Fund)
Global Fund (help against AIDS, tuberculosis and malaria)
IDA (International Development Agency/World Bank – medicines and medical supplies)

B *Sources of aid to Nigeria in 2014*

What aid does Nigeria receive?

Nigeria receives about 4 per cent of aid given to African countries. In 2013 aid represented 0.5 per cent of Nigeria's Gross National Income, a total of nearly US$5000 million. Most came from individual countries such as the UK and the USA, and some from international organisations like the World Bank (graph **B**). Charities and NGOs have also supported projects in Nigeria.

What is the impact of aid in Nigeria?

Aid has brought many benefits to people living in poverty. The most successful projects are community-based, supported by small charities and NGOs. These are often delivered directly to where help is needed. The aid is all used for the project and no money is wasted.

Maths skills

Use the data to produce a divided bar graph to show the allocation of multi-lateral aid to Nigeria in 2013.

The allocation of multilateral aid to Nigeria, 2013 (by sector)

Education	6%
Health and population	63%
Social infrastructure and services	14%
Economic infrastructure and services	7%
Production	3%
Multi-sector	5%
Programme assistance	0.5%
Humanitarian aid	0.5%
Other	1%

Aid has been used to help Nigeria in a number of ways.

In 2014 the World Bank approved a US$500 million to fund development projects and provide long-term loans to businesses. This helps reduce the over-dependence on oil exports.

The UK Department for International Development has funded a health and HIV programme, providing health education in rural areas.

How does aid benefit Nigeria?

Aid from the USA helps to educate and protect people against the spread of AIDS/HIV.

The NGO Nets for Life provides education on malaria prevention and distributes anti-mosquito nets to many households.

The USAID-funded Community Care in Nigeria project provides support packages for orphans and vulnerable children.

What prevents aid being used effectively?

Official aid to Nigeria delivered through the government has been less successful than aid delivered directly to communities. There are several reasons why aid may not be used effectively.

◆ Corruption in the government, and by individuals, is a major factor in loss of aid.

◆ The government may divert money to be used for other purposes. For example, there are claims that aid may have been used to build up Nigeria's navy.

◆ Donors may have political influence over what happens to aid.

◆ Money may be used to promote the commercial self-interest of the donor.

Another view

By receiving aid, a country may become more dependent. This could slow down its economic development.

The Aduwan Health Centre

The community of Aduwan in Kaduna State, northern Nigeria did not have a health centre. The few health workers in the area used the community's only shop as a clinic. The area has a high incidence of HIV/AIDS and high infant mortality.

With support from ActionAid they received funds from the World Bank for a new health clinic built in 2010. The new clinic:

◆ trains local women to educate mothers about the importance of immunising their children against polio and other deadly diseases (photo **C**)

◆ tests for HIV and other infections

◆ immunises children against polio.

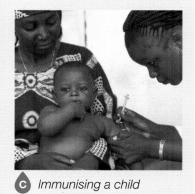

C *Immunising a child*

ACTIVITIES

1 a Describe the living conditions shown in photo **A**.

  b How do you think aid should be spent in this area?

  c How should aid be best spent to support people in Nigeria?

2 a How much aid was given to Nigeria by the UK in 2014 (graph **B**)?

  b Which NGO gave the most money to Nigeria in 2014?

  c Suggest why NGOs support health projects in Nigeria.

3 Explain why small-scale development projects are likely to be the best use of international aid for Nigeria.

Stretch yourself

Investigate examples of how UK government aid is used in Nigeria.

Practice question

Explain why aid in Nigeria may not be used effectively. *(4 marks)*

On this spread you will find out about the environmental impacts of economic development in Nigeria

How does economic growth affect the environment?

Rapid economic growth, like in Nigeria, can bring many benefits. But it can also have a negative impact on the environment.

Industrial growth

Nigeria has about 5000 registered industrial plants and 10 000 illegal small-scale industries. The fast and unregulated growth of industry has led to environmental problems.

- In Kano, Kaduna and Lagos, many harmful pollutants go directly into open drains and water channels. They are harmful to people and damage ecosystems downstream.
- Some industries dispose of chemical waste on nearby land, threatening the groundwater quality.
- Industrial chimneys emit poisonous gases that can cause respiratory and heart problems in humans (photo **A**).
- 70–80 per cent of Nigeria's forests have been destroyed through logging, agriculture, urban expansion, roads and industrial development.
- Desertification is a major problem in Nigeria, made worse by large-scale dam and irrigation schemes.

Urban growth

As Nigeria has developed, urban areas have grown rapidly. This rate of urbanisation has brought many challenges.

- Squatter settlements are common in most cities.
- Services have failed to keep pace with the rate of economic growth.
- Waste disposal has become a major issue (photo **B**).
- Traffic congestion is a major problem in most Nigerian cities, leading to high levels of exhaust emissions.

Some green belts and recreational areas are being converted into building sites. The development of Abuja has resulted in areas of rich natural vegetation being replaced by concrete. Extensive bush burning has damaged trees and wildlife species, and biodiversity has been reduced.

Commercial farming and deforestation

Commercial farming and inappropriate practices have led to land degradation. There is water pollution due to chemicals, soil erosion and silting of river channels. The building of settlements and roads has destroyed habitats and added to CO_2 emissions. Many species have disappeared because of deforestation, including cheetahs and giraffes, and nearly 500 types of plant.

A Air pollution in Lagos

B Rubbish dumped on the roadside

Mining and oil extraction

Mining and extraction of raw materials and precious metals (photo **C**) – particularly oil – can lead to serious pollution. These can damage ecosystems and affect people's jobs.

◆ Tin mining led to soil erosion. Local water supplies were also polluted with toxic chemicals.

◆ Many oil spills in the Niger Delta have had disastrous impacts on freshwater and marine ecosystems. Oil spills can cause fires, sending CO_2 and other harmful gases into the atmosphere. They cause *acid rain*, which harms plants and aquatic ecosystems.

◆ Some economic developments in the Niger Delta have caused violent conflicts with local people.

C Mining for gold in Nigeria

Another view

Shell claims that theft of crude oil, sabotage and illegal refining are the main sources of oil pollution. Do you agree with Shell? Why might they say this?

Bodo oil spills (2008/09)

In 2008 and 2009 two large oil spills devastated the livelihoods of thousands of farmers and fishermen living in the swamps around the town of Bodo in the Niger Delta (photo **D**). Leaks in a major pipeline caused 11 million gallons of crude oil to spill over a 20 km^2 area of creeks and swamps.

In 2015 Shell agreed to pay £55 million compensation to individuals and to the community of Bodo. The money will be used to build health clinics and improve schools. This is the largest compensation paid by an oil company to a local community affected by environmental damage. Shell has also agreed to clean up the swamps and fishing grounds.

D An oil-polluted fish farm in Bodo

ACTIVITIES

1 a How has rapid industrial development harmed the environment in Nigeria (photo **A**)?

b What measures could be introduced to reduce damage to the environment?

2 a Describe the waste that has been dumped in photo **B**.

b What problems might arise from this waste dump?

c Why has waste been dumped here and what could be done to solve the problem in the future?

3 What are the environmental problems associated with oil spills?

Practice question

Explain how economic growth can have harmful impacts on the environment. *(6 marks)*

Stretch yourself

Make a detailed study of either the Bodo or the Bonga oil spills.

• Find out what happened and the impacts on people and the environment.

• What has been done to clean up the area and compensate the people?

• How can oil spills be prevented?

On this spread you will find out how and why the UK economy has changed in recent years

How has the economy of the UK changed?

Before 1800 most people in the UK worked in farming, mining or related activities – the *primary* sector. But the Industrial Revolution of the nineteenth century changed all that. Many people moved to towns and cities for work – making steel, ships or textiles (the *manufacturing* sector).

In the last few decades it all changed again, with a big shift to jobs in the service (or *tertiary*) sector – health care, offices, financial services and retailing (pie chart **A**). Most recently, the *quaternary* sector has developed, with jobs in research, information technology and the media (graph **B**).

Why has the economy of the UK changed?

De-industrialisation and the decline of traditional industries

For several decades the UK has been experiencing **de-industrialisation**. This is the decline in manufacturing (secondary) industry and the subsequent growth in tertiary and quaternary employment. In the UK this has happened because:

◆ machines and technology have replaced many people in modern industries, for example in car production

◆ other countries, for example China, Malaysia and Indonesia, can produce cheaper goods because labour there is less expensive

◆ lack of investment, high labour costs and outdated machinery made UK products too expensive.

Traditional UK industries – coal mining, engineering and manufacturing – have declined (photo **C**). Instead, the UK is now a world centre for financial services, media, research and the creative industries.

Globalisation

Globalisation is the growth and spread of ideas around the world. This can involve the movement or spread of cultures, people, money, goods and information.

Globalisation has been made possible by developments in transport, communications and, in particular, the internet. Consider how easy it is to communicate with friends and family abroad, or to travel to Europe or America.

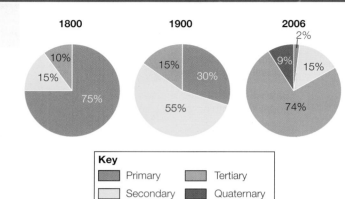

Key
- Primary
- Secondary
- Tertiary
- Quaternary

A *The UK's changing employment structure*

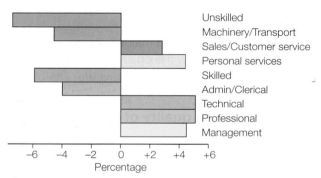

Unskilled
Machinery/Transport
Sales/Customer service
Personal services
Skilled
Admin/Clerical
Technical
Professional
Management

Percentage

B *Changes in UK employment, 1984–2014*

C *A derelict factory in Oldham, Lancashire*

Globalisation has been partly responsible for the explosion of the quaternary sector in the UK, with many people now working on global brands and products. It has also boosted world trade and enabled more imported products to the UK. This in turn has contributed to the decline in UK manufacturing.

Government policies

1945–1979

In this period the UK government created state-run industries such as the National Coal Board, British Rail and British Steel Corporation.

Government money was spent on 'propping up' declining UK industries. Ageing equipment, outdated working practices and too many employees made them unprofitable. The 1970s was a decade of strikes, social unrest, factory closures and power cuts.

1979–2010

State-run industries were sold off to private shareholders to create a more competitive business environment. This is called *privatisation*. Many older industries closed down and many jobs were lost.

Many new private companies brought innovation and change. In the UK derelict industrial areas were transformed into gleaming new financial centres with offices and modern retail outlets (photo **D**).

2010 onwards

The aim of the government from 2010 was to 'rebalance' the economy. This was to be done by rebuilding the UK's manufacturing sector and relying less on the service industries, in particular the financial sector. Government policies have included:

- improvements to transport infrastructure (for example, the London Crossrail link and plans for high-speed rail connections, such as HS2)

- more investment in manufacturing industries

- easier access to loans and finance, especially for small businesses

- encouraging global firms to locate within the UK.

D *Canary Wharf – the UK's financial centre on the site of London's docklands*

ACTIVITIES

1. **a** Define primary, secondary, tertiary and quaternary activities.

   b Suggest why tertiary activities take over from manufacturing at a time of de-industrialisation.

2. Describe and attempt to explain the employment trends shown in graph **B**.

3. **a** Describe the main features of the derelict factory in photo **C**.

   b How could buildings like this be used today given the UK's changing economic structure?

4. What is globalisation and how has it affected the structure of the UK economy?

5. Write a short summary of the government's role in the UK economy during each of the three periods in the timeline.

Maths skills

Use the percentage data to draw a pie chart of the UK's employment structure in 2015.

Year	Primary	Secondary	Tertiary	Quaternary
2015	2	10	78	10

Stretch yourself

Find out more about manufacturing in the UK today. Has the government been successful in encouraging the development of a modern manufacturing sector?

Practice question

Explain the causes and impacts of de-industrialisation in the UK. *(6 marks)*

On this spread you will find out about the development of the UK's post-industrial economy

What is a post-industrial economy?

A post-industrial economy is where manufacturing industry declines to be replaced by growth in the service sector and the corresponding development of a quaternary sector. This happened in the UK from the 1970s. By 2015, 78 per cent of UK employment was in the tertiary sector and 10 per cent in the quaternary sector. Only 10 per cent of employment was in manufacturing compared to 55 per cent in 1900.

Development of information technology

The use of information technology (IT) has transformed the way that people live and work. It is one of the main factors in the UK's move to a post-industrial economy.

◆ Computers allow large amounts of data to be stored and accessed very quickly.

◆ The internet enables people to communicate with each other instantly across the world.

◆ Technology continues to develop at a rapid pace (for example, high-speed broadband).

◆ Many people can access the internet using smart phones and tablets.

Developments in IT have affected the UK economy in a number of ways.

◆ Internet access enables many people to work from home.

◆ Many new businesses are directly involved with IT, manufacturing hardware and designing software.

◆ Over 1.3 million people work in the IT sector.

◆ The UK is one of the world's leading digital economies, attracting businesses and investment from abroad.

The government is committed to making the UK the best place in the world to start a technology business.

Service industries and finance

Individuals employed in the service sector produce services rather than products (photo **B**). The UK service sector has grown very rapidly since the 1970s. Today it contributes over 79 per cent of UK economic output, compared with 46 per cent in 1948.

Finance is an important part of the service sector. This includes banking, insurance, securities dealing and fund management. The UK is the world's leading centre for financial services, with the City of London as the UK's financial centre. The financial services sector accounts for about 10 per cent of the UK's GDP and employs over 2 million people.

A A UK high-tech company in Lancashire

Did you know?

The worldwide web was invented by British computer scientist Sir Tim Berners-Lee in 1989.

B Jobs in the UK service sector

Research

The UK research sector – part of the quaternary sector – employs over 60000 highly-qualified people and is estimated to contribute over £3 billion to the UK economy. Research is done in British universities, such as Cambridge, Manchester and Edinburgh. It is also done by private companies and government bodies (figure **C**). The research sector is likely to be one of the UK economy's main growth areas in the future.

C *Some UK research organisations*

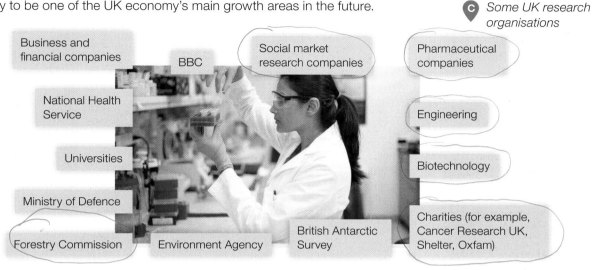

Business and financial companies

BBC

Social market research companies

Pharmaceutical companies

National Health Service

Engineering

Universities

Biotechnology

Ministry of Defence

Forestry Commission

Environment Agency

British Antarctic Survey

Charities (for example, Cancer Research UK, Shelter, Oxfam)

British Antarctic Survey

The British Antarctic Survey (BAS) employs over 500 highly-skilled people based in Cambridge (UK), Antarctica and the Arctic. It is linked to the University of Cambridge, with its well-qualified graduates and tradition of scientific research.

BAS scientific research involves the use of ships, aircraft and research stations. Its research in polar regions (photo **D**) helps our understanding of the Earth and human impact on its natural systems.

The BAS is currently investigating topics such as future changes in Arctic sea-ice, the impact of ocean acidification on ecosystems, and space weather research.

D *The BAS Halley VI Research Station*

ACTIVITIES

1 In what ways have developments in IT affected the UK's economy?

2 Work in pairs to identify a range of jobs in the service sector (photo **B**).

3 Why do you think the UK has developed into a major global financial centre?

4 Suggest what kind of research is carried out by each of the organisations listed with photo **C**. Use the internet to help you.

Stretch yourself

Find out about the research carried out by *one* of the organisations (or one of your own choice) listed with photo **C**. Where does the research take place and what are the reasons for the chosen location?

Practice question

How has the development of IT affected the growth and characteristics of the UK's economy? *(6 marks)*

On this spread you will find out about science and business parks in the UK

What is a science park?

A **science park** is a group of scientific and technical knowledge-based businesses located on a single site. There are over one hundred science parks in the UK. Most are associated with universities, enabling them to use research facilities and employ skilled graduates . Science parks may include support services such as financial services and marketing. Around 75 000 people work in the UK's science parks.

University of Southampton Science Park

Southampton Science Park opened in 1986. It has expanded since then to include a hundred small science and innovation businesses. It has strong links with the University of Southampton. Amongst the companies located in the park are:

◆ Fibrecore – manufacturer of optical fibres, set up at the University in 1982 and later moved to the science park

◆ Symetrica – specialising in gamma-ray spectrometers, established in 2002 with close links with the University's Department of Physics and Astronomy

◆ PhotonStar – specialising in lighting products, this company was founded in 2007 as another spin-off company from the University

◆ SEaB Energy Systems – founded in 2009, this company specialises in sustainable energy options and has developed an anaerobic digestion system (the 'Muckbuster') which produces heat and power from waste.

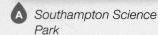

A Southampton Science Park

Benefits

For companies locating at Southampton Science Park the benefits include:

◆ excellent links with the University, providing research facilities and academic talent

◆ source of graduate employees from the University

◆ attractive location, with extensive green areas and woodlands (photo **A**)

◆ meeting rooms, a coffee shop, high-speed broadband, a nursery and a health club all available on site

◆ excellent transport links – close to M3, Southampton international airport and rail links (map **B**).

B Southampton Science Park – transport links

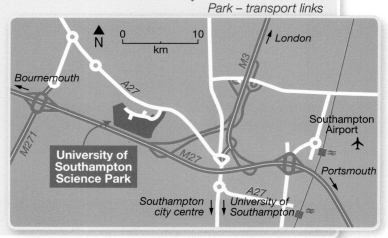

What is a business park?

A **business park** is an area of land occupied by a cluster of businesses. Business parks are usually located on the edges of towns because:

◆ land tends to be cheaper than in town centres

◆ with more land available, it may be possible to extend businesses

◆ access is better for workers and distribution, using by-passes and motorways, with less traffic congestion than in towns or cities.

◆ businesses can benefit by working together.

Cobalt Business Park, Newcastle-upon-Tyne

Cobalt Park is the UK's largest business park. There are several support facilities including retail outlets and opportunities for recreation, with a fitness centre, cycleways and green spaces. The park is close to the main A1 road and just 20 minutes from the international airport (map **D**).

The north-east of England has suffered from economic decline as traditional industries have closed down. Businesses locating in Cobalt Park therefore qualify for government assistance.

In 2015 the international energy company Siemens moved to the park (photo **C**). It is one of a number of well-known companies located in the park, including Procter & Gamble, IBM, Barnardos and Santander.

C *New Siemens offices at Cobalt Park*

D *OS map extract showing the location of the Cobalt Business Park*

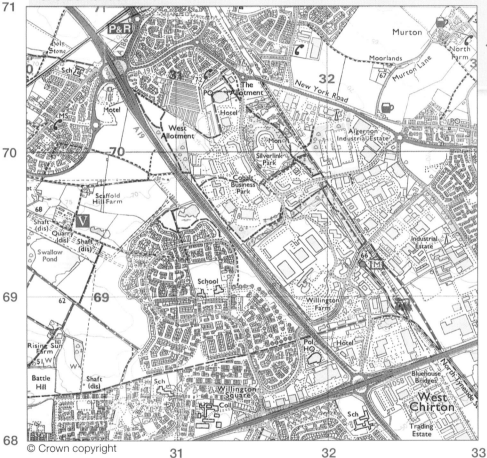

© Crown copyright

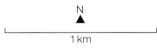
1 km

Stretch yourself

Find out about a science park or business park close to your home or school. Which companies are located there? What are the locational advantages? Produce an annotated map with photos.

Practice question

How do science and business parks provide opportunities for regional economic growth? *(6 marks)*

ACTIVITIES

1 What are the main differences between a science park and a business park?

2 Why are most UK science parks located close to universities?

3 a In which direction is photo **A** looking?

  b Which major road can be seen in the photo (refer to map **D**)?

  c Describe the environmental characteristics of the science park.

  d Why is it important to create a pleasant working environment?

4 a Give the four-figure grid reference for the Cobalt Business Park (OS map **B**).

  b Describe the layout of Cobalt Business Park.

  c Using evidence from the map, assess the importance of good transport links and an attractive working environment in the location of Cobalt Business Park.

Environmental impacts of industry

On this spread you will find out about sustainable ways of reducing the impacts of industry on the physical environment

Impacts of industry on the physical environment

Large-scale extraction industries such as mining and quarrying can have an impact on the environment. Quarries have been cut out of the countryside and huge waste tips piled up on the edges of mining settlements (photo **A**).

Modern manufacturing industries have an effect on both the landscape and the environment.

◆ Manufacturing plants can look very dull and uninteresting and can have a negative visual effect on the landscape.

◆ Industrial processes can cause air and water pollution, as well as degrading the soil.

◆ The waste products from manufacturing industry are often taken to landfill, and when in the ground they can pollute and harm the air, water and soil.

◆ The transport of raw materials and manufacturing products is usually by road, which increases levels of air pollution and damage to the environment when roads are widened or new ones built.

A   *Slate waste tips, North Wales*

How can industrial development be more sustainable?

Today there is a much greater concern about the need for industries to be environmentally sustainable. This can be achieved in a number of ways.

◆ Technology can be used to reduce harmful emissions from power stations and heavy industry.

◆ Desulphurisation can remove harmful gases such as sulphur dioxide and nitrogen oxide from power station chimneys.

◆ Stricter environmental targets put in place for industry on water quality, air pollution and landscape damage.

◆ Heavy fines imposed when an industrial pollution incident occurs.

B   *Industrial waste flowing into the River Wear at Washington, north-east England*

Quarrying in the UK

Quarrying can have harmful impacts on the environment. It can:

◆ destroy natural habitats

◆ pollute nearby water courses

◆ scar the landscape.

Today, there are very strict environmental controls on quarrying in the UK.

Sustainable development is at the heart of planning regulations and approval for mining and quarrying.

Companies are expected to restore or improve a quarry after it has been used. Examples of restoration include:

◆ landfill   ◆ housing

◆ agriculture   ◆ flood storage.

◆ habitat creation

Whilst in operation, there are strict controls on blasting, removal of dust from roads, and landscaping. Recycling is encouraged to reduce waste.

Torr Quarry, Somerset

Torr Quarry is operated by Aggregate Industries. It is one of eight limestone quarries on the Mendip Hills (map **C**). Over 100 people are employed at Torr Quarry and it is estimated that the quarry contributes more than £15 million towards the local economy each year.

Quarrying began at Torr Quarry in the 1940s. Today the quarry occupies a 2.5 km^2 site located 7 km east of Shepton Mallet. The quarry has previously produced 8 million tonnes per annum, although output is currently around 5 million tonnes. Torr Quarry is a nationally important source of construction materials – rock chippings used for a variety of construction such as roads. Three-quarters of its output is transported by rail, mostly to the south-east.

Quarrying at Torr Quarry aims to be environmentally sustainable in the following ways:

◆ The quarry is being restored to create wildlife lakes for recreation and water supply (photo **E**).

◆ Characteristic limestone features will be created to make the landscape look natural.

◆ 200 acres of the site have already been landscaped to blend in with the surrounding countryside, including planting grass and trees.

◆ Regular monitoring of noise, vibration, airborne emissions (dust) and water quality.

◆ Rail transport minimises the impact on local roads and villages.

In 2010 a planning application was made to deepen the quarry to extend operations to 2040. Deepening has less impact on the environment than extending the quarry outwards into surrounding countryside. This plan was approved in 2012.

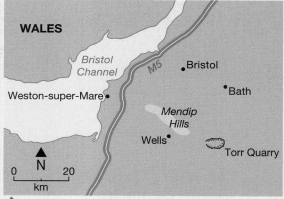

C Location of Torr Quarry, Somerset

D Torr Quarry

E Planned restoration of Torr Quarry

ACTIVITIES

1 Describe the harmful impacts that industry can have on the physical environment (photos **A** and **B**).

2 **a** What rock is quarried at Torr Quarry and what is it used for?

   b How is the rock transported away from the quarry?

   c How environmentally sustainable is this form of transport?

   d What is the importance of the quarry to the local community?

3 **a** Describe the impact of the quarry on the environment (photo **D**).

   b What is the evidence that restoration has already begun?

4 To what extent do you think the quarry is being managed in an environmentally sustainable way?

Stretch yourself

Find out more about how technology can be used to reduce harmful emissions and environmental damage.

Practice question

Use an example to demonstrate how modern industry can be more environmentally sustainable. *(6 marks)*

On this spread you will find out about the social and economic changes in two contrasting rural areas

How is the UK's rural landscape changing?

Rural landscapes in the UK are changing. For example, South Cambridgeshire is experiencing rapid population growth as people move out of Cambridge and London to enjoy a different pace of life in the countryside. However, remote rural areas such as the Outer Hebrides are experiencing population decline.

An area of population growth: South Cambridgeshire

South Cambridgeshire is the mostly rural area surrounding the city of Cambridge (photo **B**). The population of 150 000 is increasing due to migration into the area (graph **C**). Migrants have mostly come from Cambridge and other parts of the UK. However, more migrants are now arriving from Eastern Europe.

◆ In 2013, registrations for National Insurance numbers from migrant workers in South Cambridgeshire soared by 25 per cent compared to 2012.

◆ The proportion of people in South Cambridgeshire aged 65 or over is growing – by 2031 this will reach nearly 29 per cent of the population.

◆ The population of South Cambridgeshire is estimated to reach 182 000 by 2031.

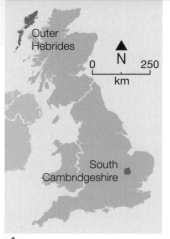

A South Cambridgeshire and the Outer Hebrides

B The landscape of Cambridgeshire

Social effects	Economic effects of a growing population
Commuters continue to use services in the places where they work, for example, Cambridge – this has a negative effect on the local-rural economy.	A reduction in agricultural employment as farmers sell their land for housing development, although this may increase jobs in construction.
80% car ownership is leading to increased traffic on narrow country roads and reducing demand for public transport.	Lack of affordable housing.
Modern developments on the edges of villages and gentrification of abandoned farm buildings can lead to a breakdown in community spirit.	This area has some of the highest petrol prices in the country due to the high demand.
Young people cannot afford the high cost of houses and move away.	The increasing number of migrants from relatively poor parts of Europe can put pressure on services and increase overall costs.

Another view

South Cambridgeshire has seen large numbers of highly-skilled and educated people move into the area. There is a high level of employment. Around 21 per cent of the workforce is employed in high-tech industries like computer software and engineering manufacture.

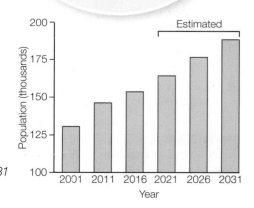

C Population growth in South Cambridgeshire, 2001–31

An area of population decline: The Outer Hebrides

The Outer Hebrides has a population of 27 400 people. Most live on the island of Lewis, one of the chain of 65 islands (photo **D**).

Despite a small population increase in recent years, there has been an overall decline of more than 50 per cent since 1901 when 46 000 people lived on the islands (graph **E**). This decline is mainly due to outward migration. With limited opportunities, younger people have chosen to move away from the area in search of better-paid employment elsewhere.

Social impacts

◆ The number of school children is expected to fall over the next few years and this may result in school closures.

◆ With many younger people moving away, there will be fewer people of working age living in the Outer Hebrides.

◆ An increasingly ageing population (graph **E**) with fewer young people to support them may lead to care issues in the future.

Economic effects

The UK and Scottish governments provide subsidies towards the costs of operating ferries and the maintenance of essential services. But it is a struggle to maintain the economy and many of these services, such as post offices, are closing.

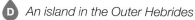

D *An island in the Outer Hebrides*

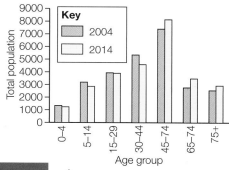

E *Population structure of the Outer Hebrides, 2004–14*

Industry	Economic impacts of a declining population
Farming	• The main farming economy is breeding sheep on small farms called crofts. • Most crofts can only provide work for up to two days per week.
Fishing	• In 1948 there were more than 900 fishing boats registered in the Outer Hebrides. By 2013 there were just a few boats catching prawns and lobsters. • Shellfish production has increased, but foreign-owned ships dominate deep-sea fishing. • The development of fish farming has been limited due to concerns about the environment.
Tourism	• Between 2007 and 2014 there was a 27% increase in visitors to the islands. • The current infrastructure is unable to support the scale of tourism needed to provide an alternative source of income.

ACTIVITIES

1 a Describe the population change in South Cambridgeshire between 2001 and 2031 (graph **C**).

  b What are the main causes for this change?

  c Identify the challenges associated with a growing rural population?

2 How might the physical landscape of the Outer Hebrides (photo **D**) contribute to migration from the islands?

3 a Describe the main population changes in the Outer Hebrides (graph **E**).

  b Suggest what social and economic impacts these changes may have in the future.

Stretch yourself

Find out more about the social and economic issues facing the Outer Hebrides. What can be done to address the rural problems associated with out-migration?

Practice question

Contrast the economic challenges associated with rural areas of population growth and decline. *(6 marks)*

On this spread you will find out about improvements and new developments to the UK's roads and railways

Transport involves the movement of people, goods and services. Improvements in transport have enabled the UK's economy to grow and develop. This remains a major issue for the future. Transport developments affect both employment and regional growth.

Road improvements

In 2014 the government announced a £15 billion 'Road Investment Strategy'. The aim is to increase the capacity and improve the condition of UK roads. New road schemes will create thousands of construction jobs and boost local and regional economies. Plans include:

◆ 100 new road schemes by 2020

◆ 1300 new lane miles added to motorways and trunk roads to tackle congestion

◆ extra lanes added onto main motorways to turn them into 'smart motorways' (photo **A**) and improving links between London, Birmingham and the north.

A Smart motorway improvements on the M25

Maths skills

The following data shows the estimated number of regional jobs that will be created by the Road Investment Strategy. Present the information in an appropriate graph.

North East and Yorkshire	1500
North West	600
Midlands	900
East of England	1000
London and South East	900
South West	1300

South-west 'super highway'

The A303 is the main route to the south-west. However, traffic flow can be 'stop-start' due to alternating stretches of dual and single carriageway. At peak times the road can become heavily congested.

The £2 billion road-widening project is the biggest to be undertaken over the next 15 years. It will create hundreds of construction jobs. Converting the route to dual carriageway, with additional improvements to connecting routes, will create a 'super highway' all the way to Plymouth and beyond (map **B**). The scheme will involve digging a 3 km tunnel beneath Stonehenge!

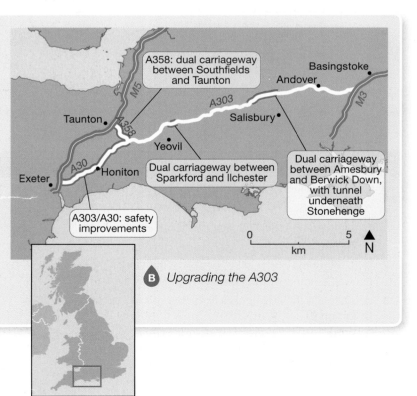

B Upgrading the A303

Railway improvements

Investment in railways is vital to improve links between regions of the UK and the rest of Europe. There are plans to stimulate economic growth in the north of the UK by improving the rail network:

◆ electrification of the Trans-Pennine Express Railway between Manchester and York by 2020, reducing journey times by up to 15 minutes and completing the electrified link between Liverpool and Newcastle

◆ electrification of the Midland mainline between London and Sheffield by 2023

◆ HS2 (High Speed 2) – a £50 billion plan for a new high-speed rail line to connect London with Birmingham and then to Sheffield, Leeds and Manchester (map **C**). It may then be extended to Newcastle and into Scotland. The scheme is due to start in 2017 for completion in 2033. Many people are against the plan, as the route passes through several stretches of countryside and close to many homes.

Key
— LGV nord (1993)
⋯⋯ Channel Tunnel (1994)
— High speed 1 (2007)
— High speed 2 (phase 1)
— High speed 2 (phase 2)
------ Future phases (proposed)

C *Plans for UK high-speed rail*

London's Crossrail

Crossrail is a new railway across the capital that links Reading and Heathrow (to the west of London), to Shenfield and Abbey Wood (to the east). Due for completion in 2018, Crossrail will involve over 32 km of new twin-bore tunnels under central London. At a cost of £14.8 billion, it is one of the most important and ambitious infrastructure projects ever undertaken in the UK.

◆ Crossrail will improve journey times across London, easing congestion and offering better connections to the Underground and to the rest of the UK and Europe.

◆ It will bring an additional 1.5 million people within 45 minutes' commuting distance of London's key business districts.

◆ Around 200 million passengers are expected to use Crossrail each year.

◆ Mainline trains operating on the lines will carry 1500 passengers.

D *The construction of Crossrail deep below London*

ACTIVITIES

1 **a** What is a 'smart motorway' (photo **A**)?

   b Why do you think the government is upgrading several motorways to become 'smart motorways'?

2 Describe and suggest reasons for the planned improvements to the A303.

3 **a** Describe the plans for high-speed rail routes in the UK.

   b Do you think it is important to have high-speed rail in the UK? Why?

4 How do road and rail improvements help to boost the national and local economy?

Stretch yourself

Find out more about the proposed HS2 (High Speed 2) rail development. Why is it so controversial? What are the arguments for and against the plan?

Practice question

How can road and rail developments improve the UK's economy? *(6 marks)*

245

On this spread you will find out about improvements and new developments to the UK's ports and airports

Developing the UK's ports

The UK ports industry is the largest in Europe, due to the length of the coastline and the UK's long trading history. Some ports specialise in handling containers whilst others service ferries and cruises. Most are all-purpose ports handling a range of goods and services. In 2014:

◆ the UK's leading port in terms of tonnage was Grimsby, followed by London (Tilbury), Milford Haven (south Wales) and Southampton

◆ Dover was the main port for freight (lorries, cars, etc.)

◆ Felixstowe was the UK's largest container port, handling 2 million containers.

About 32 million passengers travel through UK ports each year, and they employ around 120 000 people. The UK's largest ports are run by private companies which have invested heavily in the port infrastructure.

◆ Belfast – £50 million invested in facilities to service the offshore renewable energy sector. Land is available to accommodate manufacturing for wind, wave and tidal energy.

◆ Bristol (Avonmouth) – £195 million invested for bulk handling and storage facilities.

◆ Felixstowe and Harwich – a new rail terminal at Felixstowe and upgraded cruise service facilities at Harwich.

A *The location of the UK's major ports*

Did you know?

96 per cent of all UK import and export trade enters and leaves via ports.

Liverpool2

A new container terminal is being constructed at the Port of Liverpool (photo **B**). Known as 'Liverpool2', the scheme to construct a deep-water quay on the River Mersey will cost about £300 million.

The project will more than double the port's capacity to over 1.5 million containers a year to compete with other major UK ports. Due to begin operating in 2016, the new terminal will:

◆ create thousands of jobs

◆ boost the economy of the north-west

◆ reduce the amount of freight traffic on the roads.

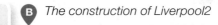

B *The construction of Liverpool2*

Airport developments

Airports are very important to the economy of the UK, creating vital global links. They provide thousands of jobs and boost economic growth both regionally and nationally.

◆ The aviation sector accounts for 3.6 per cent of the UK's GDP and employs over 300 000 people.

◆ Over two million tonnes of freight pass through the UK's airports each year.

- More than 750 000 international flights depart from the UK annually, to almost 400 airports in 114 countries (map **C**).

- Over 420 000 domestic flights provide 35 million seats annually to passengers travelling to over 60 regional airports across the UK.

Expanding London's airports

There has been much controversy about the need to expand London's airport capacity. Many options have been discussed. These include the construction of a new fourth airport in the Thames estuary and the expansion of either Heathrow or Gatwick. All the options have advantages and disadvantages.

In 2012 a government commission looked into three options and came up with a recommendation.

- a new runway at Heathrow (cost £18.6 billion)

- increasing the length of one of the existing runways at Heathrow (cost £13.5 billion)

- constructing a new runway at Gatwick (cost £9.3 billion)

The report, published in 2015, recommended a new third runway at Heathrow, although a final decision is yet to be made. This was predicted to create more jobs and make more money for the UK. Heathrow is one of the world's major airports. In 2014 it handled over 73 million passengers. Over 76 000 people work at Heathrow and the airport supports many local businesses.

People living nearby are concerned about noise from planes using the new runway. The government report recommended financial support for soundproofing homes and schools, and a ban on night-time flights.

C UK air transport routes

Key
— UK Air Links

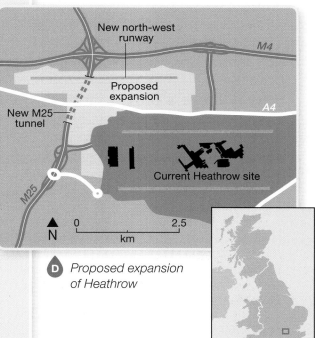

D Proposed expansion of Heathrow

ACTIVITIES

1 Why is the UK's port industry the largest in Europe?

2 **a** Mark the major UK ports on an outline map of the UK (map **A**).

   b Use the information on this spread and internet research to add annotations describing recent developments to the ports.

3 **a** Describe the recent port developments at Liverpool.

   b Give reasons for the location of Liverpool2 (photo **B**).

   c What benefits will the new developments bring to the region?

4 Describe and suggest reasons for the pattern of UK air transport routes (map **C**).

5 Comment on the decision to recommend a new runway at Heathrow.

Stretch yourself

Research the three options considered by the Davis Commission as well as the scheme to build a new airport in the Thames estuary. Consider the advantages and disadvantages of each option, and come up with your own reasoned choice.

Practice question

Discuss the arguments for and against expanding the capacity of London's airports? *(6 marks)*

The north-south divide

On this spread you will find out about strategies to address regional differences and inequalities in the UK

What is the north-south divide?

In the UK we often talk about the 'north-south divide'. This refers to the real or imagined cultural and economic differences between:

◆ the south of England (the South East, Greater London, the South West and parts of eastern England)

◆ the north of England (Yorkshire, the Humber, the North East and North West) and the rest of the UK.

In general, the south enjoys higher standards of living, longer life expectancy and higher incomes. It also has higher house prices, more congestion and, according to some measures, less 'happiness'! But what are these differences? Look at map **A** and table **B** and judge for yourself.

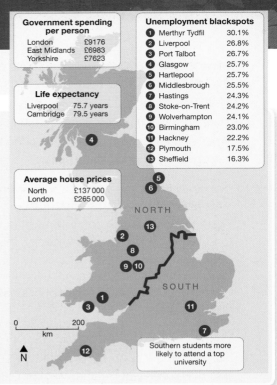

Government spending per person	
London	£9176
East Midlands	£6983
Yorkshire	£7623

Life expectancy	
Liverpool	75.7 years
Cambridge	79.5 years

Average house prices	
North	£137 000
London	£265 000

Unemployment blackspots	
1 Merthyr Tydfil	30.1%
2 Liverpool	26.8%
3 Port Talbot	26.7%
4 Glasgow	25.7%
5 Hartlepool	25.7%
6 Middlesbrough	25.5%
7 Hastings	24.3%
8 Stoke-on-Trent	24.2%
9 Wolverhampton	24.1%
10 Birmingham	23.0%
11 Hackney	22.2%
12 Plymouth	17.5%
13 Sheffield	16.3%

Southern students more likely to attend a top university

A *North and south – some facts*

Why is there a north-south divide in the UK?

During the Industrial Revolution the UK's growth was centred on the coalfields in Wales, northern England and Scotland. Heavy industries and engineering thrived in cities such as Manchester, Sheffield and Glasgow, generating wealth and prosperity.

Since the 1970s many industries (such as steel-making, ship building and heavy engineering) have declined and unemployment increased. Alternative sources of energy have reduced the importance of coalfields and modern industries have located elsewhere.

London and the South East developed rapidly due to a fast-growing service sector. London is a major global financial centre and has grown faster than the rest of the UK. This has led to high house prices across the South East.

How can regional strategies address the issue?

There have been many attempts to address the problems caused by de-industrialisation in the north. Financial support from the UK government and the EU has helped new businesses and improvements in infrastructure.

Foreign investment has been encouraged in the north: Nissan at Washington in Tyne and Wear opened in 1984, and Mitsubishi at Livingston, near Edinburgh, opened in 1975.

In 2015 the government launched a new strategy for a 'Northern Powerhouse' to help balance the wealth and influence of London and the South East. This involves developing the economies of the major cities in northern England such as Liverpool and Manchester. Tourism, food and energy are to be developed in rural areas.

	North East	North West	Yorkshire and Humberside	West Midlands	South West	East Midlands	East Anglia	South East	Scotland	Wales	Northern Ireland
Average pay	£24 000	£25 000	£24 000	£25 000	£25 000	£24 000	£26 000	£28 000	£21 000	£19 500	£18 500
% unemployed	9.9	6.9	7.3	7.1	4.9	5.5	4.9	4.4	7.7	3.4	2.1
% poverty	21	22	22	23	19	20	18	18	15	19	21
Life expectancy (years against average)	−1.3	−1.4	−0.8	−0.4	+0.9	−0.01	+1.0	+1.2	−3.0	−1.4	−1.6

B *Regional evidence for the UK's north-south divide*

Money has been invested in transport improvements including:

◆ a new high-speed rail service (HS2) between London and the north (see page 245) and the electrification of the Trans-Pennine railway

◆ upgrade of the M62 cross-Pennine motorway

◆ the new Liverpool2 deep-water container port (see page 246)

◆ the Mersey Gateway – a new 6-lane toll bridge over the River Mersey to improve access to the new deep-water port.

Local enterprise partnerships

Established in 2011, local enterprise partnerships (LEPs) are voluntary partnerships between local authorities and businesses. There are currently 39 LEPs in England. Their aim is to identify business needs in the local areas and encourage companies to invest. In this way jobs will be created boosting the local economy.

Enterprise Zones

Since 2011, 24 new Enterprise Zones have been created. Their aim is to encourage the establishment of new businesses and new jobs in areas where there were no pre-existing businesses. The government supports businesses in Enterprise Zones by:

◆ providing a business rate discount of up to £275 000 over a five-year period

◆ ensuring the provision of superfast broadband

◆ financial allowances for plant and machinery

◆ simpler planning regulations to speed up establishment of new businesses.

Lancashire LEP

Lancashire has a tradition of industry and manufacturing based on textiles and engineering. De-industrialisation led to many factory closures and job losses. Recent growth in manufacturing has been based on the development of aeronautical engineering.

The Lancashire LEP will promote new businesses and create 50 000 new jobs by 2023 (map **C**).

◆ In 2013 a Business Growth Hub was established to support small and medium-sized businesses in the area. It aims to set up 400 new businesses and create 1100 new jobs by 2016.

◆ £20 million of transport improvements are planned in cities such as Preston and Blackburn, including the major new Heysham to M6 link.

◆ A £62 million BT investment will extend superfast broadband across 97% of the region.

◆ Enterprise Zones at Samlesbury and Warton will create 6000 high-skilled jobs in advanced engineering and manufacturing (AEM) sector.

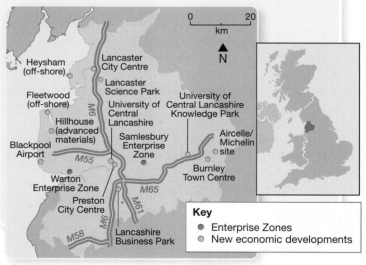

C *Economic developments in Lancashire*

ACTIVITIES

1 What is the evidence of a north-south divide (map **A**)?

2 **a** What is the evidence that the South East has the highest wealth and quality of life (table **B**)?

   b Which region appears to be most disadvantaged and why?

3 Design an information poster about Lancashire's LEP (map **C**). Include a map, text boxes and annotated photos.

4 Does the government have a responsibility to balance the economy across the UK? Why?

Stretch yourself

Use the Centre for Cities report (2015) to show how evidence on the prosperity of cities makes it difficult to locate the line of the north-south divide.

Practice question

Explain how improving transport links can help reduce the north-south divide. *(4 marks)*

On this spread you will find out about the place of the UK in the wider world

What are the UK's links with the wider world?

As globalisation increases, the world is becoming more and more inter-dependent. The UK has global links through trade, culture, transport and electronic communications. In the past the UK was one of the world's superpowers with an extensive empire in the Americas, Africa and the Asian Pacific. The UK is still an influential member of important international organisations such as the G8, NATO and the UN Security Council. Although its global position has declined, many nations see the UK as a fair, tolerant and law-abiding society with a long and rich cultural heritage.

Trade

Trade involves the movement of goods and services across the world. This usually involves transport by air, sea, road and rail. The internet is becoming more important for trade, for example in finance, communications and the creative industries.

The UK's most important trading links are with the EU (diagram **A**). As a single market, goods can be traded between member states without tariffs. The USA is an important historic trading partner, with a recent growth in trade with China.

Culture

Culture is used to describe the values and beliefs of a society or group of people. It's all about what makes a society special. Culture can include writing, painting or creativity in the form of fashion, architecture or music.

UK export destinations (£ millions)

USA 38 858	Netherlands 22 553	**Germany** 31 432	France 19 627	Switzerland 21 313
	China 15 934	Italy 8708	Rep. of Ireland 18 124	Belgium 12 597
		Spain 8774		

Sources for UK imports (£ millions)

Germany 59 365	Netherlands 31 068	**USA** 32 863	Italy 16 663
	Norway 14 958	Rep. of Ireland 11 695	**China** 33 891
	Spain 13 031	France 24 906	Belgium 20 585	

 A The UK's main trading partners

Did you know?

Peppa Pig cartoons, created in the UK, are now screened in more than 170 countries worldwide earning a total of US$1bn (£640m).

Television

Television is one of the UK's most successful media exports. In 2013–14 it accounted for over £1.28 billion of export earnings. International sales of UK television programmes has almost quadrupled since 2004. Amongst the most successful recent programmes worldwide are *Atlantis*, *Downton Abbey*, *Dr Who* and *Sherlock*.

The main markets are English-speaking countries such as the USA (47 per cent), Australia and New Zealand. The Chinese market is expanding rapidly, increasing by 40 per cent from the previous year to £17 million in 2013–14.

 **B** Dr Who – a UK export success

The global importance of the English language has given the UK strong cultural links with many parts of the world. Wherever you travel you will hear music, read books and watch films from the UK. Migrants have brought their own culture to the UK. For example:

◆ food (such as Indian, Chinese and Thai)   ◆ films ('Bollywood' from India)

◆ fashion (from France and Italy)   ◆ festivals (such as the Notting Hill Carnival).

◆ music (from America and Africa)

Today the UK is a multi-cultural society, accepting people from many other countries.

Transport

London Heathrow is one of the busiest airports in the world. It is an important aircraft hub where people transfer between flights within Europe and worldwide (see page 247).

There are important transport links between the UK and mainland Europe via the Channel Tunnel (photo **C**) and sea ferries. Southampton is a major port for cruise liners that take thousands of tourists around the world to destinations such as the Mediterranean and the Caribbean.

C The Channel Tunnel

Electronic communication

Ninety-nine per cent of all internet traffic passes along a multi-billion-dollar network of submarine high-power cables. The UK is a focus for these submarine cables (map **D**), with connections concentrated between the UK and the USA. There is a further concentration in the Far East connecting Japan, China and other countries in the region.

Electronic communication is a vital part of the global economy and fast reliable connections are essential. A project known as Arctic Fibre will lay the first cables between London and Tokyo via the Northwest Passage, linking Europe and Asia – a distance of 15000 km! The main cable will operate at speeds of 100 gigabytes. It is due to be completed in 2016.

D The global submarine cable network

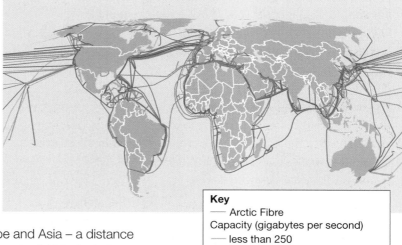

Key
—— Arctic Fibre
Capacity (gigabytes per second)
—— less than 250
—— 250-1000
—— 1000-2000
—— 2000-3000
—— more than 3000

ACTIVITIES

1 **a** Rank the UK's top five export destinations and import sources.

  b Suggest why Ireland and the USA are important trading partners of the UK.

  c Can you explain why China is a major source of imports?

2 Work in small groups to produce a collage illustrating cultural connections between the UK and the rest of the world. Use logos or photos to create your collage.

3 Suggest why UK TV programmes are such an important export worldwide.

4 Describe and suggest reasons for the pattern of submarine e-communication cables (map **D**).

Stretch yourself

Find out more about the UK's transport links with the rest of the world. Search online to find maps showing connections by air, sea, road and rail. Why is the UK such an important global transport hub in the twenty-first century?

Practice question

How does the UK benefit by having close links with the rest of the world? *(6 marks)*

The UK in the wider world (2)

On this spread you will find out about the UK's economic and political links with the European Union and the Commonwealth

What are the UK's links with the European Union?

In 1973 the UK became a member of the European Union (EU). The EU began as a small trading group of industrial countries in north-west Europe. Over the years it has expanded to its current total of 28 countries, with the latest member Croatia joining in 2013 (map **A**).

The EU is still an important trading group, but its powers have extended to exert political influence over its members and elsewhere in the world. Many in the UK feel that the EU capital in Brussels is becoming too influential in making laws which affect the UK. There was a political movement for the UK to leave the EU or at least limit its powers. In a referendum held in June 2016, the people of the UK voted to leave the EU by 52 per cent to 48 per cent. The UK will remain a member of the EU until exit negotiations are complete.

The EU has affected the UK in a number of important ways (diagram **B**).

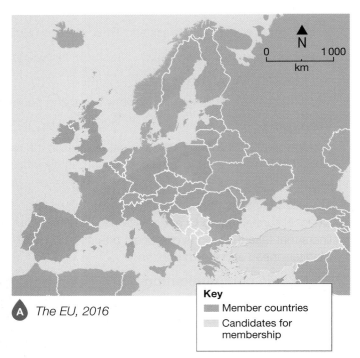

A *The EU, 2016*

Key
- Member countries
- Candidates for membership

 B *How the EU has affected the UK*

Financial support for farmers. The Single Payments Scheme, introduced in 2015, is part of the EU's Common Agricultural Policy. In 2015, £18 million was made available to support dairy farmers in England and Wales.

There are EU laws and controls on crime, pollution and consumers' rights. These rules can be restrictive for individuals and companies in the UK.

How has the EU affected the UK?

The European Structural and Investment Funds has provided support for disadvantaged regions in the UK and sectors such as fisheries.

High unemployment and low wages in poorer EU countries – particularly in Eastern Europe – may have led to mass migration of workers to the UK. In 2013 over 200 000 immigrants (about 40% of total UK immigrants) came to the UK from the EU.

The EU is the biggest single market in the world. Goods, services, capital and labour can move freely between member states and encourage trade.

What are the UK's links with the Commonwealth?

The UK is a member of the Commonwealth, a voluntary group of 53 countries, most of which were once British colonies. It is home to 2.2 billion people, 60 per cent under the age of 30. The Commonwealth includes some of the world's largest, smallest, richest and poorest countries (map **C**). Thirty-one of its members are small states, many of them island nations.

The Commonwealth Secretariat represents Commonwealth countries and provides advice on a range of issues including human rights, social and economic development and youth empowerment. The Secretariat aims to help governments achieve sustainable, inclusive and equitable development. The heads of each country meet every two years to discuss items of common interest.

There are important trading and cultural links between the UK and the Commonwealth countries. There are also sporting connections, with the Commonwealth Games – the so-called 'Friendly Games' – held every four years – (photo **D**).

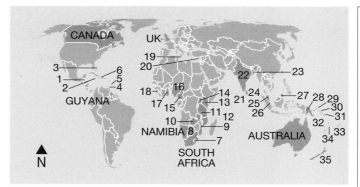

C *Some of the larger Commonwealth countries*

1. Belize
2. Jamaica
3. Bahamas
4. Trinidad & Tobago
5. Barbados
6. Dominican Republic
7. Lesotho
8. Botswana
9. Mozambique
10. Zambia
11. Tanzania
12. Seychelles
13. Kenya
14. Uganda
15. Cameroon
16. Nigeria
17. Ghana
18. Sierra Leone
19. Malta
20. Cyprus
21. Maldives
22. India
23. Bangladesh
24. Sri Lanka
25. Malaysia
26. Singapore
27. Brunei
28. Papua New Guinea
29. Solomon Islands
30. Tuvalu
31. Samoa
32. Vanuatu
33. Tonga
34. Fiji Islands
35. New Zealand

D *The Commonwealth Games in Glasgow, 2014*

ACTIVITIES

1 **a** Which European countries may wish to join the EU in the future?

   b Suggest advantages and disadvantages of continued expansion of the EU.

2 Consider how membership of the EU affects the UK (diagram **B**). Sort the list into positive and negative points. Use the internet to add to your two lists.

3 Draw a matrix like the one shown here. Write the names of Commonwealth countries into the appropriate boxes. Use the internet to help you.

	Rich	Poor
Large		
Small		

4 What are the economic, social and political benefits of events such as the Commonwealth Games? How would the city of Glasgow have benefited from the 2014 Commonwealth Games?

Maths skills

Search online for the final medals table of the 2014 Commonwealth Games. Present one aspect – say the number of gold medals – using located proportional bars on a blank world map. Draw a scattergraph (see page 338) to investigate the relationship between a country's wealth (GDP) and the number of medals won.

Stretch yourself

Investigate the pros and cons of the UK's membership of the EU. Do you think the UK should be part of the EU?

Practice question

Suggest how the UK benefits economically and politically from its membership of **either** the EU **or** the Commonwealth. *(6 marks)*

Section C The challenge of resource management

Nuclear power station at Tihange, Belgium

Unit 2 Challenges in the human environment is about human processes and systems, how they change both spatially and temporally. They are studied in a range of places, at a variety of scales and include places in various states of development. It is split into three sections.

Section C The challenge of resource management includes:

- resource management
- managing food resources
- managing water resources
- managing energy resources

You need to study resource management and one topic from food, water or energy in Section C – in your final exam, you will have to answer Question 3 and one other question.

What if...

1 all countries were equal?

2 we only ate genetically modified food?

3 everyone had clean water to drink?

4 all power was renewable?

Your key skills

To be a good geographer, you need to develop important geographical skills – in this section you will learn the following skills:

- Describing patterns of distribution in maps and graphs
- Carrying out research
- Using numerical data
- Presenting data using different graphical techniques
- Drawing and labelling diagrams.

Your key words

As you go through the chapters in this section, make sure you know and understand the key words shown in bold. Definitions are provided in the Glossary on pages 346–9. To be a good geographer you need to use good subject terminology.

Your exam

Section C makes up part of Paper 2 – a one and a half-hour written exam worth 35 per cent of your GCSE.

19.1 | The global distribution of resources

On this spread you will find out about the uneven distribution of food, water and energy

What are resources?

A **resource** is a stock or supply of something that has a value or a purpose. The three most important resources are food, energy and water. Adequate supplies of these resources are essential for countries to develop.

These resources are unevenly distributed across the world. Most HICs have plentiful supplies and enjoy a high standard of living. But many of the world's poorer countries, such as those in sub-Saharan Africa, lack resources and struggle to progress or improve quality of life for their people. As the world's population continues to grow, resource management will present many challenges.

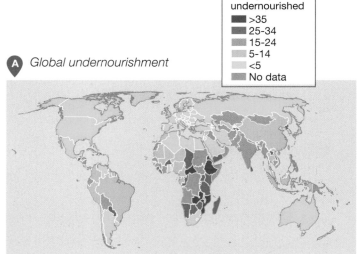

A *Global undernourishment*

Key
% of population undernourished
- >35
- 25-34
- 15-24
- 5-14
- <5
- No data

Food

Your health is affected by how much you eat and the food's nutritional value. The World Health Organisation (WHO) suggests that we need 2000–2400 calories per day to be healthy. Over one billion people in the world fall below this level and are described as *malnourished*.

A further two billion people suffer from **undernutrition** (malnutrition) – a poorly-balanced diet lacking in minerals and vitamins. This can result in a range of illnesses and diseases. It can also have economic effects. People need to be well fed to be productive at work and contribute to the economic development of their country. Obesity (being overweight) is an increasing problem.

Country	Gross National Product (GNP) per head (US$)	Human Development Index (HDI) Ranking	Water per head (m³)
Canada	22 480	1	94 000
Australia	20 210	7	185 000
Saudi Arabia	10 120	78	2176
Burkina Faso	1010	171	1535
Niger	850	173	346

Water

Think about how much water you have used and drunk today. Imagine if you had just one bucket of water to use each day, including water for drinking. Both the quantity and the quality of water are important for our well-being and for economic development (table **B**).

Water is not only essential for people and animals to drink, but is vital for crops and food supply. It is also important as a source of power for producing energy. As the world's population grows, more people are faced with a shortage of water.

The imbalance in water supply is due mainly to variations in climate and rainfall. Rainwater needs to be captured and stored in reservoirs or taken from rivers or aquifers deep underground. All of these are very expensive and require high levels of investment.

Many of the world's poorest countries, particularly in Africa, have a shortage of water. They become trapped in a cycle of poverty. The UN estimates that by 2025 there will be 50 countries facing water scarcity (map **C**).

B *How water supply relates to development*

C *Projected areas of water scarcity by 2025*

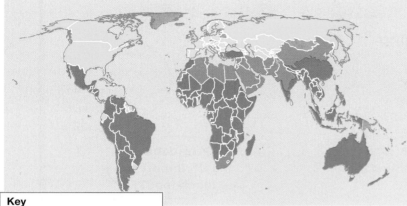

Key
- Physical water scarcity (lack of water, e.g. deserts)
- Economic water scarcity (countries that cannot afford to exploit water supplies)
- Little or no water scarcity
- No data

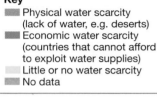

D *How water use links to a country's income*

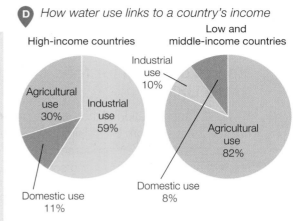

High-income countries

Agricultural use 30%
Industrial use 59%
Domestic use 11%

Low and middle-income countries

Industrial use 10%
Agricultural use 82%
Domestic use 8%

There are significant differences in water use between low/middle and high-income countries (graph **D**). Low/middle-income countries use a higher proportion of water for agriculture compared to high-income countries where most water is used for industry.

Energy

Think of the energy needed in your home and school for light and heat, and to power things like cookers, TVs and tablets. Energy is required for economic development. It powers factories and machinery and provides fuel for transport. In the past many countries could depend on their own energy resources. Today the situation is much more complex, with energy being traded worldwide.

Energy consumption is increasing as the world becomes more developed and demand increases (graph **E**). The world's richest countries use far more than poorer countries in Africa and the Middle East. The Middle East supplies much of the world's oil yet its own consumption is relatively small.

As NEEs become more industrialised, the demand for energy will increase and patterns of energy trading will change.

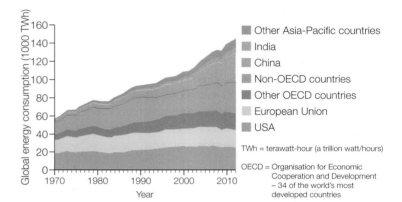

- Other Asia-Pacific countries
- India
- China
- Non-OECD countries
- Other OECD countries
- European Union
- USA

TWh = terawatt-hour (a trillion watt/hours)

OECD = Organisation for Economic Cooperation and Development – 34 of the world's most developed countries

E *Global energy consumption*

Maths skills

Calculate the percentage increase in global energy consumption between 1970 and 2012 (graph **E**).

Stretch yourself

Describe the pattern of global oil supply and demand. Where is the greatest demand for oil? How is oil transported from regions of supply to regions of demand?

Practice question

Describe the global inequality in the supply and consumption of *either* food *or* water *or* energy. (*6 marks*)

ACTIVITIES

1 **a** What is a 'resource'?
   b How is 'malnourishment' different to 'undernutrition'?
2 Which parts of the world are suffering extreme undernourishment (map **A**)?
3 Why do some countries suffer from water shortages?
4 Describe the pattern of global energy consumption (graph **E**).

On this spread you will find out about the opportunities and challenges faced by the UK in the provision of food

How is demand for food changing in the UK?

By 2037 the population of the UK is expected to rise to 73 million (from 64 million in 2015). This will increase the future demand for food. Despite the UK's efficient and productive farming sector the UK is not self-sufficient for food supplies. In fact the UK imports about 40 per cent of the total food consumed and this proportion is increasing.

Diagram **A** shows reasons why the UK imports such a high proportion of its food.

What is the impact of importing food?

Map **D** shows the distances travelled by foods imported to the UK – known as **food miles**. Transporting food by air is very expensive. Importing food also adds to our **carbon footprint** – the emission of carbon dioxide into the atmosphere. This comes from producing the energy for commercial cultivation, and from transport by planes and lorries.

UK-produced food can be expensive because of poor harvests and the price of animal feed

Availability of cheaper food from abroad imported by supermarkets who compete for low prices

Why does the UK import so much food?

Demand for greater choice and more exotic foods

UK climate is unsuitable for the production of some foods, such as cocoa, tea and bananas

Demand for seasonal produce all year round, such as strawberries and apples

 A *Why is the UK not self-sufficient in food?*

Importing high-value foods: vegetables from Kenya

The growing of vegetables such as mangetout is Kenya's biggest source of income (photo **B**). The cost of air freight to keep produce fresh is very high. UK customers are prepared to pay higher prices for vegetables when they are not in season in the UK.

Kenyan farmers only earn a fraction of the price of the vegetables in a UK supermarket (table **C**). Two-thirds are casual labourers with no job security or benefits and are paid very little.

Stage	Price per tonne (£)	% of final price
Producer	630	12
Exporter	290	6
Packaging	280	5
Air freight/handling	1040	20
Importer	620	12
Supermarket	2500	45
Total price	5360	100

C *Price breakdown for one tonne of Kenyan mangetout*

Maths skills

Draw a pie graph to show the breakdown of the price of imported mangetout.

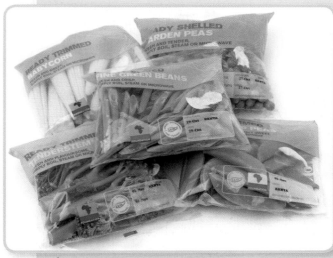

B *Vegetables produced in Kenya*

How is the UK responding to the challenges?

There is concern about the UK's dependency on foreign food imports and the need for greater food security. This has led to a growing interest in sourcing food locally to reduce carbon emissions (page 274). People are being encouraged to eat seasonal foods produced in the UK.

There are two major recent trends in UK farming:

◆ *Agribusiness* – intensive farming aimed at maximising the amount of food produced. Farms are run as commercial businesses. They have high levels of investment, and use modern technology and chemicals.

◆ **Organic produce** – grown without the use of chemicals. Organic food has become increasingly popular, although higher labour costs often make it more expensive. Organic food production is often associated with buying local produce and producing seasonal foods (page 274).

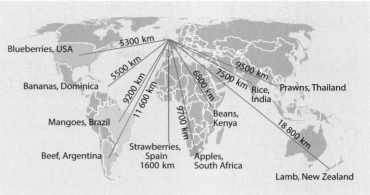

D *Distances travelled by UK imported food*

Lynford House Farm – an agribusiness

Lynford House Farm in East Anglia is a large arable farm of 570 hectares. As an agribusiness it has high inputs of chemicals, machinery and other investments.

◆ The flat, fertile land is intensively farmed to maximise productivity and profitability.

◆ The main crops are wheat, sugar beet and potatoes which are well suited to the fertile soils and a warm, sunny climate.

◆ Chemicals are widely used as pesticides and fertilisers.

◆ Machinery costs are high but make the farm efficient. It only employs a small number of workers.

◆ The farm has invested in a 54-million litre reservoir to tackle frequent water shortages in this dry area.

Riverford Organic Farms

Riverford Organic Farms began as an organic food and dairy farm in rural Devon. It supplied local people with fresh boxes of food delivered weekly. The company now delivers boxes of vegetables around the UK from its regional farms in Devon, Yorkshire, Peterborough and Hampshire. These farms help Riverford to:

◆ reduce food miles

◆ support local farmers

◆ provide local employment

◆ build a strong link between grower and consumer.

ACTIVITIES

1 Why does the UK import 40 per cent of its food?

2 **a** If a pack of mangetout costs £2 in a UK supermarket, how much money does the producer receive (table **C**)?

 b How much money does the supermarket receive?

 c Why do you think the producer receives such a small share of the retail price? Is this fair?

3 Which foods travel over 9000 kilometres to reach the UK (map **D**)?

4 How is the UK dealing with the problems of importing food?

Stretch yourself

How has foreign travel and migration affected the demand for, and supply of, food in the UK?

Practice question

Explain the UK's attempts to respond to changing demands for food. (*4 marks*)

On this spread you will find out about the opportunities and challenges faced by the UK in the provision of water

What are the demands for water in the UK?

Think about all the ways you use water, for washing, drinking, flushing the toilet, cleaning and cooking. Almost 50 per cent of the UK's water supply is used domestically. But 21 per cent is wasted through leakage (graph **A**)!

The Environment Agency estimates that the demand for water in the UK will rise by 5 per cent by 2020 because of:

◆ the growing population

◆ more houses being built

◆ an increase in the use of water-intensive domestic appliances.

How far does the UK's water supply meet demand?

The main sources of water in the UK are rivers, reservoirs and groundwater aquifers. The UK currently receives enough rain to supply the demand, but rain doesn't always fall where it is most needed (map **B**).

◆ The north and west of the UK has a **water surplus** where supply exceeds demand. There is high rainfall, lower evaporation rates and plenty of potential reservoir sites. These areas have a relatively low population density.

◆ The south and east of the country has a **water deficit** where demand exceeds supply. This is the most densely populated part of the country and has the lowest annual rainfall.

Water stress (where demand exceeds supply) is experienced in more than half of England (map **C**). The south east of England ranks very low in the world in terms of water availability. The situation is made worse in times of drought, such as in 2010–12.

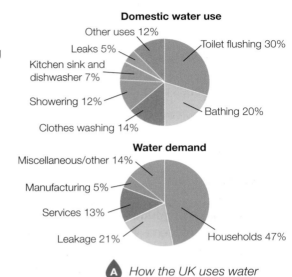

Domestic water use

- Other uses 12%
- Toilet flushing 30%
- Leaks 5%
- Kitchen sink and dishwasher 7%
- Showering 12%
- Bathing 20%
- Clothes washing 14%

Water demand

- Miscellaneous/other 14%
- Manufacturing 5%
- Services 13%
- Leakage 21%
- Households 47%

A *How the UK uses water*

Did you know?

Every person in the UK uses an average of 150 litres of water per day.

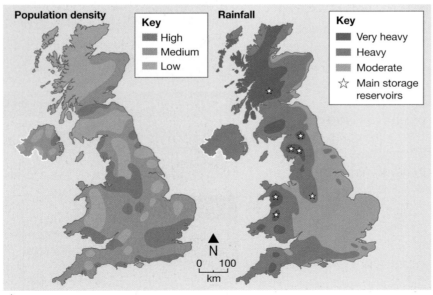

Population density

Key
- High
- Medium
- Low

Rainfall

Key
- Very heavy
- Heavy
- Moderate
- ☆ Main storage reservoirs

0   100
km
N

B *UK population density and water supply*

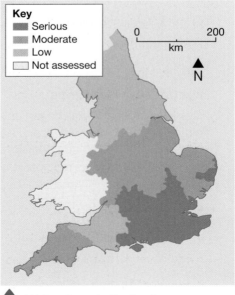

Key
- Serious
- Moderate
- Low
- Not assessed

0   200
km
N

C *Water stress in England*

Saving water can help to manage water supplies. Savings can be made by:

◆ the use of domestic water meters

◆ increasing the use of recycled water

◆ more efficient domestic appliances.

Waste water (**grey water**) from people's homes can be recycled and put to good use. It can be used to irrigate both food and non-food plants. The phosphorus and nitrogen in the water are an excellent source of nutrients.

Water transfer

In 2006 the government proposed to establish a water grid to transfer water from areas of water surplus to areas of water deficit. The enormous cost of such an engineering project has stopped it happening. Water is currently only transferred via the Rivers Tyne, Derwent, Wear and Tees to as far south as Yorkshire.

There is a growing need to increase **water transfer** in order to meet demand (map **D**). But there is opposition to large-scale water transfer because of:

◆ the effect on the land and wildlife – river habitats would need to be protected

◆ the high costs involved

◆ the greenhouse gases released in the process of pumping water over long distances.

Managing water quality

Water quality is just as important as water quantity. Much has been done to improve the quality of the UK's rivers and water sources. The Environment Agency manages water quality by:

◆ monitoring the quality of river water

◆ filtering water to remove sediment

◆ purifying water by adding chlorine

◆ restricting recreational use of water sources

◆ imposing strict regulations on the uses of water.

But some groundwater sources have deteriorated as a result of pollution due to:

◆ leaching from old underground mine workings

◆ discharge from industrial sites

◆ runoff from chemical fertilisers used on farmland

◆ water used for cooling in power stations released back into rivers.

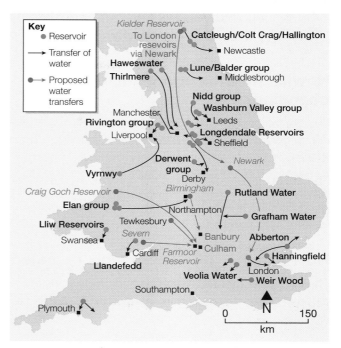

D *Water transfer schemes in England and Wales*

ACTIVITIES

1 a Name the most densely populated areas of the UK (map **B**).

   b Describe the distribution of rainfall in the UK.

2 How do the maps in **B** account for the areas of serious water stress (map **C**)?

3 a Name the proposed major water transfer schemes (map **D**).

   b Explain how these could help solve the problems of water supply and demand in the UK.

Stretch yourself

Imagine you're a farmer living in a water deficit area. Investigate how water transfer could help your business. Think about how water transfer could have an impact on the environment.

Practice question

Evaluate the issue of large-scale water transfers in the UK. *(6 marks)*

On this spread you will find out about the opportunities and challenges faced by the UK in the supply of energy

How is the UK's energy demand changing?

Despite increasing demand for electricity in the UK, energy consumption has fallen in recent years. This is due mainly to the decline of heavy industry and improved **energy conservation**. Low-energy appliances, better building insulation and more fuel-efficient cars have resulted in a 60 per cent fall in energy use by industry and a 12 per cent fall in domestic energy use.

How has the UK's energy mix changed?

The UK's **energy mix** (the range and proportions of different energy sources) has changed in the last 25 years (figure **A**). By 2020 the UK aims to meet 15 per cent of its energy requirement from renewable sources. However, in 2015 the government decided to phase out subsidies for wind and solar energy development.

How and why has the UK's energy mix changed?

The UK is no longer self-sufficient in energy. About 75 per cent of the UK's known oil and natural gas reserves have been exhausted. By 2020 the UK is likely to be importing 75 per cent of its energy. The UK's **energy security** is affected as it becomes increasingly dependent on imported energy.

Two-thirds of UK gas reserves remain, with oil remaining in less accessible oilfields. The remote Mariner oilfield (150km east of the Shetland Isles) will start producing in 2017, but UK oil production overall has declined by 6 per cent each year during the last decade.

The major change in the UK energy mix has been the decline of coal. Between 1990 and 2007 there was a steady decline because of concerns about greenhouse gas emissions and ageing coal-fired power stations.

However, fossil fuels are likely to remain important in the future because:

- the UK's remaining reserves of fossil fuels will provide energy for several decades
- coal imports are cheap – over three-quarters of the UK's coal now comes from abroad, mainly from Russia, Colombia and the USA
- existing UK power stations use fossil fuels – all coal-fueled power stations to be closed by 2025
- shale gas deposits will be exploited in the future.

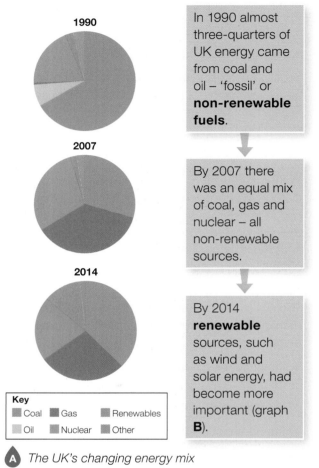

In 1990 almost three-quarters of UK energy came from coal and oil – 'fossil' or **non-renewable fuels**.

By 2007 there was an equal mix of coal, gas and nuclear – all non-renewable sources.

By 2014 **renewable** sources, such as wind and solar energy, had become more important (graph **B**).

Key
- Coal
- Gas
- Renewables
- Oil
- Nuclear
- Other

A The UK's changing energy mix

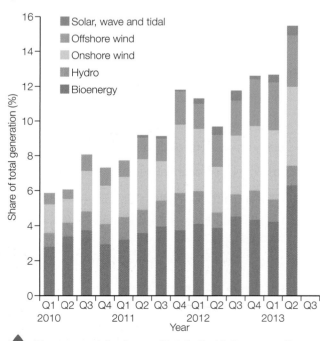

B The renewable share of total electricity generation

What is the fracking issue?

The UK has rich reserves of natural gas trapped deep underground in shale rocks. To extract the gas, high-pressure liquids (water, sand and chemicals) are introduced to fracture the shale and release the gas. This process is called **fracking**.

Fracking has become a very controversial issue (photo **C**). People are concerned about:

◆ the possibility of earthquakes

◆ pollution of underground water sources

◆ the high costs of extraction.

The impacts of energy exploitation

Energy exploitation can have both economic and environmental impacts. Table **D** focuses on two important energy sources being developed in the UK.

C An anti-fracking protest

D The economic and environmental impacts of nuclear and wind power development

	Economic	Environmental
Nuclear	• Nuclear power plants are very expensive to build. The proposed new Hinkley Point plant (page 46) could cost £18 billion, with funding from China. • High costs for producing electricity. • Decommissioning old nuclear power plants is expensive. • Construction of new plants provides job opportunities and boosts the local economy.	• The safe processing and storage of the highly toxic and radioactive waste is a big problem. • Warm waste water can harm local ecosystems. • The risk of harmful radioactive leaks.
Wind farms	• High construction costs. • May have negative impacts on local economy by reducing visitor numbers. • Some wind farms attract visitors by becoming tourist attractions. • At Delabole wind farm in Cornwall, the UK's first commercial wind farm, local homeowners benefit from lower energy bills. The wind farm has also set up a Community Fund.	• Visual impact on the landscape. In the Lake District, concerns about falling visitor numbers have resulted in several plans being rejected. • Wind farms avoid harmful gas emissions and help reduce the carbon footprint. • Noise from wind turbines. • Construction of a wind farm and access roads can impact on the environment.

ACTIVITIES

1 **a** What percentage of the UK's energy came from oil in 1990 and 2014 (figure **A**)?

   b Which energy source saw the greatest reduction between 1990 and 2014?

   c Describe the UK's changing energy mix.

2 Work out the percentage electricity generation from wind energy in the second quarter of 2013 (graph **B**).

3 Suggest reasons why 'solar, wave and tidal' contribute so little to the renewable share of total electricity generation.

4 Look at table **D**. Sort the economic and environmental impacts for each energy source into advantages and disadvantages. Debate which energy source would be best for your local area.

Stretch yourself

Research the controversial issue of fracking. What are the arguments for and against exploiting shale gas? How might it affect the UK's energy mix in the future?

Practice question

Explain why the UK's energy mix will include both renewable and non-renewable sources in the future.
(6 marks)

On this spread you will find out that demand for food is rising globally but supply is not spread evenly

Global patterns of food consumption

The level of food consumption varies across the world (map **A**). Canada, the USA and Europe consume the most, with an average daily intake of over 3400 calories. Most countries consume closer to the recommended daily 2000–2400 calories. However, in some parts of world such as sub-Saharan Africa, daily calorie intake per head is below this level.

Global food consumption is increasing for several reasons. For example:

◆ increasing levels of development and higher standards of living mean that people can afford to buy more food

◆ there are growing populations, particularly in India, Indonesia, China and much of Africa

◆ there is greater availability of food due to improved transport and storage.

Global patterns of food supply

The global supply of food is also uneven (map **B**). Countries with vast human resources like China and India have high agricultural outputs. The USA, Brazil and UK also achieve high outputs due to intensive farming methods and high capital investment. But countries in sub-Saharan Africa produce less food because they have unreliable rainfall, drought, low investment and a lack of education and training.

What is meant by food security?

The term **food security** means having access to enough safe affordable nutritious food to maintain a healthy and active life.

Map **C** shows global food insecurity as measured by the Food Security Index (FSI). This is calculated using indicators including a country's level of nutrition, food stocks and political stability. The highest concentration of countries at risk of food insecurity is in sub-Saharan Africa. Other countries with food insecurity include Afghanistan, Haiti and Bangladesh.

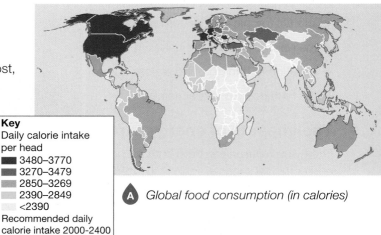

Key
Daily calorie intake per head
- 3480–3770
- 3270–3479
- 2850–3269
- 2390–2849
- <2390
Recommended daily calorie intake 2000-2400

A *Global food consumption (in calories)*

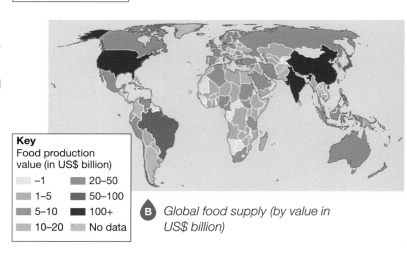

Key
Food production value (in US$ billion)
- –1
- 1–5
- 5–10
- 10–20
- 20–50
- 50–100
- 100+
- No data

B *Global food supply (by value in US$ billion)*

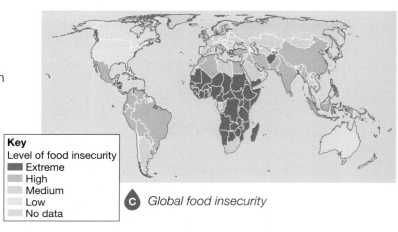

Key
Level of food insecurity
- Extreme
- High
- Medium
- Low
- No data

C *Global food insecurity*

Countries which produce more food than is needed by their population have a **food surplus**. Most countries in the world do not produce enough food to feed their people and have to rely on imported food. Many countries which have a **food deficit** also experience **food insecurity**.

What factors affect food supply?

There are several factors that affect food supply, both now and in the future.

D *Factors affecting food supply*

Climate affects productivity and the types of food that can be grown. Regions experiencing extreme temperatures and rainfall struggle to produce food.

Climate change affects global farming patterns and productivity (graph **E**). Weeds and pests will thrive in warmer conditions.

Without technology, food yields tend to remain low. Unskilled use of technology, like the poor use of **irrigation**, can lead to waterlogging and salinisation. In HICs, mechanisation and agribusiness give high levels of productivity.

Rising global temperatures are causing pests and diseases to spread north and south from the Tropics.

Lack of water affects many areas that suffer food scarcity, particularly in sub-Saharan Africa. These areas are likely to become drier and more desertified in the future as temperatures rise.

Conflicts can lead to the destruction of crops and livestock, to food insecurity, and possibly even famine and death.

Poverty – the poorest people cannot afford any form of technology, irrigation or fertilisers.

ACTIVITIES

1 **a** Name *one* country in *each* of the highest and lowest categories on map **A**.

   b Describe the global pattern of food consumption.

2 What is meant by the term 'food security'? Why is it an important issue?

3 To what extent are the areas of high food productivity the same as the areas of high food security (maps **B** and **C**)?

4 How and why does climate change affect yields of maize and wheat (graph **E**)?

Maize

China	Brazil	France
−7%	−8%	−3%

Global (−4%)

Wheat

China	Russia	France
−2%	−14%	−5%

Global (−5%)

E *The effects of climate change on crop yields*

Stretch yourself

The USA supplies 30 per cent of the world's wheat, maize and rice. Investigate how climate change is affecting American agriculture and how this may affect global food security.

Practice question

Explain why there is increasing global food insecurity. (*6 marks*)

On this spread you will find out about the impacts of food insecurity

Food insecurity occurs when a country can't supply enough food to feed its population. This can have significant economic, social and environmental impacts.

Famine

Famine is a widespread shortage of food often causing malnutrition, starvation and death. There have been some devastating famines resulting from food insecurity:

◆ Former Soviet Union – droughts and crop failures resulted in the death of nine million people in the 1920s and 1930s.

◆ China – droughts and political decisions led to serious famines when millions died. There were famines in 1928–30, 1942–43, and 1959–60.

◆ Ethiopia – in the 1980s an estimated 400 000 people died of starvation due to drought and political conflict.

A *Somalian refugees in Kenya*

Famine in Somalia (2010–12)

The UN estimates that 258 000 people died in Somalia as a result of food insecurity during the famine of 2010–12. In southern Somalia an estimated 18 per cent of the child population died due to lack of food or because they were too weak to resist disease (photo **A**). At the height of the crisis 30 000 people were dying each month.

The famine was the result of two successive seasons of low rainfall, poor harvests and the death of livestock. The worst-hit areas, in southern and central Somalia, were under the control of the al-Shabab militant group. They blocked aid from humanitarian agencies, making the crisis worse.

Undernutrition

Undernutrition is the lack of a balanced diet, and deficiency in minerals and vitamins. The Food and Agricultural Organisation (FAO) estimates that 805 million people suffered from undernutrition between 2012 and 2014. It is a major public health problem, particularly in southern Asia and sub-Saharan Africa. Diets in these regions are frequently deficient in protein, carbohydrates, fats, minerals and vitamins. This causes around 300 000 deaths per year and contributes to half of all child deaths.

Soil erosion

Soil erosion involves the removal of fertile top soil layers by wind and water. There are several causes (diagram **B**).

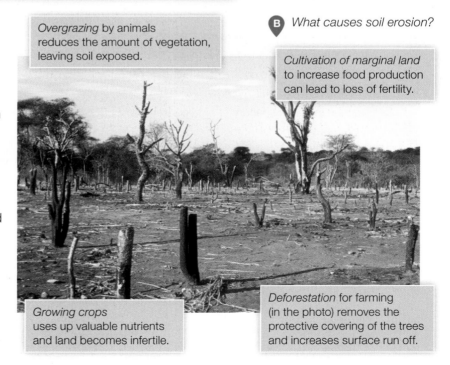

Overgrazing by animals reduces the amount of vegetation, leaving soil exposed.

B *What causes soil erosion?*

Cultivation of marginal land to increase food production can lead to loss of fertility.

Growing crops uses up valuable nutrients and land becomes infertile.

Deforestation for farming (in the photo) removes the protective covering of the trees and increases surface run off.

Rising prices

Food prices are rising across the world. This is mainly due to increased prices for fertilisers, animal feed, food storage, processing and transportation. LICs and the poorest people in NEEs are hit hardest by higher food costs. This is because food represents a larger share of their spending.

Table **C** shows recent price increases for main staple foods, using a base index of 100 in 2004. For example, a value of 200 means the price has doubled since then.

	2007	2008	2009	2010	2015
Maize	141	179	186	176	155
Wheat	157	219	211	204	157
Rice	132	201	207	213	192
Soya beans	121	156	150	144	127
Soya oil	138	170	162	153	119
Sugar	135	169	180	190	185

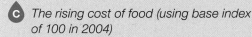

C *The rising cost of food (using base index of 100 in 2004)*

(Base index = 100 in 2004)

The quantity of rice the same amount of money could buy in 2007 (right) and 2008 (left).

Social unrest

The 21st century has seen social unrest, especially in North Africa and the Middle East. These incidents – sometimes called 'food riots' – correspond with high rises in the price of food.

The food price index (graph **D**) increased dramatically in 2008 (the start of the global economic recession) and again in 2011. The graph shows how both these 'spikes' in food prices coincide with outbreaks of social unrest. Most of these incidents occurred in LICs or NEEs in Africa and the Middle East. For example, in 2011 the price of cooking oil and flour doubled. In Algeria this price rise led to five days of rioting, with four people killed.

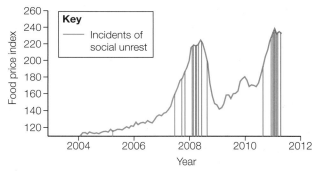

D *The link between food prices and social unrest*

ACTIVITIES

1 What is famine and why is it often a consequence of food insecurity?

2 What is soil erosion and what are its main causes?

3 a Describe the pattern of food prices between 2004 and 2011 (graph **D**).

 b Is there evidence that high food prices lead to civil unrest?

Maths skills

Draw a series of line graphs using different colours to show the trends in food price rises in table **C**. Start your graph in 2004 using the base index value of 100.

Stretch yourself

Why are decreasing yields and the increasing price of wheat, maize and rice significant?

Practice question

Explain how both physical and human factors affect the world's supply of food. (*6 marks*)

On this spread you will find out about different strategies used to increase food supply

How can food supply be increased?

Globally there is sufficient food to feed the world's population. Food supply is unevenly distributed and food consumption varies greatly from one region to another. As the world's population grows, traditional methods and modern technology are being used to increase food production.

A *Drip irrigation*

Irrigation

Irrigation is the artificial watering of land. Most methods involve the extraction of water from rivers or underground aquifers. Irrigation is needed where there are water shortages during the growing season.

In some countries irrigation projects have involved the construction of expensive dams and reservoirs such as in the Indus Valley of Pakistan. These high-profile projects often benefit larger commercial farming rather than small-scale local farmers.

In contrast, there are many smaller schemes, such as in Makueni County in eastern Kenya. Here a 35 km pipeline and use of water storage tanks has enabled drip irrigation (photo **A**) to support domestic food cultivation. This has increased Kenya's food security.

Aeroponics and hydroponics

Most plants obtain nutrients (plant foods) from the soil. Modern techniques are used to deliver these direct to the plants. This speeds up plant growth, enabling seasonal produce to be grown throughout the year and reducing the use of chemicals. But some think foods grown this way don't taste as good, and the cost of heating and lighting can be high.

- **Aeroponics** – plants are sprayed with fine water mist containing plant nutrients. Excess water can be collected and re-used. In Vietnam disease-resistant potatoes are produced. This enables small-scale farmers to increase yields and lower production costs.

- **Hydroponics** – plants are grown in gravel or mineral-rich water (diagram **B**).

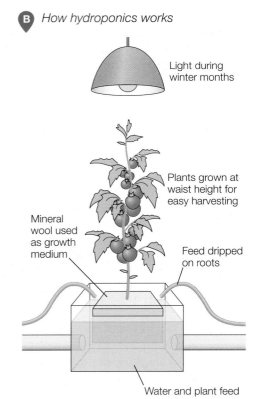
B *How hydroponics works*

Light during winter months

Plants grown at waist height for easy harvesting

Mineral wool used as growth medium

Feed dripped on roots

Water and plant feed

The 'new' green revolution

The term '**green revolution**' was first used in the 1950s and 1960s. Modern farming techniques – such as the use of machines, chemicals and new strains of plants – were used to increase food production in poorer parts of the world.

Today, there is a 'new' green revolution focusing on sustainability and community. In 2006, the Indian government began a 'second' Green Revolution, using techniques such as:

- water harvesting
- irrigation
- soil conservation
- improving seed and livestock quality using science and technology.

Along with improved rural transport and affordable credit, these innovations have enabled the Indian state of Bahir to double its rice output.

Biotechnology

Biotechnology uses living organisms to make or modify products or processes. In farming this includes the development of genetically modified (GM) crops. These produce higher yields, use fewer chemicals and reduce carbon dioxode emissions. In the UK there is opposition to GM crops because of the possible effects on the environment and human health.

Despite these concerns, GM crops are grown elsewhere in the world. For example, over half the world's soya beans are GM. In the Philippines GM maize has given a 24 per cent increase in yield. GM oilseed rape is widely cultivated in Canada.

Appropriate technology

Appropriate technology means using skills or materials that are cheap and easily available to increase output without putting people out of work. This form of technology typically involves small-scale water harvesting equipment, irrigation methods or farming techniques. It is particularly appropriate for people living in the poorer countries of the world.

Photo **C** shows how a bicycle can produce the power needed to remove the outer shell of coffee beans. Farmers can then roast their own beans in the sun and add value to their product.

C *Appropriate technology: a bicycle being used to de-husk coffee beans*

Think about it

Some people think that the benefits of growing GM crops are greater then the risks. Do you agree?

Maths skills

The table shows data for the increase in the total global area of GM crops (millions of hectares) between 1996 and 2009. Use the data to drawn a line graph. What trends does your graph show?

2000	2001	2002	2003	2004	2005	2006	2007	2008	2009	2010	2011	2012	2013	2014
44.2	52.6	58.7	67.7	81.0	90.0	102.0	114.3	125.0	134.0	148.0	160.0	170.3	175.2	181.5

ACTIVITIES

1 What is irrigation and how can it increase food supply?

2 **a** Copy diagram **B** and add labels to describe the advantages and disadvantages of hydroponics.

   b How is hydroponics different from aeroponics?

3 **a** Describe what is happening in photo **C**.

   b Why does it benefit small coffee farmers to be able to roast their own beans?

Stretch yourself

Research more about the 'new' green revolution as it is being applied in India.

Practice question

'Strategies for increasing global food supply will be both large-scale and small-scale.' Evaluate the effectiveness of this approach to improving global food supply.
(9 marks)

On this spread you will find out about the advantages and disadvantages of a large-scale agricultural development to increase food supply

Example

The source of the Indus River is high in the Tibetan Plateau. From there it flows roughly north to south through the length of Pakistan to reach the Arabian Sea (map **A**). The Indus basin covers about one million km² and includes parts of India, China and Afghanistan.

What is the Indus Basin Irrigation System? (IBIS)

The Indus River is an important water source for the two NEEs India and Pakistan. The Indus Water Treaty (1960) gave India control of the eastern rivers – but Pakistan gained control of the Indus and the western tributaries, the Jhelum and Chenab. These mountain rivers, fed by heavy rain and snowmelt, supply the water used to irrigate the drier agricultural land further south.

The IBIS is the largest continuous irrigation scheme in the world. It began as a system of irrigation canals built during the period of British rule (1847–1947) and has been developed since. Today it consists of three large dams and over a hundred smaller dams that regulate water flow. Twelve link canals enable water to be transferred between rivers. Over 64 000 km of smaller canals distribute the water across the countryside. In total, over 1.6 million km of ditches and streams provide irrigation for Pakistan's agricultural land.

Map **B** is a topological map of the Indus irrigation system. Just like the map of the London Underground, it does not show distances, but links and connections. There are three large reservoirs in the north of Pakistan, at Tarbela, Mangla and Chashma. Tarbela is Pakistan's largest reservoir, with a capacity of 11 billion m³ (photo **C**).

Did you know?
India takes its name from the Indus River.

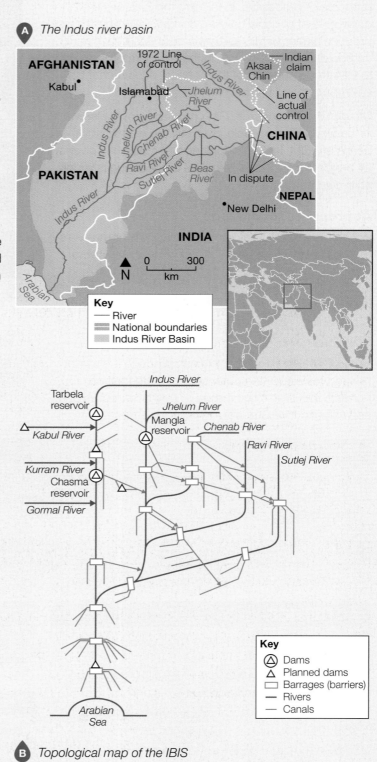

A *The Indus river basin*

Key
— River
▓ National boundaries
▓ Indus River Basin

B *Topological map of the IBIS*

Key
△ Dams
△ Planned dams
▭ Barrages (barriers)
— Rivers
— Canals

C *The Tarbela reservoir*

D *Farmland ruined by salinisation, Pakistan*

What are the advantages and disadvantages of the IBIS?

Advantages	Disadvantages
• Improves food security for Pakistan, making 40% more land available for cultivation.	• Some farmers take an unfair share of water, depriving others downstream.
• Over 14 million ha of land is now irrigated.	• High summer temperatures result in high water loss through evaporation.
• Irrigation has increased crop yields: wheat (36%), rice (39%) and fruit (150%).	• Poor irrigation techniques mean water is wasted. Salinisation (page 76) can cause long-term damage to the soil.
• Diets have improved as a greater range of food products are available.	• Population growth will increase the demand for water in the future.
• Fish farming in storage reservoirs provides a source of protein.	• High costs to maintain reservoir capacity, barrages and canals.
• Agriculture-based industries have developed, providing value-added goods for export.	• High water levels have led to waterlogging in places.
• HEP is generated by main dams.	

ACTIVITIES

1 a Describe the Indus drainage system in Pakistan (map **A**).

  b Why was the Indus River Treaty such an important agreement?

2 Describe the structure and layout of the IBIS (map **B**).

3 a What are the disadvantages for the farmer in photo **D**?

  b How has Pakistan benefited from the scheme?

  c Identify two disadvantages of the IBIS scheme. Think of one to the individual farmer and one to the country as a whole.

Stretch yourself

Research the process of salinisation. Describe how it can be caused by poor methods of irrigation. Find out what effect it has on soil fertility in India.

Maths skills

Draw a pie chart to show land use in the Indus Valley using the data below.

Wheat 44%
Sugar cane 7%
Animal food crops 15%
Rice 15%
Cotton 19%

Practice question

Use a named example to evaluate the effects of a large-scale irrigation scheme. *(6 marks)*

On this spread you will find out about the potential for sustainable food supplies

What is sustainable food supply?

A **sustainable food supply** ensures that fertile soil, water and environmental resources are available for future generations. In the long term this principle must be at the heart of all strategies to increase food supply.

Organic farming

Organic farming is growing crops or rearing livestock without the use of chemicals. Production and labour costs may be higher than in other forms of farming. However, many people choose to pay higher prices for **organic produce**.

Permaculture

The word 'permaculture' is from 'permanent agriculture' and 'permanent culture'. It is a system of food production which follows the patterns and features of natural ecosystems. It aims to be sustainable, productive, non-polluting and healthy. It is a philosophy of living as well as a system of farming.

Permaculture includes harvesting rainwater, composting waste and re-designing gardens to include a wide variety of plants and trees (photo **B**). These provide a range of wildlife habitats. Permaculture practices include:

- organic gardening
- use of crop rotation
- keeping animals like sheep and pigs, and bees
- managing woodland.

Urban farming

Urban farming is the cultivation, processing and distribution of food in and around settlements. It is becoming more popular because:

- a greater choice of fresh foods is available for a healthier diet
- new jobs are created in deprived urban areas
- it brightens up urban environments
- it attracts wildlife such as birds and butterflies
- there are social benefits from communities working together on joint farming projects.

A What is organic farming?

GM crops (see page 269) are banned

Wildlife is encouraged as a natural pest control

Crop rotation is used to keep soil fertile and reduce pests and diseases

Clover is grown to maintain nitrogen – an important nutrient – in the soil

Crops grown in soil that has been chemical-free for five years

B Permaculture in a city garden

Fish from sustainable sources

Almost 90 per cent of the world's fisheries are fully or over-exploited. Increasing demand for food and technological improvements has resulted in greater catches of fish. Commercial trawlers using close-meshed nets catch all fish large and small, discarding smaller ones. Nets can damage marine ecosystems and fish breeding grounds. These practices are unsustainable.

Intensive fish farming (salmon, trout and prawns) using chemicals has boomed in recent years. Ecosystems can be harmed and diseases spread to wild fish populations.

Sustainable fishing involves setting catch limits (quotas) and monitoring fish breeding and fishing practices. The European Union sets standards as part of its Common Fishery Policy, with limits on fish catches. In Norway, salmon farms are spread out along the coastline to reduce the possible spread of disease. Public awareness has increased in recent years and sales of fish from sustainable sources have increased.

Meat from sustainable sources

Intensive livestock production often involves practices that are unsustainable.

◆ Large amounts of energy (heat and light) are used for indoor rearing.

◆ Chemicals are used to maximise production.

◆ Large volumes of waste need to be removed and safely disposed of without polluting the environment.

◆ High concentrations of animals can damage the soil.

Sustainable meat production involves small-scale livestock farms, using free-range or organic methods (photo **D**). Stocking levels are low and there is minimal impact on the environment. Prices may be higher in the shops but quality and animal welfare standards are higher.

ACTIVITIES

1 **a** What are the characteristics of permaculture shown in photo **B**?

  b Explain why this is an example of sustainable food production.

2 What are the social, economic and environmental benefits of urban farming?

3 Should we only ever buy fish and meat from sustainable sources? Justify your answer.

Stretch yourself

Investigate urban farming in the UK. Find an example and describe its characteristics and benefits to the local community.

Practice question

Evaluate how far it is possible to increase food supply sustainably. (*6 marks*)

The Michigan Urban Farming Initiative

The Michigan Urban Faming Initiative in the USA aims to address problems of urban decay, poor diet and food insecurity in Detroit, Michigan. Urban communities are encouraged to work together to turn wasteland into productive farmland, providing jobs and easier access to healthy food. A community resource centre and 'demo' farm have been set up (photo **C**). Over 150 raised garden beds have been created on waste ground for use by local community groups.

C *Urban farming in Detroit*

D *Cattle grazing on a biodynamic organic farm in Wales*

On this spread you will find out more about how we can manage food supplies in a sustainable way

Seasonal food consumption

In the past, food was bought from local sources – farms or markets. Fruit and vegetables were only available when 'in season', for example, strawberries in the summer and apples in the autumn. With better storage and faster transport around the world, it is now possible to eat every type of food throughout the year!

Local food sourcing is more sustainable. It reduces both 'food miles' (see page 258) and our carbon footprint. Local farmers' markets and other initiatives make fresh locally produced seasonal foods more readily available (photo **A**).

Reducing food loss and waste

Around 32 per cent of all food produced is lost or wasted each year (diagram **B**), most in richer countries. Almost half of this is fruit and vegetables. This is because they have to be kept fresh and stored in cool conditions, with careful packaging and transportation. By halving the amount of food waste, the gap between food supply and demand could be reduced by 22 per cent by 2050 (diagram **C**).

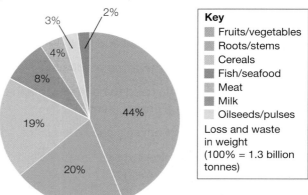

A A local farmers' market

B The food we waste (figures for 2011)

3% 2% 4% 8% 19% 20% 44%

Key
- Fruits/vegetables
- Roots/stems
- Cereals
- Fish/seafood
- Meat
- Milk
- Oilseeds/pulses

Loss and waste in weight (100% = 1.3 billion tonnes)

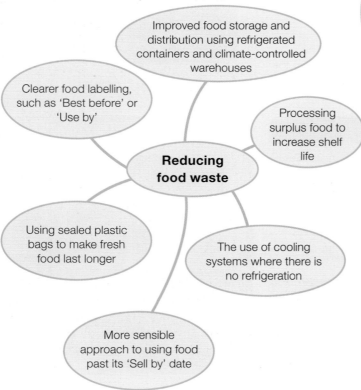

Improved food storage and distribution using refrigerated containers and climate-controlled warehouses

Clearer food labelling, such as 'Best before' or 'Use by'

Processing surplus food to increase shelf life

Reducing food waste

Using sealed plastic bags to make fresh food last longer

The use of cooling systems where there is no refrigeration

More sensible approach to using food past its 'Sell by' date

C How can we reduce the amount of food we waste?

Why eat seasonal produce?

There are good reasons for increasing the amount of locally produced food we eat.

- It reduces the energy needed to grow and transport food, and reduces CO_2 emissions.
- It gives support to the local economy.
- Food that has been transported a long way may be more expensive.
- Food production follows the natural seasonal cycle.
- Locally-produced food often tastes better and is likely to be more nutritious.

Increasing sustainable food supplies in Makueni, Kenya

Makueni County in eastern Kenya is 200 km south east of Nairobi (map **D**). It has a population of 885 000 with most living in small isolated rural communities. The average annual rainfall is just 500 mm.

What food is grown in Makueni?

The main crops grown in the county are maize, beans, millet, sorghum, cassava and sweet potatoes. The area has rich, dark volcanic soils which are high in nutrients. Low and unreliable rainfall affects agricultural output with frequent crop failures.

D *The location of Makueni*

E *A sand dam in Makueni*

The Makueni Food and Water Security Programme

In April 2014 the charity organisation 'Just a Drop', together with the African Sand Dam Foundation, provided direct help to two small villages, Musunguu and Muuo Wa Methovini (population 800), and to Kanyenyoni Primary School (463 pupils).

The programme included:

◆ improving access to a clean and safe water supply by building sand dams for each village

◆ a rainwater harvesting tank on the school roof

◆ increasing food security by providing a reliable source of water for crops and keeping livestock

◆ a training programme to support local farmers

◆ growing trees to reduce soil erosion, increase biodiversity and provide medicinal products.

Sand dams store water in the ground, filtering and cleaning the rainwater as it soaks into the soil (photo **E**). With minimal operation and maintenance costs, sand dams provide a cost-effective and sustainable way to provide a water supply in rural areas.

The project has been very successful:

◆ crop yields and food security have increased

◆ water-borne diseases have been reduced

◆ less time is wasted fetching water – more time is available for work or education

◆ the school now has a safe and clean water supply.

ACTIVITIES

1 What kinds of seasonal foods are available all year in local shops and supermarkets?

2 **a** Which food group in diagram **B** shows the greatest loss?

   b Suggest reasons for this.

   c What can be done to reduce food waste?

3 What is a sand dam (photo **E**) and why is it a sustainable solution to the water supply problem?

4 How has the Makueni project increased the community's food security and benefited the community?

Stretch yourself

Carry out more research about the Makueni Food and Water Security Programme. (Try visiting the 'Just a Drop' website.)

Practice question

The involvement of local people is important in sustainable schemes to increase food security. Describe how the Makueni scheme achieves this. *(4 marks)*

On this spread you will find out that demand for water is rising globally but supply is not spread evenly across the world

Water surplus and deficit

Map **A** shows global patterns of **water surplus** and **water deficit**. Regions with a water surplus have a plentiful supply of water with supply exceeding demand. These regions include North America, Europe and parts of Asia. Other regions have a water deficit, where demand exceeds supply and supplies are under pressure.

Regions with high rainfall usually have a water surplus. Areas with low rainfall, such as hot deserts, are more likely to have a water deficit.

Areas of high population density and high concentrations of industry have the highest demand for water. Without sufficient water supply these areas may experience a water deficit. Areas with low rainfall but a lower demand may have a water surplus!

Water security/insecurity

Water security means having access to enough clean water to sustain well-being, good health and economic development. Regions which do not have access to sufficient safe water supplies are described as being in a situation of **water insecurity**.

Water security is very important for improving quality of life because it:

◆ reduces poverty

◆ helps to improve education

◆ increases living standards.

What is water stress?

Many countries face high water stress (map **B**). This means that more than 80 per cent of available water is used every year, leaving the threat of water scarcity.

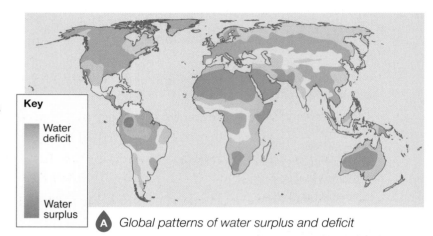

Key

| | Water deficit |
| | Water surplus |

A Global patterns of water surplus and deficit

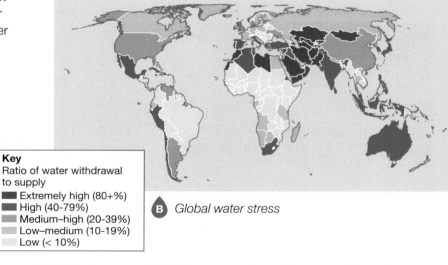

Key
Ratio of water withdrawal to supply

- ■ Extremely high (80+%)
- ■ High (40-79%)
- ■ Medium–high (20-39%)
- ■ Low–medium (10-19%)
- ■ Low (< 10%)

B Global water stress

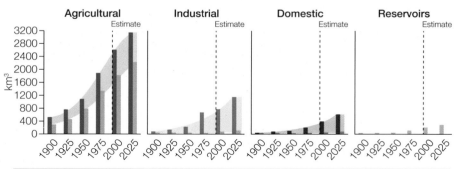

Key

■ Withdrawal	■ Withdrawal	■ Withdrawal	■ Evaporation
■ Consumption	■ Consumption	■ Consumption	
■ Waste	■ Waste	■ Waste	

C Global water consumption since 1900

Countries begin to experience water stress when less than 1700m³ is available per person per year. Below 1000m³, water stress may damage economic development – and human health and well-being. Regions with high water stress include several Caribbean islands, Bahrain, Cyprus, Malta and the Middle East.

Why is water consumption increasing?

The steady growth of the world's population, by roughly 80 million each year, means that more water is needed. Water consumption is increasing because economic development results in greater demand (figure **C**).

There are other reasons why we are all using more water.

◆ Changes in lifestyle and eating habits have increased the average use of water per head.

◆ Global demand for food is expected to increase by 70 per cent by 2050 – water is used to irrigate crops and in food processing.

◆ All sources of energy require water in their production. Global energy consumption is expected to increase by 50 per cent by 2035.

◆ As urbanisation increases, more water is needed for drinking, sanitation and drainage.

Water availability

There are a number of factors affecting the availability of water supplies.

What is water stress?

Water stress takes into account several physical factors that are related to water resources. These include:

◆ water scarcity

◆ water quality

◆ accessibility of water

◆ environmental flows (the quality, quantity and timing of water flow needed to maintain healthy ecosystems in streams, rivers, and the estuaries they feed).

Geology – infiltration of water (as in the Sahara Desert) through permeable rock builds up important groundwater supplies. Much of London's water comes from the chalk underlying the city.

Climate – regions with high rainfall usually have surplus water. Those with drier climates have less water available.

Over-abstraction – pumping water out of the ground faster than it is replaced by rainfall. This can cause wells to dry up, sinking water tables and higher pumping costs. Lower water tables mean that rivers are not fed by springs in the dry season.

Pollution – increasing amounts of waste and growing use of chemicals in farming have led to higher levels of pollution. In some LICs and NEEs water sources are often used as open sewers leading to **waterborne diseases**.

Limited infrastructure – poorer countries may lack the infrastructure for transporting water to areas of need (for example, pumping stations and pipes).

Poverty – many poorer communities lack mains water or only have access to shared water supplies.

D *What affects the availability of water?*

ACTIVITIES

1 a Define the meaning of 'water surplus' and 'water deficit'.

 b Which continent has the highest water deficit (map **A**)?

 c Which are the main areas of water surplus?

 d Explain the different patterns of water surplus and deficit.

2 Describe and suggest reasons for the pattern of global water stress (map **B**).

3 a Describe the trends in each of the four graphs in figure **C**.

 b Which shows the most waste? Why?

4 'Poverty is the main factor affecting water supply.' To what extent do you agree with this statement?

Stretch yourself

Explain how access to safe water can improve people's standard of living.

Practice question

Explain how both physical and human factors can influence the availability of water. (*6 marks*)

The impact of water insecurity

On this spread you will find out about the impacts of water insecurity

What are the impacts of water insecurity?

Water insecurity can cause social, economic and environmental problems. It is experienced in some richer, more developed countries, as well as LICs.

Waterborne disease and water pollution

In countries where water supply infrastructure is limited, there may be little or no sanitation. There may be open sewers and high levels of pollution in rivers and other water sources. Contaminated drinking water can cause outbreaks of life-threatening disease such as cholera and dysentery.

With a shortage of clean water, people may have to queue for a long time to obtain a supply from standpipes (photo **A**). This wastes time and reduces levels of productivity.

A Queuing for water at a standpipe

Water pollution: The River Ganges, India

The River Ganga (Ganges) is 2520 km long and flows through northern India and Bangladesh. It is the most polluted river in the world, with both human and industrial waste.

- Over one billion litres of raw sewage enter the river each day from the cities, towns and villages along its banks.

- Hundreds of factories discharge 260 million litres of untreated wastewater from factories is discharged into the river daily.

- The major polluting industry along the Ganges is the leather industry, because toxic chemicals leak into the river.

- Run off from pesticides and fertilisers is another major source of pollution.

Pollution of the Ganges has become so serious that bathing in the river and drinking its water have become very dangerous.

Food production

Agriculture uses 70 per cent of global water supply and suffers the most from water insecurity. Drier regions of the world with unreliable rainfall are most at risk. The USA supplies 30 per cent of the world's wheat, maize and rice. Droughts and water shortages are serious issues across much of the USA (graph **B**) and can have a global impact on food production and supply.

Water shortages in Egypt

Shortage of water is affecting Egypt's food security. The River Nile is Egypt's primary source of water. Climate change and the demands of countries upstream are expected to reduce its flow by 90 per cent by the end of the century. Although 80 per cent of Egypt's water supply is used in agriculture, food production is likely to decline by 30 per cent over the next 30 years. Egypt currently has to import 60 per cent of its food.

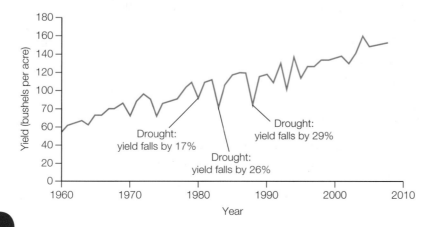

Drought: yield falls by 17%

Drought: yield falls by 26%

Drought: yield falls by 29%

B Crop yields in the USA

Industrial output

Growth of manufacturing industry, particularly in NEEs, is making increasing demands on water supplies.

Water conflict

In the past, wars were fought over oil supplies – in the future, they may be fought over water. This is because water sources, such as rivers and groundwater aquifers, cross national and political borders. Many of the world's great rivers, such as the Nile and Danube, flow through several countries. Issues such as reservoir construction and pollution can impact on more than one country, and create conflict (map **C**).

Did you know?

About one fifth of the world's population live where there is not enough water to meet demand.

Chinese industry

By 2030 Chinese industry will use 33 per cent of the country's available water. Water shortages cost China US$40 billion in lower industrial production. Some factories have closed temporarily due to water shortages. Also, China is depends on its coal resources to drive its economic growth. Coal mining and power stations use 20 per cent of China's water.

Turkey built a large number of dams on the Tigris and Euphrates Rivers, causing an angry response from Iraq and Syria. Water is sold from the Manavgat River.

The River Jordan flows through Jordan and Israel. Israel draws water from the Sea of Galilee. Groundwater is polluted and in short supply. Israel buys water from Turkey.

Lake Chad has shrunk to 5% of its previous size, due to climate and over-abstraction. This affects the whole population.

Egypt's population of 160 million relies on the Nile for its water. The river flows through seven other countries. Egypt will not allow those countries to do anything to affect its flow (for example, to build dams). This causes great tension in the region.

The River Ganges flows through Northern India and Bangladesh. India has built barriers to control the flow, and this affects the water supply to Bangladesh.

C *The world's potential water conflict zones*

The Rogun Dam, Tajikistan

Several rivers flow from mountainous Tajikistan to Uzbekistan in the west. In 2014 the World Bank agreed to finance the Rogun Dam Project on the Vakhsh River (map **D**). The construction of the dam has been hit by economic and political problems. Once completed, electricity generated by the dam will support industrial development in Tajikistan.

The project is very controversial in neighbouring Uzbekistan, which suffers from a water deficit. Irrigated cotton is Uzbekistan's main export and there is concern about the impact of reduced water supplies on its economy. The Nurek Dam (map **D**), built in the 1970s, already affects the flow of the river. The new dam could lead to further tensions.

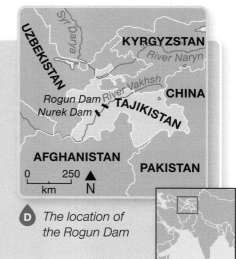

D *The location of the Rogun Dam*

ACTIVITIES

1 Suggest the social and economic impacts of polluted water supplies in LICs.

2 **a** In which year was the greatest fall in yields due to drought?

   b Which decade saw the worst drought?

3 What are the causes of the potential conflicts identified on map **C**?

Stretch yourself

Research more information on one of the potential conflicts in map **C**. Do you agree that future wars may be fought over water resources? Why?

Practice question

Explain how human actions can contribute to water insecurity. *(6 marks)*

How can water supply be increased?

On this spread you will find out about strategies to increase water supply

The amount of water available is limited. To make more water available means finding new sources or moving it from areas of surplus to areas of deficit.

Diverting supplies and increasing storage

Water supplies can be artificially diverted and stored for use over longer periods. For example, in some parts of the world surface water evaporates rapidly and is lost. This water can be stored in deep reservoirs or in permeable rocks (aquifers) underground.

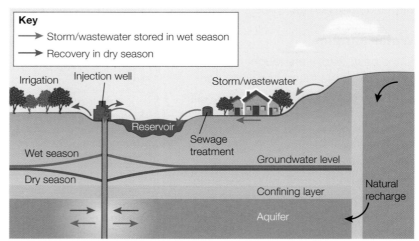

Key
→ Storm/wastewater stored in wet season
→ Recovery in dry season

Aquifer storage and recovery

In Oklahoma, USA, rainfall is infrequent but heavy. Surface water quickly evaporates. So it is collected and diverted into underlying alluvial soils where it can be stored (diagram **A**). Alluvial soils are loose at the surface with good water-holding capacity.

Dams and reservoirs

Dams control water flow in rivers by storing water in reservoirs. Rainfall can be collected and stored when it is plentiful and then released gradually during drier periods. The control of water flow enables it to be transported and used for irrigation. It helps to prevent flooding.

Dams range widely in size. There are huge, multi-purpose dams like the Three Gorges Dam in China. Small earth or cement dams a few metres high are common in sub-Saharan Africa. Large dams are expensive to construct and maintain. They can lead to the displacement of large numbers of people. Also, they may reduce the flow of water downstream. In hot and arid regions, reservoirs with a large surface area can lose a lot of water through evaporation.

Kielder Water, Northumberland

Kielder Water is the UK's largest reservoir in terms of its storage capacity (photo **B**). The dam, 1.2 km long and 50 m high, was built in the late 1970s at North Tyne Valley near Falstone. The valley is relatively narrow, reducing the cost of building. The 10 km-long reservoir took two years to fill.

The reservoir regulates flow in the North Tyne, making up for water abstracted (taken) further downstream. Water is also used to generate electricity.

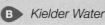 Kielder Water

Water transfers

Water transfer schemes aim to redistribute water from areas of water surplus to areas of water deficit. They often involve elaborate systems of canals and pipelines to take water from one river basin to another. In the UK the Kielder transfer scheme carries water south to the rivers Wear and Tees.

China's south–north water transfer scheme

China is spending over US$79 billion on an ambitious project to transfer water from the Yangtze River in the south to the Yellow River Basin in the arid north (map **C**). The water will be transferred through three canal systems. The eastern and central routes were completed in 2015. The western route is due for completion by 2020.

The controversial western route involves building several dams and hundreds of tunnels through the Bayankela Mountains. The entire project could take 50 years. However, it is still uncertain whether the scheme will actually be completed.

Key
→ Waterway routes
  Dry region
  Semi-dry region
  Semi-humid region
  Humid region

C *China's water transfer project*

Desalination

Desalination involves removing salt from seawater to produce fresh water. This is a very expensive process. It is used only where there is a serious shortage of water with few alternatives to increase water supply. Both Saudi Arabia and UAE have developed desalination plants.

There are several issues linked with the process of desalination, such as:

◆ environmental impacts on ecosystems when salt waste is dumped back into the sea

◆ the vast amount of energy required, adding to carbon emissions

◆ the high cost of transporting the desalinated water to inland areas.

Future technological improvements may reduce costs and make this process more economically viable for NEEs or even LICs.

D *Global desalinisation*

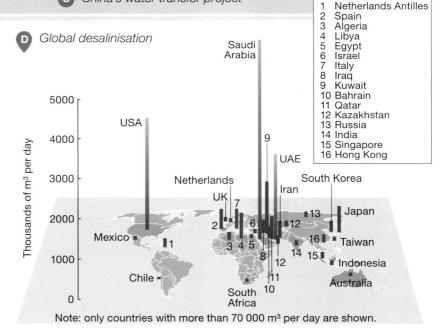

1	Netherlands Antilles
2	Spain
3	Algeria
4	Libya
5	Egypt
6	Israel
7	Italy
8	Iraq
9	Kuwait
10	Bahrain
11	Qatar
12	Kazakhstan
13	Russia
14	India
15	Singapore
16	Hong Kong

Note: only countries with more than 70 000 m³ per day are shown.

ACTIVITIES

1 Copy diagram **A** and add labels to describe how water can be collected and stored underground for future use.

2 What are the advantages and disadvantages of creating reservoirs such as Kielder Water (photo **B**)?

3 Explain how China's climate has influenced the Chinese plan to transfer water (map **C**).

4 Describe the distribution of the countries where large amounts of water are desalinised (diagram **D**).

Stretch yourself

Imagine you're a journalist sent to investigate the western route of China's south–north water transfer scheme. Investigate the advantages and disadvantages of this route. Write a front-page report for tomorrow's paper.

Practice question

Explain the costs and benefits involved in strategies to increase water supply. (*6 marks*)

The Lesotho Highland Water Project

On this spread you will find out about a large-scale water transfer scheme in Lesotho

Example

Lesotho is a highland country in southern Africa surrounded by the country of South Africa (map **A**). It has few resources, high levels of poverty, and is unable to feed its growing population. Most farms are for subsistence and productivity is low. Lesotho is heavily dependent economically on South Africa.

Despite experiencing food insecurity, Lesotho has a water surplus. The mountains receive high rainfall (graph **B**) and the demand for water is low.

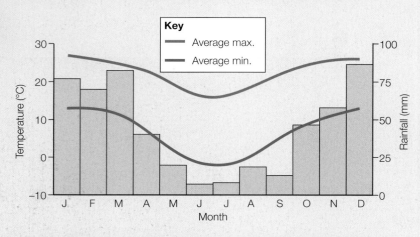

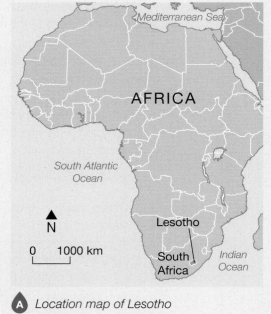

A *Location map of Lesotho*

B *The climate of Lesotho*

What is the Lesotho Highland Water Project?

The Lesotho Highland Water Project is a huge water transfer scheme aimed to help solve the water shortage in South Africa. On completion, 40 per cent of the water from the Segu (Orange) River in Lesotho will be transferred to the River Vaal in South Africa. It is a massive scheme involving the construction of dams, reservoirs and pipelines as well as roads, bridges and other infrastructure developments (map **C**). It will take 30 years to complete.

The main features of the scheme include:

◆ The Katse and Mohale Dams (completed in 1998 and 2002) store water that is transferred through a tunnel to the Mohale Reservoir.

◆ Water is then transferred to South Africa via a 32 km tunnel enabling HEP to be produced at the Muela plant.

◆ The Polihali Dam will hold 2.2 billion m^3 of water with a 38 km transfer tunnel.

◆ The Tsoelike Dam will be built at the confluence of the Tsoelike and Senqu rivers. It will have a storage capacity of 2223 million m^3 and a pumping station.

◆ The Ntoahae Dam and pumping station will be built 40 km downstream from Tsoelike Dam on the Senqu River.

By 2020 there will be 200 km of tunnels and 2000 million m^3 of water will be transferred to South Africa each year.

C *Map of the project*

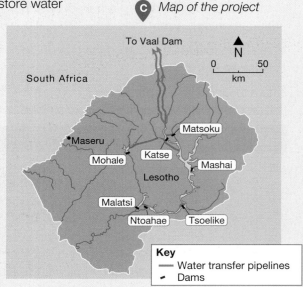

What are the advantages and disadvantages of the scheme?

Advantages for Lesotho

- Provides 75% of its GDP
- Income from the scheme helps development and to improve standard of living
- Supplies the country with all its hydro-electric power (HEP) requirements
- Improvements to transport infrastructure with access roads built to the construction sites
- Water supply will reach 90% of the population of the capital, Maseru
- Sanitation coverage will increase from 15 to 20%

Advantages for South Africa

- Provides water to an area with an uneven rainfall pattern and regular droughts
- Provides safe water for the 10% of the population without access to a safe water supply
- Fresh water reduces the acidity of the Vaal River Reservoir. Water pollution from industry, gold mines and sewage was destroying the local ecosystem
- The influx of water from Lesotho is restoring the balance

 The Katse Dam

Disadvantages for Lesotho

- Building of the first two dams meant 30 000 people had to move from their land
- Destruction of a unique wetland ecosystem due to control of regular flooding downstream of the dams
- Corruption has prevented money and investment reaching those affected by the construction
- Construction of the Polihali Dam will displace 17 villages and reduce agricultural land for 71 villages

Disadvantages for South Africa

- Costs are likely to reach US$4 billion
- 40% of water is lost through leakages
- Increased water tariffs to pay for the scheme are too high for the poorest people
- Corruption has plagued the whole project

ACTIVITIES

1. Describe the location of Lesotho (map **A**).
2. **a** How can you work out from graph **B** that Lesotho is in the southern hemisphere?
   b Describe the seasonal distribution of rainfall.
   c During which months would you expect there to be a water surplus?
3. Using photo **D**, draw an annotated field sketch of the Katse Dam.
4. Discuss in pairs the advantages and disadvantages of this scheme. Who are the winners and losers?

Stretch yourself

Use the internet to research the impact of the scheme on the people of Lesotho.

Maths skills

Use climate graph **B** to estimate the total annual rainfall for Lesotho.

Practice question

Evaluate whether the Lesotho Highland Water Project is worth the enormous costs involved. (*6 marks*)

On this spread you will find out about strategies for a sustainable water supply

What is sustainable water supply?

Population growth and economic development will lead to greater demand for water in the future. In 2005 the United Nations began a 10-year 'Water for Life' campaign worldwide. The aim was to ensure that water resources are managed in a sustainable way in the long term.

Sustainable approaches to water supply focus on the careful management of water resources and the need to reduce waste and excessive demand.

Water conservation

Conserving water is about reducing waste and unnecessary use. This can include strategies such as:

◆ reducing leakages (25–30 per cent of global water supply is lost through leakage)

◆ monitoring illegal and unmetered connections

◆ water tariffs, with charges increasing sharply after a certain level of usage

◆ improving public awareness of the importance of saving water

◆ water meters to encourage people to use water wisely

◆ preventing **pollution**.

Did you know?
Turning the tap off while you brush your teeth saves 6 litres of water per minute!

A *How to use water sustainably*

Plant drought-tolerant plants

Wash vegetables in a container, not under a running tap

Only use washing machines and dishwashers for a full load

Live with a dirty car rather than washing it frequently

How can you save water at home?

Install low-flow shower heads

Collect rainwater for use in the garden

Water gardens early morning or late evening to reduce water loss

Install a twin-flush toilet system

Turn off tap when brushing teeth

Groundwater management

Groundwater stored in underground aquifers has to be managed to maintain the quantity and **quality** of the water. To ensure supplies are sustainable, water abstraction (loss) must be balanced by recharge (gain). If groundwater levels fall, water can become contaminated, making expensive water treatment necessary.

In many LICs individual families or groups own wells. National laws are ineffective and often ignored. So effective community-based management is needed.

B *Repairs to a well in Sierra Leone*

Recycling

Water recycling involves re-using treated domestic or industrial wastewater for useful purposes like irrigation and industrial processes. For example:

◆ Large quantities of recycled water are used for cooling in electricity-generating and steel-making plants. In some Australian power stations recycled water replaces enough fresh water to fill an Olympic-sized swimming pool.

◆ In Kolkata, India, sewage water is re-used for fish farming and agriculture. Sewage is pumped into shallow lagoons where sunlight helps algae to photosynthesise. This oxygenates the water so that it can be re-used.

◆ Some nuclear power plants, like the Palo Verde nuclear generating station in Arizona, USA, use recycled water for cooling (photo **C**).

Using grey water

Grey water is taken from bathroom sinks, baths, showers and washing machines. It may contain traces of dirt, food, grease, hair and some cleaning products. If used within 24 hours it contains valuable fertiliser for plants. Water from toilets is considered to be 'black' water and cannot be used in the same way.

Grey water is mainly used for irrigation and watering gardens (diagram **D**). In Jordan 70 per cent of the water used for irrigation and gardens is grey water.

Participatory Groundwater Management (PGM), India

In rural India, 50 per cent of water for irrigation and 85 per cent of drinking water is groundwater. Communities are encouraged to conserve water from their wells through the PGM scheme. Without careful management, the future of some rural communities is at risk. The PGM scheme involves:

◆ training local people to record rainfall and to monitor groundwater levels and water abstraction (photo **B**)

◆ helping farmers to plan when and how much water to use for irrigation

◆ encouraging farmers to plant crops to fit in with annual periods when water is available.

Through PGM, rural communities have used scientific monitoring to balance water supply and demand using sustainable practices.

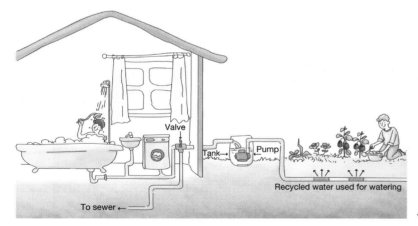

Recycled water used for watering

Valve

Tank → Pump

To sewer ←

C Palo Verde nuclear generating station

D Re-using grey water

ACTIVITIES

1 Explain whether you think your household makes sustainable use of water (diagram **A**).

2 Why is the involvement of local communities so important in the sustainable management of groundwater in rural areas?

3 What is the difference between 'grey' water and 'black' water?

Stretch yourself

Investigate why it is not only in areas of low rainfall where sustainable use of water is needed.

Practice question

Use examples to explain why both demand and supply affect the sustainable use of water. (*6 marks*)

The Wakel River Basin project

On this spread you find out about a local scheme to increase sustainable water supply

Example

Rajasthan is a region in north west India (map **A**). It is the driest and poorest part of India, and largely covered by the Thar Desert (see page xx). Summer temperatures can reach 53°C. Rainfall is less than 250 mm per year with 96 per cent between June and September. There is little surface water, as rain quickly soaks away or evaporates.

What are the issues with water supply?

Water management in the region has been poor. Over-use of water for irrigation has led to waterlogging and salinisation. Over-abstraction from unregulated pumps has resulted in falling water tables in aquifers and some wells have dried up. With access to wells controlled by households or villages, there has been little coordination of water management.

The Wakel River Basin Project

The Wakel river basin is located in the south of Rajasthan. The United States Agency for International Development has funded a project called The Global Water for Sustainability Program (2004–14). This NGO (non-government organisation) has been working with local people in the Wakel river basin to improve their water security and overcome the problems of water shortages. Local people needed to be actively involved in the decision-making process to make the water management successful.

The two main aims of the scheme are to:

◆ increase water supply and storage using appropriate local solutions

◆ raise awareness in local communities of the need for effective water management.

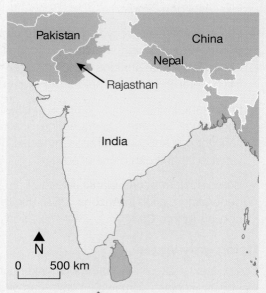

A *Location map of Rajasthan*

Did you know?

Each taanka can hold 20 000 litres – enough to supply a family for several months.

Increasing water supply

The project has encouraged greater use of rainwater harvesting techniques to collect and store water. This benefits villages and individual families. The methods used include the following.

◆ *Taankas* – underground storage systems about 3 m in diameter and 3–4 m deep (photo **B**). They collect surface water from roofs.

◆ *Johed* – small earth dams to capture rainwater. These have helped to raise water tables by up to 6 m. Five rivers that used to dry up following the monsoon now flow throughout the year.

◆ *Pats* – irrigation channels that transfer water to the fields.

B *Collecting water from a concrete* taanka

How does the *pat* system work?

In the *pat* system, a small dam called a bund diverts water from the stream towards the fields. Bunds are made of stones, and lined with leaves to make them waterproof.

Villagers take turns to irrigate their fields using water controlled in this way. The irrigation channels need regular maintenance to avoid them breaking or silting up. This is done by the villager whose turn it is to receive the water.

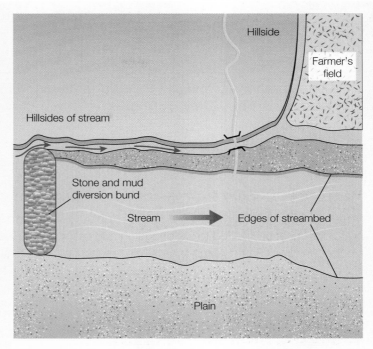

C *The* pat *irrigation system*

Increasing public awareness

Education is used to increase awareness of the need for communities to work together to conserve water (photo **D**). By conserving water, water security is increased and problems such as soil erosion, desertification and groundwater pollution are reduced.

D *A cycle campaign to spread the message about the importance of water conservation*

ACTIVITIES

1 Summarise the physical and human factors causing water insecurity in the region.

2 **a** What are the key design features of a *taanka* (photo **B**)?

   b How does this form of water harvesting benefit local communities?

3 **a** Copy diagram **C** and add labels to describe how the *pat* system works.

   b How does this system demonstrate the importance of communities working together to reduce water security?

Stretch yourself

Give reasons for working at a local rather than a national level when developing an effective sustainable water scheme.

Practice question

Evaluate the success of a local scheme for increasing sustainable water supplies. (*4 marks*)

On this spread you find out that global demand for energy is rising but supply is not evenly distributed

Global energy consumption and supply

Energy consumption per person is very high in countries like the USA, Canada, Australia, much of Europe, and parts of the Middle East (map **A**). It is low across most of Africa and parts of south east Asia. In regions of high energy consumption there is a growing demand for industry, transport and domestic use.

Some regions have energy resources such as coal, oil or gas. Some areas are also able to produce electricity (map **B**), for example by using nuclear power.

The balance between energy supply (production) and demand (consumption) determines the level of **energy security**. If supply exceeds demand then a country has an *energy surplus*. If demand exceeds production, there is an energy deficit and the country suffers from *energy insecurity* (table **C**).

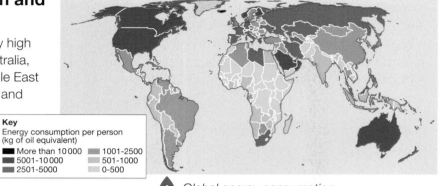

Key
Energy consumption per person (kg of oil equivalent)

- More than 10 000
- 5001-10 000
- 2501-5000
- 1001-2500
- 501-1000
- 0-500

A *Global energy consumption*

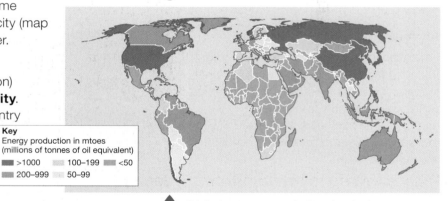

Key
Energy production in mtoes (millions of tonnes of oil equivalent)

- >1000
- 200–999
- 100–199
- 50–99
- <50

B *Global energy supply (production)*

C *Energy security status, by region*

Region	Energy security	Energy sources
Russia and Eastern Europe	Energy surplus	• Large reserves of natural gas and oil. • Uranium resources which can be used for nuclear energy.
Western Europe	Energy insecurity (deficit)	• Dependent on energy imports, particularly oil and gas. • Low energy efficiency.
Middle East	Energy surplus	• Large oil reserves. • Unstable political regimes affect fuel supply.
North America	Energy insecurity	• Large coal reserves. • Opportunity to exploit oil reserves in sensitive areas like the Arctic. • Huge energy consumption. • Deficit in energy until technological advances allowed exploitation of oil shale.
Asia	Energy insecurity	• Large coal and uranium deposits. • Rapidly increasing demand outstrips supply.
Sub-Saharan Africa	Energy insecurity	• Depends on foreign TNCs to exploit reserves, for example Nigerian oil. • Limited energy supplies with rising rates of consumption.

What factors affect energy supply?

Costs of exploitation and production

Some energy sources are costly to exploit. Oil rigs and pipelines require huge investment. Nuclear power stations are expensive to build.

Physical factors

The geology of an area determines the location and availability of **fossil fuels**. Coal is formed from vegetation laid down and altered by pressure and heat over millions of years. Natural gas and oil is trapped in folded layers of rocks. **Geothermal energy** is produced in areas of tectonic activity like Iceland and the Pacific Rim.

Technology

Technological advances have allowed energy sources in remote or difficult environments, such as the North Sea and the Arctic, to be exploited. They can also reduce costs. Technology has made possible the exploitation of shale gas by fracking (see page 263).

Political factors

Political factors affect decisions about which energy sources to exploit and from which countries energy can be obtained.

◆ Political instability in the Middle East has meant that many oil-consuming countries are looking for alternative sources of energy.

◆ Some Western countries and Israel currently want to stop Iran developing nuclear power. They fear it will be used for non-peaceful purposes.

◆ The German government is planning to stop generating nuclear power by 2020.

◆ The UK government has decided to cut subsidies for renewable energy such as solar and wind.

Climate

The amount of sunshine and wind influence the availability of **solar energy** and **wind energy**. Tidal power needs a large tidal range in order to be effective. HEP needs a suitable dam site, often in sparsely populated mountainous areas with high rainfall.

Why is energy consumption increasing?

Economic development

As countries develop their demand for energy supplies rises. NEEs will account for more than 90 per cent of the growth in demand for energy to 2035. Recent growth in Asia's energy demand has been led by China, but this has now started to slow down. Greater energy demand is expected to accompany rapid economic growth in India and other parts of south east Asia. This is due to industrialisation and greater wealth.

Rising population and technology

In 2015 the world's population was 7.5 billion. By 2050 it is predicted to rise to 9 billion. All these extra people will use more energy. Many will grow up in an increasingly energy-thirsty world.

The increasing use of technology, like computers and other electrical equipment, means a greater demand for energy. As quality of life improves and prosperity increases, the demand for vehicles, lighting and heating also increases.

ACTIVITIES

1 a Describe the pattern of global energy consumption (map **A**).

 b Compare map **A** with map **B**. What are the main similarities and differences?

2 What are the reasons for the growth in global energy consumption?

3 How have technological advances affected the consumption and supply of energy?

Stretch yourself

Investigate the rising energy demand in Nigeria. Why is its demand for energy increasing?

Practice question

Explain why many countries are experiencing energy insecurity. (*6 marks*)

On this spread you will find out about the costs and impacts associated with energy insecurity

What can be done about energy insecurity?

Many countries experience energy insecurity. In order to secure their future energy needs, they must consider a range of options. To increase its energy supply a country may:

◆ try to further exploit its own energy sources

◆ reach agreements with other countries to import energy

◆ reduce its energy consumption through new technologies or greater energy saving.

Energy insecurity can have **economic, social and environmental impacts**.

Exploiting resources in difficult and sensitive areas

In the past, energy resources were relatively easy to exploit. For example, coal seams have been exposed at the Earth's surface. Today, complex techniques and expensive equipment are needed to extract oil and gas reserves in sensitive areas, such as deep below the North Sea.

Energy resources exist in some of the world's most hostile, dangerous and environmentally sensitive regions. These include the Amazon and Antarctica. Exploiting these resources in the future will depend on:

◆ the development of technologies that make exploitation cost-effective

◆ the environmental implications of **energy exploitation** in areas that are extremely sensitive and could easily be damaged.

Exploiting energy resources in the Arctic

Map **A** shows energy resources in the Arctic. This region holds an estimated 13 per cent (90 billion barrels) of the world's undiscovered oil resources and 30 per cent of its unexploited natural gas.

This region has great potential to supply energy in the future, but exploitation is difficult and expensive. The environmental consequences of an oil spill, for example, would be catastrophic for the fragile Arctic ecosystem. Recovery from damage would be slow given the low temperatures and short growing season.

To exploit energy resources in the Arctic, several economic and environmental factors need to be addressed (figure **B**).

A Oil and natural gas resources in the Arctic (yellow shading)

B Economic and environmental costs of oil and gas exploitation in the Arctic

People demand higher wages to work there

Drilling equipment may sink during the summer thaw

Political issues develop because the territory north of the Arctic Circle is claimed by eight countries

Strict environmental controls are needed to prevent damage

Long distances and limited transportation increases transport costs

Special equipment is needed to withstand the extreme temperatures

Impacts of energy insecurity on food production

Food production uses 30 per cent of global energy. Energy is used to power farm machinery, store farm produce, and to manufacture fertilisers and chemicals.

Agriculture is also an energy generator. Use of biofuels has increased in response to concerns about carbon dioxide emissions. Use of biofuels like maize and sugar cane have contributed significantly to increased food prices. In addition, biofuels are often grown on land previously used for growing food crops (cartoon **C**).

In some LICs such as Tanzania and Mali firewood is the main source of energy. Instead of working on the land, people – often women – have to spend hours walking to collect the wood. This impacts on food production in regions with high food insecurity.

Impacts of energy insecurity on industry

Energy is essential for industry as a source of power and a raw material. Oil, for example, has many uses in manufacturing chemicals, fuels, plastics and pharmaceuticals.

Some countries suffer from shortfalls in electricity production, resulting in frequent power cuts. In Pakistan, regular power cuts can last for 20 hours a day. This costs the country an estimated 4 per cent of its GDP. Energy shortages have led to the closure of more than 500 companies in the industrial city of Faisalabad alone. Pakistan relies heavily on imported oil which makes energy expensive as well as insecure.

Potential for conflict

Shortages of energy can lead to political conflict when one state holds a bigger share of an energy resource. For example, Russia controls 25 per cent of the world's natural gas supplies. It could put pressure on its customers – mostly in Western Europe – by raising prices or even cutting off supplies.

The Middle East produces 40 per cent of the world's gas and 56 per cent of its oil. The Gulf and Iraq wars in the 1990s and 2000s were driven by the West's fear of a global oil shortage and rising prices.

There are flashpoints where the transport of oil is at risk from political conflicts, terrorism, hijack or collision (map **D**).

FOOD NOT FUEL

C *ActionAid's 'Food not fuel' campaign*

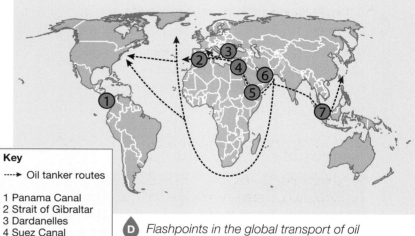

Key

---▶ Oil tanker routes

1 Panama Canal
2 Strait of Gibraltar
3 Dardanelles
4 Suez Canal
5 Bab el Mandeb
6 Strait of Hormuz
7 Strait of Malacca

D *Flashpoints in the global transport of oil*

ACTIVITIES

1 a What are the main economic and environmental costs and impacts associated with exploiting oil and gas from the Arctic (figure **B**)?

b Do you think the Arctic should be exploited for energy resources? Justify your answer.

2 What are the benefits and costs of growing crops for fuel (**C**)?

3 Which flashpoints on map **D** do you think are the most vulnerable and why?

Stretch yourself

Investigate recent international incidents which have been linked to energy insecurity.

Practice question

Explain how physical and human factors can contribute to energy insecurity. (*6 marks*)

Strategies to increase energy supply

On this spread you will find out about how energy supplies can be increased

What are the options for increasing energy supplies?

There are two main options for increasing energy supplies.

◆ Develop and increase the use of renewable (sustainable) sources of energy, such as **wind**, **solar** and HEP.

◆ Continue to exploit non-renewable fossil fuels such as oil and gas and develop the use of nuclear power.

Graph **A** shows the trend for global energy consumption up to 2015. Notice the balance between renewables, including **hydroelectric power (HEP)** and non-renewables. To achieve a sustainable **energy mix** countries need to develop the use of renewable energy sources (map **B**).

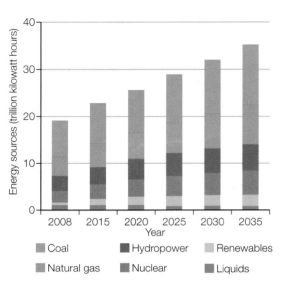

| Coal | Hydropower | Renewables |
| Natural gas | Nuclear | Liquids |

A *Global energy sources for electricity production*

Renewable energy sources

Renewable energy source	How does it work?	Can it increase energy supplies?
Biomass	Energy produced from organic matter includes: • burning dung or plant matter • the production of biofuels, by processing specially grown plants such as sugar cane.	• Using land to grow biofuels rather than food crops is very controversial. • Burning organic matter can create smoky unhealthy conditions. • Fuelwood supplies are limited.
Wind	Turbines on land or at sea are turned by the wind to generate electricity.	• In 2014, wind power met 10% of the UK's electricity demand. • Unpopular, but considerable potential.
Hydro (HEP)	Large-scale dams and smaller micro-dams create enough water to turn turbines and generate electricity.	• Large dams are expensive and controversial. • Micro-dams are becoming popular options at the local level. • An important energy source in several countries. It currently contributes 85% of global renewable electricity.
Tidal	Turbines within barrages (dams) built across river estuaries use rising and falling tides to generate electricity.	There are few tidal barrages (the largest is the Rance in France) due to high costs and environmental concerns.
Geothermal	Water heated underground in contact with hot rocks creates steam that drives turbines to generate electricity.	Limited to tectonically active countries: • the USA (has the most geothermal plants – 77) • Iceland (provides 30% of the country's energy) • the Philippines and New Zealand.
Wave	Waves force air into a chamber where it turns a turbine linked to a generator.	• Portugal has built the world's first wave farm, which started generating electricity in 2008. • There are many experimental wave farms but costs are high and there are environmental concerns.
Solar	Photovoltaic cells mounted on solar panels convert sunlight into electricity.	• Energy production is seasonal. • Solar panel 'farms' need a lot of space. • Great potential in some LICs with high levels of sunshine.

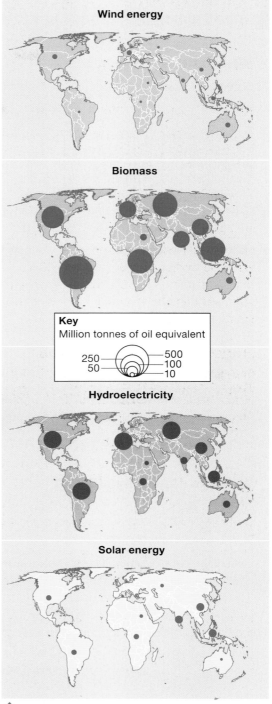

Non-renewable energy sources

Non-renewable energy resources are unsustainable. At some point the economic and environmental costs of these resources will become too high. Or they will run out.

Fossil fuels

Fossil fuels are sources of energy formed from organic matter millions of years ago. They include coal, gas and oil. Although limited, there are still plenty of resources left in the world. Despite high carbon dioxide emissions they remain important for electricity production. Carbon capture techniques (page 46) can help overcome the environmental impact.

Nuclear power

Nuclear power stations are very expensive to build. However, the cost of the raw material uranium is relatively low because small amounts are used.

The main problem with nuclear power is disposal of the radioactive waste. It can remain dangerous for longer than 100 years. Despite the good safety record, there is considerable opposition. There is fear of further accidents like those at Chernobyl, or Fukushima after the Japanese tsunami in 2011.

ACTIVITIES

1 a How much electricity will be produced from coal in 2020 (graph **A**)?

   b Describe the pattern shown on the graph.

2 Describe and suggest reasons for global variations in hydro-electricity potential (map **B**).

3 a Which form of renewable energy has the most potential as an energy source?

   b Suggest advantages and disadvantages of focusing on this form of renewable energy.

Stretch yourself

Produce a case study of one type of renewable energy in the UK. Outline its location, when it started and the short-term and long-term advantages and disadvantages. Do you think your choice is a viable option for the future?

Maths skills

Use an appropriate presentation to show the global use of energy, using these figures:

78% fossil fuels
19% renewables (of which traditional biomass 13%; hydro 3%; rest 3%)
3% nuclear.

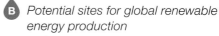 Potential sites for global renewable energy production

Practice question

Explain why the contribution of renewable energy to world energy production is likely to remain less than fossil fuel production.
(6 marks)

Example

On this spread you will find out about the advantages and disadvantages of extracting a fossil fuel

What is natural gas?

Natural gas is a hydrocarbon. Like oil, natural gas forms from the decomposition of organisms deposited on the seabed millions of years ago (diagram **A**). This is why it is called a fossil fuel. The organic matter was buried by layers of sediment and heated by compression. Lack of oxygen produced thermal reactions that converted the organic material into hydrocarbons.

The colourless and odourless natural gas rises up through cracks and pores (holes) in the overlying rocks. It then collects in concentrations called reservoirs. It is from these reservoirs that natural gas is extracted.

A *The formation of oil and natural gas*

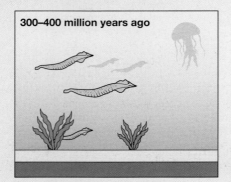

300–400 million years ago

Remains of tiny sea plants and animals buried on ocean floor. Over time these are covered by sand and sediment.

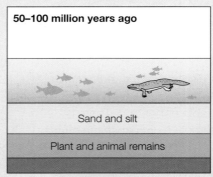

50–100 million years ago

Sand and silt

Plant and animal remains

Over millions of years the remains are buried deeper. Enormous pressure and heat turns them into hydrocarbons (oil and gas).

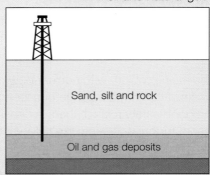

Sand, silt and rock

Oil and gas deposits

Today, oil and gas deposits are reached by drilling down through layers of sand, silt and rock.

Where is natural gas found?

Nearly 60 per cent of known natural gas reserves are in Russia, Iran and Qatar (table **B**). These reserves are sufficient to last for the next 54 years at the current rate of production.

Recently technology has allowed **shale gas** to be extracted (map **C**). Shales are black sedimentary rocks formed from the same organic matter that is the source of oil and gas. To extract shale gas, the rock is broken up by a process called fracking. This process is very controversial and there is a lot of opposition (see page 263).

Country	Production in million m³
Russia	48
Iran	36
Qatar	25
Saudi Arabia	9
Turkmenistan	8
United Arab Emirates	7
Venezuela	6
Nigeria	5.5
Réunion	5.5
China	4.5

B *The top ten gas-producing countries in 2015*

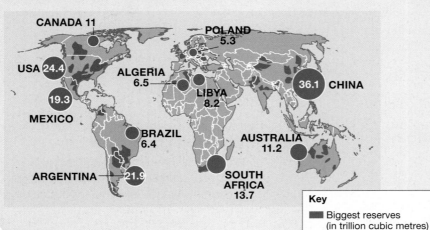

CANADA 11
POLAND 5.3
USA 24.4
ALGERIA 6.5
LIBYA 8.2
CHINA 36.1
MEXICO 19.3
BRAZIL 6.4
AUSTRALIA 11.2
ARGENTINA 21.9
SOUTH AFRICA 13.7

Key
■ Biggest reserves (in trillion cubic metres)

 C *Global shale gas deposits*

Extracting natural gas

Advantages

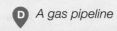 *A gas pipeline*

Disadvantages

Cleanest of the fossil fuels with 45% less CO_2 emissions than other non-renewable sources and less toxic chemicals like NO and SO_2.

Less risk of environmental accidents than oil.

Provides employment for 1.2 million people.

Can be transported in a variety of ways, i.e. through pipelines or by tankers over land and sea.

Relatively abundant compared to other fuels. This is increasing as technology makes exploitation of shale gas more economic.

Provides electricity during peak demand periods. Unlike nuclear and HEP, gas-fired power stations can be shut off and turned back on as required.

Dangerous if handled or transported carelessly.

Some gas reserves are in countries that are politically unstable or prepared to use gas supply as a political weapon.

Contributes to global warming by producing CO_2 and methane emissions.

Fracking is controversial. Lots of water is needed. Wastewater and chemicals could contaminate groundwater and minor earthquakes are possible.

Pipelines are expensive to build and maintain.

Extracting natural gas in the Amazon

The Camisea project began in 2004 to exploit a huge gas field in the Amazonian region of Peru. The project has brought both advantages and disadvantages for Peru.

Advantages

◆ It could save Peru up to US$4 billion in energy costs.

◆ Peru could make several billion dollars in gas exports – up to US$34 billion over the 30-year life of the project.

◆ It provides employment opportunities and helps boost local economies.

◆ Improved infrastructure could bring benefits to local people. Agriculture could become more productive.

Disadvantages

◆ Deforestation associated with the pipeline and other developments will affect natural habitats.

◆ The project could impact on the lives of several indigenous tribes, affecting their traditional way of life and their food and water supplies.

◆ Local people have no immunity to diseases introduced into the area by developers.

◆ Clearing routes for pipelines has led to landslides and pollution of streams resulting in decline of fish stocks.

ACTIVITIES

1 Using diagram **A**, describe the formation of natural gas.

2 **a** Why is fracking an important technological development?

b Describe the global distribution of shale gas basins (map **C**).

c Why is fracking such a controversial process?

3 Suggest advantages and disadvantages of transporting gas by pipeline (figure **D**).

Maths skills

Present the figures in table **B** as bars or proportional circles on an outline map of the world. Take care to work out an appropriate scale. Use a single colour to shade the bars or circles.

Stretch yourself

Evaluate the importance of shale gas in the natural gas market. Present your findings as a news report.

Practice question

'The advantages of exploiting natural gas outweigh the disadvantages.' Do you agree with this statement? Justify your decision. *(9 marks)*

On this spread you will find out how it is possible to move towards a sustainable energy supply

What is a sustainable energy supply?

A sustainable energy supply involves balancing supply and demand. It also involves reducing waste and inefficiency. Moving towards a more sustainable future needs individual actions and decisions made by businesses, councils and national governments.

To increase energy supply, renewable sources of energy can be developed and fossil fuels can be exploited more efficiently. Energy demand can be reduced by increasing **energy conservation** and designing more energy-efficient homes and workplaces (diagram **A**). Reducing the use of fossil fuels and increasing efficiency will help reduce carbon dioxide emissions and our **carbon footprint**.

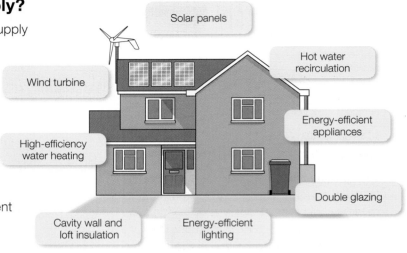

Solar panels

Hot water recirculation

Wind turbine

Energy-efficient appliances

High-efficiency water heating

Double glazing

Cavity wall and loft insulation

Energy-efficient lighting

A *Energy conservation in the home*

Sustainable energy developments in Malmo, Sweden

To improve energy conservation, we need to consider sustainable approaches to the design of homes, workplaces and transport systems. These can be seen in innovations in Malmo, an industrial city of about 300 000 people on Sweden's west coast.

Malmo's Western Harbour (figure **B**) is one of the best examples of sustainable urban redevelopment in the world. The houses have been designed to generate and conserve energy and the transport system aims to reduce car usage.

Creation of green spaces and roof gardens.

All 1000 buildings in the district use 100% renewable energy.

From 2019 all buses will run on a mixture of biogas and natural gas.

Solar tubes on the outside of buildings produce hot water which can be stored in aquifers 90 m below ground and used to heat buildings during the winter. The water is pumped using electricity from wind power.

Cyclists have priority at crossroads. A sensor system turns lights green when a cyclist approaches.

Energy comes from photovoltaic panels on the roofs of houses and workplaces, a 2 MW wind turbine, and biogas from local sewage and rubbish.

Frequent buses and water taxis offer public transport options for local people. This has reduced car usage and people's carbon footprint. A car share scheme has also been introduced.

B *Malmo's Western Harbour*

Reducing energy demand

There are several ways of reducing energy demand. These can include:

◆ financial incentives

◆ raising awareness of the need to save and use energy more efficiently

◆ greater use of off-peak energy tariffs

◆ using less hot water for domestic appliances.

How can technology increase efficiency of fossil fuels?

Vehicle manufacturers are using technology to design more fuel-efficient cars to reduce oil consumption and their carbon footprint. These developments include the use of carbon fibre, which is lighter than conventional steel, improved engines and aerodynamic designs to increase fuel efficiency.

The recent development of electric and hybrid cars will increase the efficient use of fossil fuels (photo **D**). In the USA the growth in the use of electric cars could reduce the use of oil for transport by up to 95 per cent.

The development of biofuel technology in car engines can reduce the use of oil. Brazil has reduced its petrol consumption by 40 per cent since 1993 by using sugar cane ethanol (photo **E**). Around 90 per cent of all new cars in Brazil can run on both ethanol and petrol. Brazilians are increasingly choosing environmentally-friendly ethanol because it is cheaper than petrol. However, growing biofuels rather than food crops is a controversial issue in Brazil.

Reducing energy demand at Marriott hotels

The chain of Marriott hotels in the UK and Europe spends £60 million a year on energy. An automated system places the hotel chain on energy-saving standby if the national electricity supply grid needs to reduce demand.

Everything from air conditioning to ice coolers in the corridors can be turned down at a moment's notice without customers noticing. Not only will this reduce energy demand but the hotel chain is paid a supplement for reducing its energy use.

C *An energy-saving Marriot hotel*

D *An electric car*

E *Ethanol fuel at a gas station in São Paulo, Brazil*

ACTIVITIES

1 **a** In what ways is your own home energy-efficient?

   b What are the least energy-efficient aspects of your home?

2 How is Malmo's Western Harbour a model for sustainable urban development?

3 Choose a new model *either* of a car *or* aircraft. State how the manufacturers have designed it to be more energy efficient.

Stretch yourself

Investigate how energy ratings on electrical appliances contribute to energy sustainability.

Practice question

Evaluate why changes in individuals' actions and in the built environment are necessary if energy use is to become sustainable. (*6 marks*)

On this spread you will find out about a local sustainable energy scheme

Where is Chambamontera?

Chambamontera is an isolated community in the Andes Mountains of Peru (map **A**). It is more than two hours' drive on a rough track from Jaén, the nearest town.

Why does Chambamontera need a sustainable energy scheme?

Most people in the area are dependent on subsistence farming with some small-scale coffee growing and rearing of livestock (photo **B**). Development has been severely restricted by a lack of electricity for heat, light and power. Despite farming being efficient, nearly half the population survive on just US$2 a day.

The steep slopes rise to 1700m and the rough roads are impassable in winter. This makes Chambamontera a very isolated community. Due to the low population density it was uneconomic to build an electricity grid to serve the area.

What is the Chambamontera micro-hydro scheme?

The solution to Chambamontera's energy deficit involved the construction of a micro-hydro scheme supported by the charity Practical Action. The high rainfall, steep slopes and fast flowing rivers make this area ideal for exploiting water power as a renewable source of energy (diagram **C**).

The total cost of the micro-hydro scheme was US$51 000. There was some government money and investment from Japan, but the community had to pay part of the cost. The average cost per family was US$750. Credit facilities were made available to pay for this.

A Location map of Chambamontera

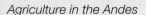 **B** Agriculture in the Andes

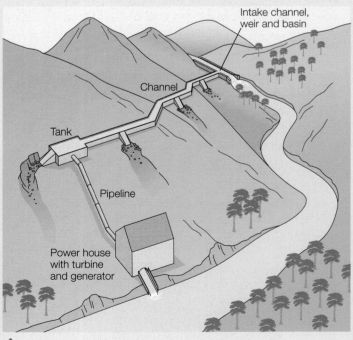

 C How the scheme works

How has the scheme benefited the local community?

The micro-hydro scheme has benefited this isolated community in many ways.

The scheme has provided local people with a sustainable source of energy. It enables them to make a more productive living. The scheme:

◆ provides renewable energy

◆ has low maintenance and running costs

◆ has little environmental impact

◆ used local labour and materials.

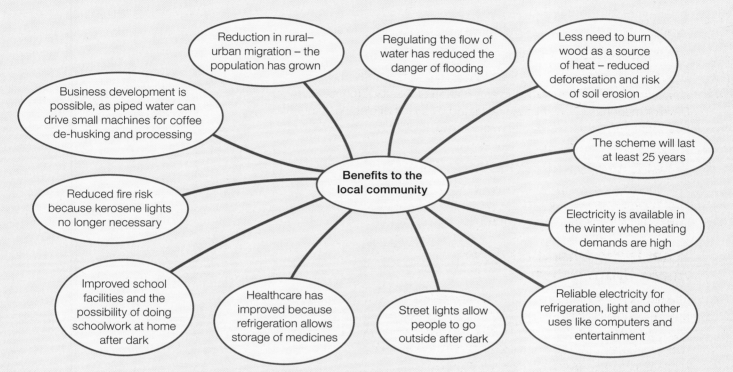

Reduction in rural–urban migration – the population has grown

Regulating the flow of water has reduced the danger of flooding

Less need to burn wood as a source of heat – reduced deforestation and risk of soil erosion

Business development is possible, as piped water can drive small machines for coffee de-husking and processing

Benefits to the local community

The scheme will last at least 25 years

Reduced fire risk because kerosene lights no longer necessary

Electricity is available in the winter when heating demands are high

Improved school facilities and the possibility of doing schoolwork at home after dark

Healthcare has improved because refrigeration allows storage of medicines

Street lights allow people to go outside after dark

Reliable electricity for refrigeration, light and other uses like computers and entertainment

ACTIVITIES

1 Describe the location of Chambamontera (map **A**).

2 **a** Describe the landscape shown in photo **B**.

  b What evidence is there to suggest that farming is difficult?

  c Explain why this is likely to be a subsistence farming area.

3 Evaluate the success of the micro-hydro scheme.

Stretch yourself

Why is a 'bottom-up' scheme like the Chambamontera micro-hydro an effective way of providing sustainable energy?

Practice question

What makes the Chambamontera scheme a sustainable way of providing energy? (*4 marks*)

D *Local people carry the turbine to the micro-hydro site*

299

About Paper 3

Unit 3 Geographical applications and skills is about identifying, understanding and appreciating the interrelationships between the different aspects of your geography study. It is split into *three* chapters.

◆ **Chapter 23 Issue evaluation** includes a critical thinking and problem-solving task based on a current issue.

◆ **Chapter 24 Fieldwork** includes two contrasting enquiries to help you understand each stage of the enquiry process in a fieldwork context, and the interrelationships between them.

◆ **Chapter 25 Geographical skills** includes a summary of the cartographic, graphical, numerical, statistical, and data skills required to become a good geographer. Remember – geographical skills will be assessed in **all** three written exams.

Your exam

Chapters 23 and 24 make up Paper 3 – a one hour 15 minutes written exam worth 30 per cent of your GCSE. The marks for this paper are divided equally between the issue evaluation and fieldwork.

What is the Issue evaluation?

The theme will come from one of the core topics in the specification and not from one of the options. It will be *synoptic* – it will be likely to draw on both the human and physical aspects of your GCSE Geography course.

A resource booklet will be published 12 weeks before the exam to help you become familiar with the issue evaluation theme and resources. You will not be allowed to take this booklet into the exam, but will be given a new copy.

In the exam you'll be asked to make decisions and judgements based on the resources provided, using evidence to support your decisions. Remember – you must focus on the *evidence from the resources* to support your decision or choice.

How to approach a decision-making question

When you are tackling the decision-making questions in the Issue evaluation, remember the following steps:

Whether or not the decision is an ideal solution, it is sometimes necessary to consider:

◆ the short-term and long-term costs and benefits

◆ whether anyone is inconvenienced by the decision

◆ how that inconvenience might be reduced.

Sources of geographical evidence

Sources in the exam booklet might include maps (on several different scales), diagrams, graphs, statistics, photos, satellite images, sketches, extracts from published materials (including management plans), and quotes from different interest groups.

You may need to consider several sources before making a balanced decision. Take time to read and interpret the sources carefully before answering any questions. *Do not ignore any sources* – all the sources provided should help you to argue for or against a point of view, or provide a context for the issue.

1 Read the question carefully and make sure you fully understand the issue.

2 Consider all the evidence and make your decision. Take time to make sure you have made a sensible choice.

3 Remember there is no wrong answer — just a poor answer that is not supported by the evidence! Always use evidence from the resources to support your decisions.

4 Read through your answer and check that you have backed up your decisions with relevant facts and figures taken from the resources.

How is fieldwork assessed?

Geographical enquiry

Before sitting your exam you will need to carry out two contrasting geographical enquiries. One of the enquiries should show *the interaction between human and physical geography*. For example:

◆ how flood defences have been used on a stretch of river

◆ how the physical environment influences land use in a place.

Both enquiries must include the use of primary data collected as part of fieldwork exercises. Chapter 24 uses investigations of river processes and urban quality of life as model enquiries. The exam could cover other areas of geography and locations.

Fieldwork in the exam

In the exam, you should be prepared to discuss fieldwork that shows *interaction between aspects of physical and human geography*.

Some of the questions will require you to make connections between different elements of the enquiry process. For example:

◆ Assess the appropriateness of your data collection methods.

◆ To what extent can the fieldwork results be deemed to be reliable?

◆ Evaluate the accuracy and reliability of your results/conclusions.

Your understanding of the *enquiry* process will be assessed in the following ways:

1 Questions based on the use of fieldwork data from an unfamiliar context.

2 Questions based on your own individual enquiry work. For these questions you will have to write the titles of your own individual enquiries.

How to approach a fieldwork question

The specification requires you to understand the process of geographical enquiry. This includes:

◆ choosing a suitable question for a geographical enquiry

◆ selecting, measuring and recording data appropriate for your chosen enquiries

◆ selecting appropriate ways of processing and presenting fieldwork data

◆ describing, analysing and explaining fieldwork data

◆ reaching conclusions

◆ evaluating geographical enquiries.

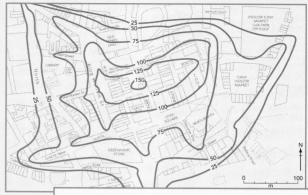

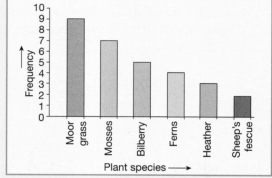

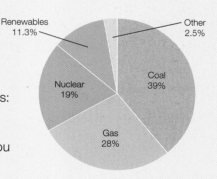

In this chapter you will attempt a sample Issue evaluation

Should a tunnel be constructed under Stonehenge?

In December 2014 the government announced plans to construct a 2.9 km tunnel beneath Stonehenge as part of a plan for upgrading the busy A303 trunk road to the West Country. This was part of government plans to improve the UK's road and railway infrastructure.

Study Figure 1.

1 In which county is Stonehenge located? *(1 mark)*

2 What is the straight line distance between Basingstoke and Honiton? *(1 mark)*

3 Describe the location and route of the proposed tunnel. *(2 marks)*

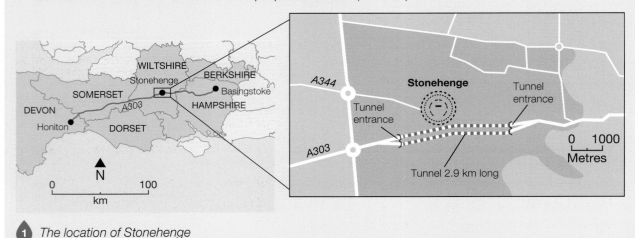

1 *The location of Stonehenge*

Study Figure 2.

4 Describe the physical landscape in the photo. *(2 marks)*

5 Traffic congestion is common on this stretch of the A303. Describe one *economic* and one *environmental* problem resulting from traffic congestion. *(2 marks)*

2 *Stonehenge and congestion on the A303*

Study Figure 3.

6 Give the four-figure grid reference for Stonehenge. *(1 mark)*

7 Assume that the eastern end of the tunnel would be at the end of the dual carriageway at 132420. Give the six-figure grid reference of the point where it would re-emerge on the A303 west of Stonehenge. *(2 marks)*

8 What is the evidence that there are many historic monuments in the area? *(2 marks)*

3 *1:50 000 OS map extract of Stonehenge*

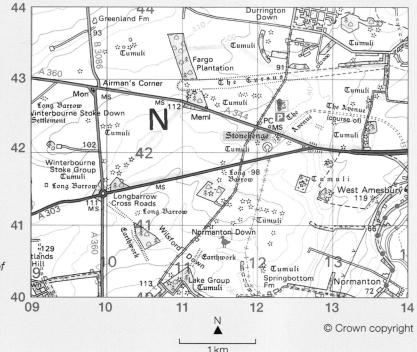

© Crown copyright

Study the table (Figure 4) which shows the visitor numbers to Stonehenge since 2004.

9 Use the data to plot a graph of visitor numbers for 2004–14 *(3 marks)*

10 Describe and suggest reasons for the trend in visitor numbers at Stonehenge between 2004 and 2014. *(4 marks)*

11 Calculate the percentage increase in visitor numbers between 2008 and 2009. *(1 mark)*

12 Use the graph to extrapolate the trend. Suggest what the visitor numbers might be in 2020. *(1 mark)*

Year	Number of visitors	Percentage +/-
2014	1 346 177	+8
2013	1 241 296	+18.9
2012	1 043 756	−5.1
2011	1 099 656	+9
2010	1 009 973	+2
2009	990 705	
2008	883 603	+2
2007	869 432	−1
2006	879 393	+4
2005	833 617	+3
2004	817 924	+6.8

4 *Stonehenge visitor numbers*

Study Figure 5, a cartoon produced by Heritage Action.

13 What is the evidence in the cartoon that this campaign group does not support the construction of a tunnel beneath Stonehenge? *(4 marks)*

14 Suggest how construction of the tunnel might effect the local economy. *(4 marks)*

5 *Cartoon produced by Heritage Action*

FATHER TIME FACES DEATH AT STONEHENGE

News articles about the proposed Stonehenge tunnel

A

Plans for new Stonehenge tunnel

The government has announced plans to construct a tunnel to take a congested main road past Stonehenge. The announcement is part of the government's infrastructure plan. The 2.9 km tunnel is part of a £2 billion plan to turn the main A303 road into a dual carriageway. The busy road links London and the South West.

A previous plan to build a tunnel on the route was dropped seven years ago because of the high cost. But local councils have continued to lobby for a tunnel and the widening of the A303. Chancellor George Osborne said the plan would transform the A303 and help to boost the economy in the South West.

The Stonehenge Traffic Action Group has campaigned for action to be taken to relieve congestion on the A303 in this area. They said the new plan was good news.

English Heritage, which runs the Stonehenge site, has previously described the bottleneck road as 'highly detrimental' to the ancient monument.

The Campaign to Protect Rural England (CPRE) is campaigning for a longer tunnel. It argues that the proposed tunnel is too short and would create two huge holes near Stonehenge. They say this would affect the landscape around the World Heritage site.

B

Impact on the World Heritage Site

The heritage organisations English Heritage and the National Trust have been arguing for a short tunnel for the A303 at Stonehenge. They argue that the negative impacts of the tunnel entrances on the World Heritage Site (WHS) would be greater than the benefits

If the construction of the tunnel were to go ahead, it is possible that both Stonehenge and Avebury could lose WHS status. This would cause economic, aesthetic, cultural and social losses. The loss of WHS status for would be a national disgrace. The future reputation of Britain's heritage would be at risk. Would we want that on our conscience?

C

Views of Senior Druid

There are plans are for a tunnel to be bored through the chalk to the south of the existing A303. This would move the road further away from Stonehenge. The Senior Druid said that he will support the proposal only if he can be sure that any human remains excavated as a result of the construction are reinterred as closely as possible to what should have been their final resting place.

Without such guarantees he has said that he will protest and take non-violent action against the proposed tunnel. He is therefore undecided whether he will support or oppose the plan.

D

Archaeologists against Stonehenge tunnel

Archaeologists have discovered an ancient site which could help to unlock the mysteries about Stonehenge. They are calling on the government to rethink plans for a 2.9 km tunnel under the 5000 year-old World Heritage Site in Wiltshire. The aim of the tunnel is to help relieve one of Britain's most congested roads.

A series of digs at nearby Amesbury, next to the A303, has unearthed evidence of a settlement that pre-dates Stonehenge by thousands of years. Thousands of tools have been discovered at the site as well as evidence of giant structures.

Archaeologists warn that the chance to find out more about the earliest chapter of Britain's history could be ruined if the tunnel goes ahead.

Read the news articles in Figure 6. Using these articles and the other resources, attempt the following question:

15 Do you think the tunnel should be constructed beneath Stonehenge? Justify your answer using evidence from the resources. *(9 marks +3 SPaG)*

On this spread you'll find out how to prepare a fieldwork enquiry to investigate river processes and management

Enquiry questions

Rivers are popular places for people to live near and to enjoy. Photos **A** and **B** shows two contrasting river locations. Think about the kind of questions geographers might ask when they see places like this. For example, they might ask:

◆ What has happened here?

◆ How did it happen (short and longer-term reasons)?

◆ What might happen to this place in future and why?

Many river (or fluvial) locations are good places for fieldwork since there are lots of questions like these to investigate. This is the starting point for any enquiry. An enquiry is a series of stages that start with a question (diagram **C**, stage 1) and end up with an answer or conclusion (stage 5). You will probably have completed an enquiry in geography (or science) before and have used fieldwork and practical work in the same way.

Each stage is equally important, right from the initial question, to the research and context, through to the overall evaluation. Only at the end can you have an opportunity to reflect on what you have found and what it means.

A Floods in York, 2015

B An area of the Peak District, Derbyshire

(6) Evaluating and reflecting on the enquiry

(1) Setting up a suitable enquiry question

(2) Selecting, measuring and recording primary and secondary data appropriate to the enquiry

(3) Selecting appropriate methods (e.g. graphs, charts, maps) of processing and presenting fieldwork data

(4) Describing, analysing and explaining fieldwork data

(5) Reaching conclusions and considering their significance

C The route to enquiry – a planning 'pathway' for your investigation

Developing an enquiry question

A good enquiry depends upon having a good question. A good question must be directly linked to the overall theme – in this case, rivers! Consider the following question:

How do different drainage basin and river channel characteristics influence flood risk for people and property along the River Exe?

This question is too broad. It needs to be broken down into sub-questions that are simpler and more workable. For example:

1 What are the main characteristics of the river's channel in its upper course?

2 What places are most at risk from flooding along the River Exe?

3 What impacts does channel shape have on flood risk?

4 What impacts does valley shape and land use have on flood risk?

To complete the enquiry you must use primary data. You can use secondary data if you wish.

Primary data

Fieldwork data which you collect yourself (or as part of a group) are called 'primary' data – first-hand information that comes from you and people you have worked with. There are many different types of primary data and information. You can find out more about these on page 308.

Secondary data

Secondary data are information that another person, group or organisation has collected. Secondary data are very important in providing background information and a context for the enquiry. It helps you to understand more about places and the kinds of questions that might be relevant. Past river flow data are one example of secondary data (see Figure **D**).

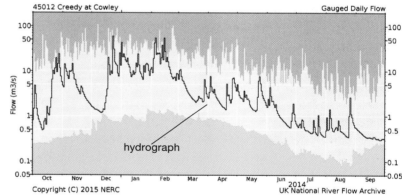

D *An annual hydrograph showing river data for the River Creedy, near Exeter (2013) – this information helps you get a context for the enquiry*

ACTIVITIES

1 Make a copy of this table, which refers to the stages shown in diagram **C**. In pairs, complete the table using each stage.

Stage	Your chosen response
(1) An enquiry question we could investigate for either photo **A** *or* photo **B**.	
(2) Three methods we could use to collect primary data.	
(3) Two sources of secondary data we could use.	
(4) Three ways in which we could present and process our data.	

2 Explain how the data in graph **D** could help you in your enquiry.

Practice questions

1 Study the OS map of the River Tees on page 127. Identify *one* question or aim that could be used to investigate physical processes in this river landscape. *(2 marks)*

2 For your own fieldwork enquiry, explain *two* reasons why particular aims or questions were developed. *(4 marks)*

On this spread, you'll find out about different techniques for collecting primary data in a rivers enquiry

Collecting primary data

The purpose of fieldwork is to collect your own data. The student in photo **A** is using a measuring instrument – a 'pebbleometer'– to find out the size of stones collected in a river. He is investigating river processes and wants to find out about stone size changes along the river.

Data are essential! They help you to understand what is happening in a drainage basin. You can also compare your river fieldwork to what textbooks tell you. It makes good teamwork.

It's very important to consider what data you need when you design your investigation. Any data collected should be as reliable and accurate as possible. In particular, you should think about:

A Measuring pebble size from a river using a homemade pebbleometer

◆ *Sample size* – How many measurements will you be taking, and why? More measurements – for example, measuring more stones – will generally produce more reliable data. But this is time-consuming, and group collection of data can save time.

◆ *Survey locations/sites* – Where will you collect the data and how? Will you collect data along a line (called a *transect*) and how far apart will these locations be?

◆ *Accuracy* – How can you ensure that your data are accurate? Will you need to measure several samples of stones and calculate an average?

Quantitative and qualitative data

Most data you'll collect on a river investigation involves numbers. There are two types of data – quantitative and qualitative. Whatever data you collect must link directly to the enquiry question that you have set yourself (see page 307).

Quantitative data

River studies include a number of quantitative fieldwork techniques. Table B shows data that are commonly collected in river investigations. All quantitative techniques need equipment, like the metre stick shown in photo **B**.

Sample size needs careful consideration. Three types of sampling are used in collecting quantitative data:

◆ *Random* – where samples are chosen at random, for example every pebble has an equal chance of being selected.

◆ *Systematic* – means working to a system to collect data, for example every 50 cm across the river.

◆ *Stratified* – deliberately introducing bias to ensure a sample addresses the question. For example, deliberately selecting samples of different pebble sizes from a point in the river so that the whole range of pebble sizes is included within the sample.

B Using a metre stick to measure the depth of a stream at intervals across its width

Data required	Equipment needed	Brief description and reasons for doing this
River gradient (in degrees)	Clinometer, tapes and ranging poles	The gradient is measured at sites along the river over length of 10 or 20 m. The gradient of a river can, for example, help us to understand more about the processes operating and the influence of geology.
River speed (velocity)	Flow meter or float	Measures how fast the river is flowing. This tells us about the amount of energy in a river and we can investigate whether it conforms to a model.
Pebble size in cm or mm	Ruler or pebbleometer/ calliper (photo **A**)	Measures the length of the long axis of a sample of stones. This helps to link processes of river erosion with position within a river catchment.

 Examples of quantitative data used in river investigations

Qualitative data

Qualitative data include a number of techniques that don't involve numbers or counting. They are subjective and involve the judgment of the person collecting the data. Techniques for collecting qualitative data with rivers areas include:

◆ written site descriptions

◆ taking photographs

◆ recording videos

◆ field sketches (shown in photo **D**).

These techniques can be used to record the use of fieldwork equipment as well as capturing river landscapes and management. When taking a photo, always think carefully about the frame of your picture. Where necessary use an object such as a coin for scale if you are taking close-up pictures of things such as river sediment.

Think about it

What should you consider before taking a photograph to support your fieldwork enquiry?

 A good field sketch. Remember to use annotations to explain the geography of the landscape around you.

Practice questions

1 For your physical geography enquiry, explain two ways that you collected quantitative fieldwork data. *(4 marks)*

2 Explain one way in which you attempted to make your data collection reliable. *(2 marks)*

ACTIVITIES

1 Explain the differences between quantitative and qualitative data.

2 Using the headings in table **C**, complete a table for the four types of qualitative information.

3 In pairs, draw up a table of the advantages and disadvantages of quantitative and qualitative data.

4 Look at the OS map of the River Tees on page 123.

   a Choose one location in which you could investigate river processes.

   b Outline three quantitative techniques that you would use to collect data.

   c For each technique, decide whether you would use random, stratified or systematic sampling to collect your data. Explain your reasons.

On this spread, you'll find out how to present your data using graphs, photos or maps

Pulling it all together

Managing and organising your data is very important. Make sure that you organise both your own and any group data that you have collected. A spreadsheet which you can complete and share is a good way of doing this.

You might find it helpful to note down how and why you selected your data, and how they link to the enquiry aim or focus.

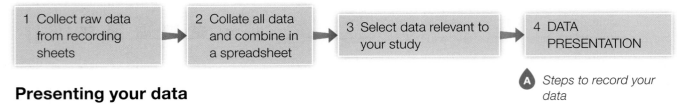

| 1 Collect raw data from recording sheets | → | 2 Collate all data and combine in a spreadsheet | → | 3 Select data relevant to your study | → | 4 DATA PRESENTATION |

A *Steps to record your data*

Presenting your data

When considering how best to present your data, think more widely than just bar charts, histograms and pie charts. Table **B** shows a range of other ways to present your information.

Maps	GIS and photos	Tables	Graphs and charts
• Used to show locations and patterns. • Mini-graphs and charts can be located on maps. • This makes it easier to compare patterns at specific locations. • Consider using isolines or choropleth maps.	• Used to show historic maps or sites which have been lost to erosion. • Useful for aerial shots of rivers to show land use. • Helps to show how places have changed after being affected by storms.	• Can be used to present raw data that you and your group collected. • Useful to highlight patterns and trends. • Can be highlighted and annotated, and can help to identify anomalies (any data which look unusual).	• There is a wide range of graphs and charts available. (Hint: make sure you choose the right chart, e.g. do you know when to use a pie chart or bar chart?) • Can show data and patterns clearly – easier to read than a table of data.

B *A range of data presentation techniques*

Consider which presentation technique is most appropriate for the data. For example, are you dealing with *continuous data* or *categories*? Are you dealing with numbers or percentages? How can you present your data spatially?

◆ *Continuous data* show change along a line of study or over a period of time. River gradients, for example, are continuous, so are best presented using a line graph. An example is shown in graph **C**, which has been created with GIS (ArcGIS Online).

◆ *Categories* show classifications – for example, measuring pebble size or long axes and grouping them into sizes. A bar chart would be the best method for presenting this information. An example is shown in graph **D**.

◆ Where your *sample sizes* are different (e.g. 15 pebbles at one location, 17 at another), turn raw numbers into percentages of different sizes. Then you should use a bar chart.

◆ Instead of just presenting graphs, locate them on a map or aerial photo (e.g. using Google Maps or GIS, see map **E**). This makes change easy to spot, and turns simple data into a geographical display.

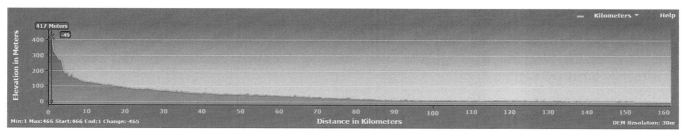

Think about it

What are the most appropriate techniques to present the data that you have collected?

You will find other techniques to use in presenting your fieldwork data. For example:

◆ *annotated photographs* show evidence of river processes, e.g. erosion on the outer edge of a meander

◆ *field sketches* highlight the way in which people and property are vulnerable to river flooding.

GIS is another good way of presenting information since it allows you to overlay different types of data (as in map **E**) as well as begin to do some analysis of more complex data. GIS has a number of geo-processing tools that allow you to create specialised maps, as well as look for patterns and relationships.

C *A line graph showing a river valley long profile helps you to understand the drainage basin*

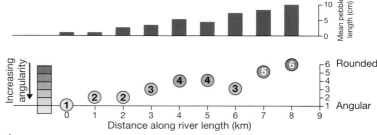

D *These river pebble data show pebble length (top graph) and shape (bottom graph). This is category data (where there are gaps between plot points). The data are plotted on two graphs, one above the other, to aid comparison.*

E *The data are plotted on a simple base map to aid comparison between the different sites*

Practice questions

1 Explain *one* advantage of using a line graph to show the long profile of a river. *(2 marks)*

2 Describe *one* technique that you used to present your river sediment data. *(3 marks)*

ACTIVITIES

1 In pairs, list some examples of when you should use the following: bar charts, pie charts, line graphs and histograms.

2 In pairs, research and identify *two* ways in which you could use GIS to:

   a research river processes

   b present fieldwork data collected along the River Tees (pages 122–3).

3 Study the data in graphs **C** and **D**. Explain the reasons why the methods used to present the data in each graph are **a** effective, and **b** the most mathematically accurate ways of showing the data.

4 Look at the data in graphs **C** and **D**.

   a Identify any anomalies which do not seem to fit the general pattern.

   b Explain possible reasons for these.

On this spread you'll find out about how to analyse and draw conclusions

What is analysis?

To analyse your data, you need to:

◆ identify patterns and trends in your results, and describe them

◆ make links between different sets of data – for example, how sediment size and roundness seem to change at the same time

◆ identify *anomalies* – unusual data which do not fit the general pattern of results

◆ explain reasons for patterns you are sure about – for example, data that might show a process operating along a river, such as deposition

◆ suggest possible reasons for patterns you are unsure about – for example, why results suddenly change in a way that you can't explain.

Cause and effect	Emphasis	Explaining	Suggesting
as a result of…	above all…	this shows…	could be caused by…
this results in…	mainly…	because…	this looks like…
triggering this…	mostly…	similarly…	points towards…
consequently…	most significantly…	therefore…	tentatively…
the effect of this is…	usually…	as a result of…	the evidence shows…

A *The language of analysis – these words and shorts phrases are useful to use in analysis*

Writing your analysis

When you write your analysis you should have a clear and logical format. Start with an introductory statement and then write about each point in more detail. Good analysis writing also:

◆ uses the correct geographical terminology

◆ uses the past tense

◆ is written in the third person

◆ avoids the use of 'I' or 'we'.

Table **A** has some useful phrases you can use, depending on your results.

Analysing data

You need to be able to use both quantitative and qualitative techniques when analysing your data.

Using quantitative techniques

Quantitative techniques are about handling numerical data from different sites, like that shown in table **B**. These can be analysed using statistical techniques – for example, you should be able to calculate the mean (the average of the values in the data).

Site A – upstream section	Site B – middle section	Site C – downstream section
95	24	10
68	19	12
48	16	64
49	15	32
90	29	34
82	18	55
86	6	37
56	10	18
80	19	19
49	20	19
69	13	12
42	9	8
68	15	63
57	18	62
70	19	15
59	21	9

 B *River sediment data from three sites A, B and C, measuring the long axis in mm*

You can also use a dispersion diagram to help you find the following values (see pages 342–4 for more about these values):

◆ *Median* – to find the median you need to order the data (like the dispersion diagram) and then find the middle value. This divides the data set into two halves.

◆ *Mode* – the number that appears most frequently in a data set.

◆ *Range* – the difference between the highest and lowest values.

◆ *Quartiles* – dividing a list of numbers into four equal groups – two above and two below the median.

Using qualitative techniques

Just like sketches, photos are far more than mere space-fillers. They are examples of qualitative techniques that provide vital clues and evidence about the fieldwork experience, as well as help with the analysis of information. Photo **C** is an example that could be analysed using annotations. Remember – annotations are good for explaining and can also be numbered to show a sequence. Field sketches can also be used to analyse processes as well as change over time.

Writing a good conclusion

A conclusion is almost the end point in your enquiry, with several important features. It is shorter than the analysis, because it is more focused. Remember the following points when writing your conclusion.

◆ Refer back to the main aim of your investigation (see page 307). What did you find out? Make sure you answer the question!

◆ State the most important data that support your conclusion – both primary and secondary.

◆ Is your hypothesis supported/substantiated by the data, or does it need to be modified in the light of your evidence?

◆ Comment on any anomalies and any unexpected results.

◆ Comment on the wider geographical significance of your study. Think about why it might be important, whether your results could be useful to others, or whether you think all rivers are like the one you have studied.

c *A section of river with many physical features*

Practice question

Explain *two* ways in which you analysed your physical geography data. *(6 marks)*

ACTIVITIES

1 Using table **B**, calculate the mean sediment size for each of the sites, A, B and C.

2 **a** Draw a dispersion diagram for sites B and C (see pages 342–4).

   b Use your dispersion diagrams for sites B and C to calculate the median, mode and range for each site.

   c Divide each diagram into quartiles. Calculate the mean of the upper quartile and of the lower quartile.

3 Which site has **a** the largest sediment overall, **b** the smallest sediment, and **c** the most variation in size?

4 Using the data in table **B**, explain the overall changes in sediment sizes from upstream to downstream.

5 Are there any anomalies in the river data across the three sites? What possible suggestions could there be for this?

6 Draw an annotated field sketch of photo **C**.

On this spread you'll find out how to evaluate, reflect and think critically about the different parts of the enquiry process

The importance of the evaluation

The evaluation is the last part of the enquiry process. It is much more than just a list of things that might have gone wrong. It is the part of the enquiry which aims to both evaluate and reflect on:

◆ the process of collecting data

◆ the overall quality of the results and conclusion.

Many approaches to fieldwork and research have limitations and errors which can affect the results. It is very important to remember that no study is perfect – even those carried out by university academics and professionals! It is sensible to highlight where you think the shortcomings of your work might be (figure **A**).

What might have affected your results?

Reliability is the extent to which your investigation has produced consistent results. In other words, if you were to repeat the enquiry, would you get the same results? Are your results valid – that is, did your enquiry produce an outcome that you can trust?

Several factors can influence the reliability, validity and therefore the overall quality of your enquiry, as shown in table **B**.

1 What part of my fieldwork design caused errors to be introduced?

2 How might the problems introduced affect the reliability and validity of outcomes?

3 How do my results help me reflect more about geographical knowledge gained?

 A Some key questions to ask in an evaluation

Think about it

If I were to repeat my investigation, what would I do to ensure that I obtained more reliable results?

Sources of error	Impacts on quality
Sample size	Smaller sample sizes usually means lower quality data.
Frequency of sample	Fewer sites reduces frequency, which then reduces quality.
Type of sampling	Sampling approaches may create 'gaps' and introduce bias in the results.
Equipment used	The wrong / inaccurate equipment can affect overall quality by producing incorrect results.
Time of survey	Different times of the year will significantly influence the amount of water in the river and may not be representative – may create problems with measurement.
Location of survey	Big variations in river channel depth and width, as well as sediment characteristics, can occur in locations close to each other.
Quality of secondary data	Age and reliability of secondary data affect their overall quality.

B *Possible sources of error in a geographical enquiry*

Being critical of your work

Being critical means thinking about any errors that may have affected the results of your enquiry. In a geographical enquiry researchers generally try to produce the most reliable outcomes possible. They accept the limitations of their study. They may consider the following questions as part of their evaluation:

◆ How much do I trust the overall patterns and trends in my results?

◆ What is the chance that these outcomes could have been generated randomly (or by chance)?

◆ Which of my conclusions are the most reliable compared to other conclusions?

◆ Which part of my enquiry produced the most unreliable results?

Questions like these, although complex, can be useful when writing an evaluation.

How do your results affect your understanding of river processes?

Another important part of the evaluation process is to link any knowledge you gained from the enquiry back to a theoretical model or idea. It may be helpful to think about the key factors (photo C) and then try to develop your own model.

C ▸ *Factors that may influence river processes*

Nature of river course – straight or meandering?

Length of river course?

Rock type and rock structure?

Local settlements and changes of land use?

Changes in gradient or slope profile?

Local flood management?

Practice questions

1 Explain *one* factor about your own primary data which could have affected your results. *(3 marks)*

2 Evaluate the reliability of your river fieldwork conclusions. *(8 marks)*

ACTIVITIES

1 Explain why an evaluation is the last part of the enquiry process.

2 Explain why an evaluation of one enquiry can be helpful in planning a second enquiry.

3 a In pairs, list the factors in table **B** in order of importance, based on your own fieldwork

b Explain your final rank order.

c Add an extra column to table **B** with the title 'The main sources of error in our enquiry', and add points about your own fieldwork.

4 For either the River Tees (page 123) or your own rivers fieldwork, use photo **C** to add annotations to explain how these factors affect the river.

In this section you'll learn how to investigate differences in urban quality of life using fieldwork and research

Enquiry questions

Urban areas are popular for fieldwork as they are familiar and often close to us. Photos **A** and **B** shows two very different urban locations. Think about the kind of questions geographers might ask when they see places like this. For example, they might ask:

◆ Which location offers a higher quality of life?

◆ Why are they so different?

◆ What might happen to each place in future, and why?

Many urban locations are good places for fieldwork since there are lots of questions like these to investigate. This is the starting point for any enquiry. An enquiry is a series of stages that start with a question (diagram C, stage 1) and end up with an answer or conclusion (stage 5). You will probably have completed an enquiry in geography (or science) before and have used fieldwork and practical work in the same way.

Each stage is equally important, right from the initial question, to the research and context, through to the overall evaluation. Only at the end can you have an opportunity to reflect on what you have found and what it means.

A *A street in central Sheffield*

B *An area of Birmingham*

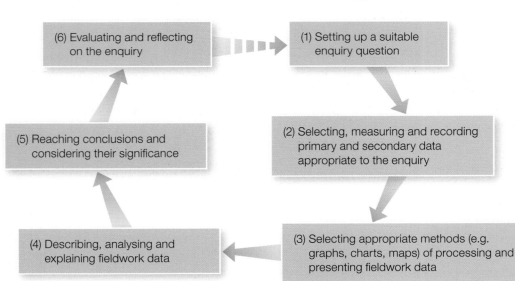

(6) Evaluating and reflecting on the enquiry

(1) Setting up a suitable enquiry question

(2) Selecting, measuring and recording primary and secondary data appropriate to the enquiry

(3) Selecting appropriate methods (e.g. graphs, charts, maps) of processing and presenting fieldwork data

(4) Describing, analysing and explaining fieldwork data

(5) Reaching conclusions and considering their significance

Enquiry means the process of investigation to find an answer to a question.

Fieldwork means work carried out in the outdoors.

C *The route to enquiry – a planning 'pathway' for your investigation*

Developing an enquiry question

A good enquiry depends upon having a good question. A good question must be directly linked to the overall theme – in this case, urban environments! An example is:

How and why are there variations in quality of life for different census output areas within Leeds?

This question is too broad. It needs to be broken down into sub-questions that are simpler and more workable. For example:

1 What are the challenges facing some areas of Leeds?

2 What is the primary evidence of differences in quality of life between areas?

3 Does environmental quality affect urban quality of life?

4 What impacts have these differences had on communities?

To complete the enquiry you must use primary data. You can use secondary data if you wish.

Primary data

Fieldwork data which you collect yourself (or as part of a group) are called 'primary' data – first-hand information that comes from you and people you have worked with. There are many different types of primary data and information. You can find out more about these on page 318.

Secondary data

Secondary data are information that another person, group or organisation has collected. Secondary data are very important in providing background information and a context for the enquiry. It helps you to understand more about places and the kinds of questions that might be relevant. The government's Index of Multiple Deprivation (IMD) is one example of secondary data because it is compiled using census data (chart **D**).

D *Different weightings given to the elements of the IMD. This information helps you get a context for the enquiry.*

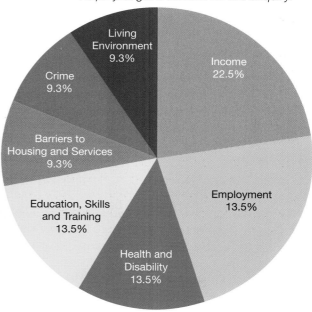

ACTIVITIES

1 Make a copy of this table, which refers to the stages shown in diagram **C**. In pairs, complete the table using each stage.

Stage	Your chosen response
(1) An enquiry question we could investigate for either photo **A** or photo **B**.	
(2) Three methods we could use to collect primary data.	
(3) Two sources of secondary data we could use.	
(4) Three ways in which we could present and process our data.	

2 Explain how IMD data (chart **D**) could help you in your enquiry.

3 Design *three* enquiry questions that could be investigated in an urban area like Bristol or Rio de Janeiro.

Practice questions

1 Identify *one* question or aim that could be used to investigate variations in quality of life. *(2 marks)*

2 For your chosen enquiry, give *two* reasons why particular aims or questions were developed. *(4 marks)*

On this spread you'll find out about different techniques for collecting primary data in an urban enquiry

Enquiry design

The purpose of fieldwork is to collect your own data. The student in photo A is using a camera to record the buildings and architecture in a coastal resort. She is investigating urban deprivation and wants to use photographic evidence to document the change between different parts of the town.

Data are essential! They help you to understand what is happening in an urban area. You can also compare your fieldwork in a town or city to what textbooks tell you. It makes good teamwork.

A *Using a camera to record the built environment at the coast*

It's very important to consider what data you need when you design your investigation, so that any data collected are as reliable and accurate as possible. In particular you should think about:

- *Sample size* – How many measurements will you be taking, and why? More measurements – for example, more questionnaires – will generally get more reliable data. But doing so takes time. This is where group collection of data helps.

- *Survey locations/sites* – Where will you collect the data and how? Will you collect data along a line (called a *transect*) and how far apart will these locations be?

- *Accuracy* – How can you ensure that your data are accurate? Will you need to complete more questionnaires and calculate an average, median or mode?

Data required	Equipment needed	Brief description and reasons for doing this
Environmental quality survey	Environmental quality survey (**C**)	Measures different characteristics of a place based on personal judgements, with a simple scoring system or using a bi-polar chart.
Annotated photos and sketches	Camera or phone and paper for a sketch	Take photos or draw sketches and annotate them to identify features and characteristics that are pertinent to your study

B *Examples of quantitative data used in urban investigations*

Quantitative and qualitative data

There are two types of primary data - quantitative and qualitative. You could use both types in an urban enquiry. Whatever data you collect must link directly to the enquiry question that you have set yourself (page 317).

Quantitative data

Urban studies include a number of quantitative fieldwork techniques. Table B shows data that are commonly collected in urban investigations. Most quantitative techniques need equipment or recording sheets (example **C**).

Sample size needs careful consideration. Three types of sampling are used in collecting quantitative data:

- *Random* – where samples are chosen at random so that every person in a questionnaire survey has an equal chance of being selected.

- *Systematic* – means working to a system to collect data, for example, every 20 metres or paces along a road to record land use.

- *Stratified* – deliberately introducing bias to ensure a sample addresses the question. For example, deliberately selecting samples of different people within the town or city so that the whole range of people is included in the sample.

Think about it

What should you consider before taking a photograph to support your fieldwork enquiry?

Qualities being assessed		High + 2	Good +1	Average 0	Fairly poor –1	Very poor – 2	
Building design and quality	1 Well designed / pleasing to the eye						Poorly designed / ugly
	2 In good condition – e.g. paintwork, woodwork						In poor condition
	3 Houses well maintained or improved						Poorly maintained / no improvement
	4 Outside, gardens are kept tidy / in good condition						Outside gardens, or land / open space in poor condition
	5 No vandalism, or any graffiti has been cleaned up						Extensive vandalism or graffiti in large amounts
Traffic noise and parking	6 Roads have no traffic congestion						Streets badly congested with traffic
	7 Parking is easy; garages or spaces provided						Parking is difficult; no parking provided / on the street
	8 No road traffic, rail or aircraft noise						High noise volume from road, rail, and air traffic
	9 No smell from traffic or other pollution						Obvious smell from traffic or other pollution

C *An example of an environmental quality survey that could be used in an urban area*

Qualitative techniques

Qualitative data include a number of techniques that don't involve numbers or counting. They are subjective and involve the judgment of the person collecting the data. Techniques for collecting qualitative data in urban areas include:

◆ written site descriptions

◆ taking photographs and videos

◆ field sketches

◆ interviews.

These techniques can be used to record use of fieldwork equipment as well as capture different aspects of the urban environment. Always think carefully about the frame of your photo. In urban areas, 360° panoramas work well to illustrate contrasts between areas.

Practice questions

1 For your chosen enquiry, explain *two* ways that you collected quantitative fieldwork data. *(4 marks)*

2 Explain *one* way in which you attempted to make your data collection reliable. *(2 marks)*

ACTIVITIES

1 Explain the differences between quantitative and qualitative data.

2 Using the headings in table **B**, complete a table for the four types of *qualitative* information.

3 In pairs, draw up a table of the advantages and disadvantages of quantitative and qualitative data.

4 Use an OS map of your local area.

   a Choose one location in which you could investigate issues of urban deprivation.

   b Justify your choice of location.

   c Outline *three* quantitative techniques that you would use to collect data there.

   d For each technique decide whether you would use random, stratified or systematic sampling to collect your data. Explain your reasons.

On this spread, you'll find out how to present your data using graphs, photos or maps

Pulling it all together

Managing and organising your data is very important. Make sure that you organise both your own and any group data that you have collected. A spreadsheet which you can complete and share is a good way of doing this.

You might find it helpful to note down how and why you selected your individual and group data, and how they link to the enquiry aim or focus.

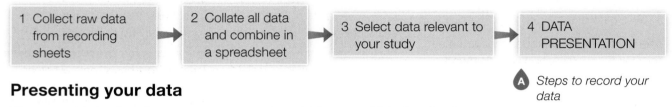

1 Collect raw data from recording sheets → 2 Collate all data and combine in a spreadsheet → 3 Select data relevant to your study → 4 DATA PRESENTATION

A Steps to record your data

Presenting your data

When considering how best to present your data, think more widely than just bar charts, histograms and pie charts. Table **B** shows a range of other ways to present your information.

Maps	GIS and photos	Tables	Graphs and charts
• Used to show locations and patterns. • Mini-graphs and charts can be located on maps. • This makes it easier to compare patterns at locations. • Consider isolines or choropleth maps.	• Used to show historic maps to show change in an urban area. • Useful for aerial photos of the town / city to show land use. • Helps to show deprivation and / or 'health' of a place.	• Can be used to present raw data that you and your group collected. • Useful to highlight patterns and trends. • Can be highlighted and annotated, and can help to identify anomalies (any data which look unusual).	• There is a wide range of graphs and charts available. (Hint: make sure you choose the right chart, e.g. do you know when to use a pie chart or bar chart?) • Can show data and patterns clearly – easier to read than a table of data.

Consider which presentation technique is most appropriate for the data. For example, are you dealing with *continuous data* or *categories*? Are you dealing with numbers or percentages? How can you present your data spatially?

B A range of data presentation techniques

◆ *Continuous data* show change along a line of study or over a period of time. Pedestrian flows might be continuous, for example, so are best presented using a line graph.

◆ *Categories* show classifications – for example, putting environmental quality scores into classified groups (circles have been used on map **C**).

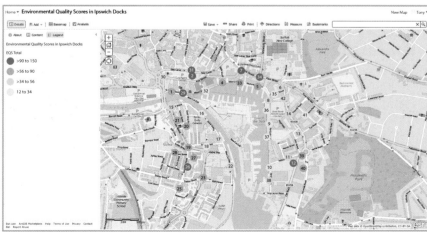

C Locations in Ipswich on a GIS map where EQA (Environmental Quality Assessment) data have been displayed as circles. Circles are colour coded with a 'colour-ramp' (values are in the centre).

◆ Where your *sample sizes* are different (e.g. 15 quality scores at one location, 17 at another), turn raw numbers into percentages of different sizes. Then you should use a pie chart.

◆ Instead of presenting graphs individually, locate them on a map or aerial photo (e.g. using Google Maps or GIS (Geographical Information Systems) (map **C**)). This makes differences easy to spot, and turns simple data into a geographical display.

There are other techniques that you may find helpful in presenting your fieldwork data. For example:

◆ *annotated photographs* show evidence of dereliction and decay, possibly indicating a lower quality of life

◆ *field sketches* highlight the way in which people and property are influenced by areas of changing environmental quality.

We have already seen that GIS is another good way of presenting information since it allows comparisons (as on map **C**). It also allows more sophisticated presentation tools to be used, such as digital choropleth maps. GIS also has a number of geo-processing tools that allow you to create specialised maps, as well as look for patterns and relationships.

Practice questions

1 Give reasons for the choice of data presentation techniques in your enquiry. *(2 marks)*

2 Describe *one* technique that you used to present your secondary IMD data. *(3 marks)*

Respondent: 65-year-old retired person, discussing changes in their local town.

'I have lived in Tiverton, Devon all my life and there has been a lot of change. For a start, many of the smaller local shops have gone; the big supermarket near the river was to blame. I'm less mobile than I was and there are also fewer bus services so I need to reply on my car (but parking is free for a short period in the town). But the town has been improved I think. It's better for people as they have stopped cars driving through the middle like they used to, plus I like the coffee shops where I can relax and they have outside seating. I don't like the fact that there are fewer banks and book shops but that's probably just an age thing. I don't really do internet shopping!'

D *A coding technique is useful for a variety of text-based data, whether primary or secondary information. Colour highlighting can be used to show positive (yellow) and negative (blue) comments. This example helps to analyse results from a questionnaire about attitudes to quality of life.*

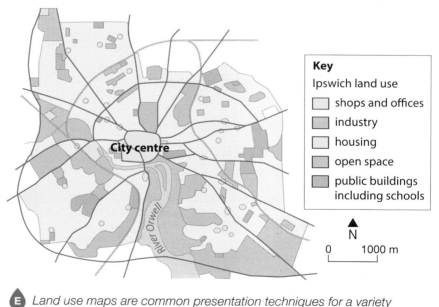

Key
Ipswich land use
☐ shops and offices
☐ industry
☐ housing
☐ open space
☐ public buildings including schools

E *Land use maps are common presentation techniques for a variety of urban studies and can be linked to urban quality of life.*

ACTIVITIES

1 In pairs, explain the differences between bar charts, pie charts, line graphs and histograms.

2 In pairs, research and identify *two* ways in which you could use GIS to **a** research urban quality of life, **b** present fieldwork data.

3 Choose *two* ways of presenting your urban fieldwork data. In pairs, make a table of the advantages and disadvantages of each of these methods.

4 Study collated data from your class fieldwork. How far are there any anomalies or data which do not seem to fit the general pattern?

What is analysis?

To analyse, you need to:

◆ identify patterns and trends in your results, and describe them

◆ make links between different sets of data – for example, how sediment size and roundness seem to change at the same time

◆ identify *anomalies* – unusual data which do not fit the general pattern of results

◆ explain reasons for patterns you are sure about – for example, data that might show a process operating in a town or city, such as spatial change in land use

◆ suggest possible reasons for patterns you are unsure about – for example, why results suddenly change in a way that you can't explain.

Cause and effect	Emphasis	Explaining	Suggesting
as a result of…	above all…	this shows…	could be caused by…
this results in…	mainly…	because…	this looks like…
triggering this…	mostly…	similarly…	points towards…
consequently…	most significantly…	therefore…	tentatively…
the effect of this is…	usually…	as a result of…	the evidence shows…

 The language of analysis – these words and shorts phrases are useful to use in analysis

Writing your analysis

When you write your analysis you should have a clear and logical format. Start with an introductory statement and then write about each point in more detail. Good analysis writing also:

◆ uses the correct geographical terminology

◆ uses the past tense

◆ is written in the third person

◆ avoids the use of 'I' or 'we'.

Table **A** has some useful phrases you can use, depending on your results.

Analysing data

You need to be able to use both quantitative and qualitative techniques when analysing your data.

Using quantitative techniques

Quantitative techniques are about handling numerical data from different sites, like that shown in table **B**. These can be analysed using statistical techniques – for example, you should be able to calculate the *mean* (the average of the values in the data).

Site A	Site B	Site C
95	24	10
68	19	12
48	16	64
49	15	32
90	29	34
82	18	55
86	6	37
56	10	18
80	19	19
49	20	19
69	13	12
42	9	8
68	15	63
57	18	62
70	19	15
59	21	9

 Totalled environmental quality scores (/100) for three different urban locations: A, B and C. A variety of different indicators have been measured in each location; higher scores indicate a better environmental quality.

You can also use a dispersion diagram to help you find the following values (see pages 344–5 for more about these values):

◆ *Median* – to find the median you need to order the data (like the dispersion diagram) and then find the middle value. This divides the data set into two halves.

◆ *Mode* – the number that appears most frequently in a data set.

◆ *Range* – the difference between the highest and lowest values.

◆ *Quartiles* – dividing a list of numbers into four equal groups – two above and two below the median. You could use quintiles (five groups) as well.

Using qualitative techniques

Just like sketches, photos are far more than mere space-fillers. They are examples of qualitative techniques that provide vital clues and evidence about the fieldwork experience, as well as help with the analysis of information. Photo **C** is an example that could be analysed using annotations. Remember – annotations are good for explaining and can also be numbered to show a sequence. Field sketches can also be used to analyse processes as well as change over time.

Writing a good conclusion

A conclusion is almost the end point in your enquiry, with several important features. It is shorter than the analysis, because it is more focused. Remember the following points when writing your conclusion.

◆ Refer back to the main aim of your investigation (see page 317). What did you find out? Make sure you answer the question!

◆ State the most important data that supports your conclusion – both primary and secondary.

◆ Comment on any anomalies and any unexpected results.

◆ Comment on the wider geographical significance of your study. Think about why it might be important, whether your results could be useful to others, or whether you think all urban locations are like the one you have studied.

C *Photo of urban environment showing problems typical for some town centres*

Practice question

Explain *two* ways in which you analysed your fieldwork data. *(6 marks)*

ACTIVITIES

1 Draw an annotated field sketch of photo **C**.

2 Use table **B** to calculate the mean environmental quality survey (EQS) scores for each of the locations, A, B and C.

3 **a** Draw a dispersion diagram for sites B and C.

   b Use the diagrams to calculate the mean, median, mode and range for each site.

   c Divide each diagram into quartiles. Calculate the mean of the upper quartile and of the lower quartile.

4 Which site has **a** the highest overall EQS score, **b** the smallest EQS score, and **c** the most variation?

5 Using the data in table **B**, suggest possible reasons for the variations in EQS between locations.

6 Are there any anomalies in the EQS data between the three locations? What possible explanations could there be for this?

On this spread you'll find out how to evaluate, reflect and think critically about the different parts of the enquiry process

The importance of the evaluation

The evaluation is the last part of the enquiry process. It is much more than just a list of things that might have gone wrong. It is the part of the enquiry which aims to both evaluate and reflect on:

◆ the process of collecting data

◆ the overall quality of the results and conclusion.

Many approaches to fieldwork and research have limitations and errors which can affect the results. It is very important to remember that no study is perfect – even those carried out by university academics and professionals! It is sensible to highlight where you think the shortcomings of your work might be (figure **A**).

What might have affected your results?

Reliability is the extent to which your investigation has produced consistent results. In other words, if you were to repeat the enquiry, would you get the same results? Are your results valid – that is, did your enquiry produce an outcome that you can trust?

Several factors can influence the reliability, validity and therefore the overall quality of your enquiry, as shown in table **B**.

> 1 What part of my fieldwork design caused errors to be introduced?

> 2 How might the problems introduced affect the reliability and validity of outcomes?

> 3 How do my results help me reflect more about geographical knowledge gained?

 Some key questions to ask in an evaluation

Think about it

If I were to repeat my investigation, what would I do to ensure that I obtained more reliable results?

Sources of error	Impacts on quality
Sample size	Smaller sample sizes usually means lower quality data.
Frequency of sample	Fewer sites reduces frequency, which then reduces quality.
Type of sampling	Sampling approaches may create 'gaps' and introduce bias in the results.
Equipment used	The wrong / inaccurate equipment can affect overall quality by producing incorrect results.
Time of survey	Different days or times of day might influence perceptions and pedestrian floes, for example.
Location of survey	Big variations in environmental quality can occur between places very close to each other.
Quality of secondary data	Age and reliability of secondary data affect their overall quality.

 Possible sources of error in a geographical enquiry

Being critical of your work

Being critical means thinking about any errors that may have affected the results of your enquiry. In a geographical enquiry researchers generally try to produce the most reliable outcomes possible. They accept the limitations of their study. They may consider the following questions as part of their evaluation:

◆ How much do I trust the overall patterns and trends in my results?

◆ What is the chance that these outcomes could have been generated randomly (or by chance)?

◆ Which of my conclusions are the most reliable compared to other conclusions?

◆ Which part of my enquiry produced the most unreliable results?

Questions like these, although complex, can be useful when writing an evaluation.

How do your results affect your understanding of urban deprivation?

Another important part of the evaluation process is to link any knowledge you gained from the enquiry back to a theoretical model or idea. It may be helpful to think about the key factors (photo **C**) and then try to develop your own model.

C *Factors that may influence urban deprivation and quality of life*

Access to services?

Ethnicity?

Availability of good employment?

Access to employment?

Quality of green spaces?

Availability of affordable housing?

Practice questions

1 Explain *one* factor about your own primary data which could have affected your results. *(3 marks)*

2 Evaluate the reliability of your fieldwork conclusions. *(8 marks)*

ACTIVITIES

1 Explain why an evaluation is the last part of the enquiry process.

2 Explain why an evaluation of one enquiry can be helpful in planning a second enquiry.

3 a In pairs, list in order of importance the factors that may be possible sources of error in your own fieldwork.

 b Explain your final rank order.

c Create a table similar to table **B** with the title 'The main sources of error in our enquiry', and add points about your own fieldwork.

4 For your own fieldwork, use photo **C** to add annotations to explain how these factors influence quality of life.

In this unit you will find out how to use and interpret the information on different kinds of maps and photos.

Atlas maps

Latitude and longitude

Any place on the Earth's surface can be located by its latitude and longitude (map **A**).

◆ *Lines of latitude* run *parallel* to the Equator. This divides the world into the northern and southern hemispheres. Latitude increases north and south of the Equator to reach 90° at the north and south poles.

◆ *Lines of longitude* run between the north and south poles. The Prime Meridian, 0° longitude, passes through Greenwich in London. Values are given east and west of this line.

Using atlas maps

Atlas maps are useful sources of information for geographers.

◆ Basic maps of countries and regions of the world show physical relief, settlements and political information.

◆ Thematic maps show factors like climate, vegetation, population, tourism and tectonics.

◆ Maps can show global issues such as pollution, global warming, desertification and poverty.

◆ Atlases also include tables of statistics and useful data.

Different thematic maps can be used to find links between patterns, such as those between physical and human factors.

> ### Remember!
> Both latitude and longitude are measured in degrees (using the symbol °). Each degree is subdivided into 60 minutes (using the symbol '). So, '30 minutes' (30') equates to half a degree.
>
> The location of a place is expressed as follows:
>
> Manchester: 53° 30' N 2° 15' W
>
> On map **A**, Mumbai is located at 18° 56' N 72° 51' E

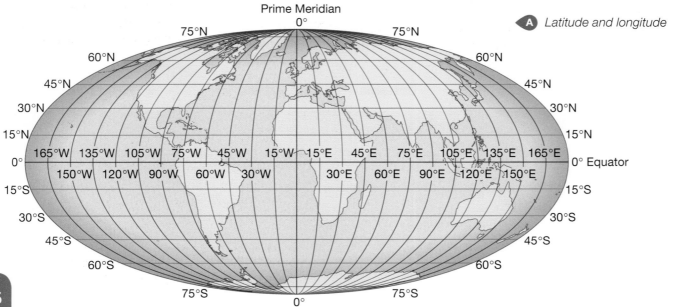

A *Latitude and longitude*

You will make use of a range of atlas resources while studying Geography and you may be asked about an atlas map in your exam. You may be asked to identify patterns or distribution on maps, so make sure you practise these skills.

◆ A *pattern* means there is some regularity or connection between things, for example population concentrated along the coast or industry concentrated along a river valley. Terms such as *radial* (spreading outwards from a central point) or *linear* can be used to describe a pattern. For example, earthquakes tend to form a linear pattern in the North Atlantic by following the North American-Eurasian plate margin.

◆ *Distribution* is a term used more broadly to describe where things are. There may or may not be a regular pattern. For example, the distribution of population in Kenya shows that most people live in the highlands where the climate is less extreme and soils are good for farming. Fewer people live on the lower ground in eastern Kenya because it is very hot and dry.

Atlas maps include a range of physical and human features such as:

◆ population distribution

◆ population movements

◆ transport networks

◆ settlement layout

◆ relief

◆ drainage.

Practice identifying and describing these features on an atlas map. You may be asked to make use of two different maps to consider links and inter-relationships between physical and human factors, for example between population distribution and relief in Kenya.

Ordnance Survey maps

You need to be confident using OS maps at a range of different scales, including 1:50 000 and 1:25 000. You should be able to identify and describe both physical and human features and to use key mapwork skills.

Four-figure and six-figure grid references

Ordnance Survey maps have numbered gridlines drawn on them. The lines that run up and down and increase in value from left to right (west to east) are called *eastings*. Those that run across the map, and increase in value from bottom to top (south to north), are called *northings* (diagram **B**).

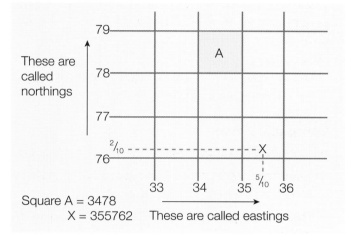

Square A = 3478
X = 355762 These are called eastings

B *How to use grid references*

To locate a point rather than a square, each square is split into tenths to give a six-figure reference. Look at diagram **B** and notice that point X is at grid reference 355762. Notice how the eastings value is represented by the three digits 355 and the northings value is represented by the digits 762. It is the third digit of each set that gives the 'tenths' value. Thus the eastings value is 35 and 5/10ths and the northings value is 76 and 2/10ths.

To locate a grid square on a map, we use a four-figure reference:

◆ the first two digits refer to the *easting* value

◆ the second two digits give the *northing* value.

Remember to give the *eastings* value *first* and then the northings. Think of the phrase 'along the corridor and up the stairs'!

For example: The four-figure reference for grid square A on diagram **B** is 3478; grid square A is the square after the values 34 and 78.

Scale

OS maps are drawn to scale. This means they are an accurate representation of the real world, but reduced to fit onto a sheet of paper! Scale can be shown using a *ratio* or a *linear scale*.

◆ A scale of 1:25 000 means that 1 unit on the map equals 25 000 units on the ground.

◆ 1:50 000 means that 1 unit on the map equals 50 000 units on the ground.

Distance

Distance can be measured as a 'straight-line' or 'curved' distance (for example, along a road or a river). You need to be able to measure both straight and curved distances.

Straight-line distance

Every map has a scale, usually in the form of a linear scale – a straight line with distances written alongside. To calculate a straight-line distance, simply measure the distance on the map between any two points, using a ruler or the straight edge of a piece of paper. Then line up your ruler or paper alongside the linear scale to find out the actual distance on the ground in kilometres (diagram **C**).

Curved distance

A curved distance takes longer to work out. Use the straight edge of a piece of paper to mark off sections of the curved line, converting the curved distance into a straight-line distance (diagram **C**).

Remember!

On a 1:25 000 map, 1 cm equals 25 000 cm on the ground, or 250 metres. 1 km on the ground equals 4 cm on the map.

On a 1:50 000 map, 1 cm on the map represents 0.5 km and 2 cm equals 1 km.

The distance between gridlines on any OS map is 1 km. On a 1:25 000 map the gridlines are 4 cm apart. On a 1:50 000 map they are just 2 cm apart.

 Measuring distance

Straight-line distance

1 Use a ruler to measure the distance between two places on the map, in centimetres.

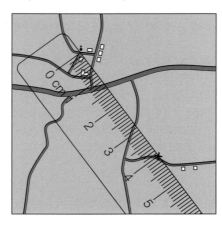

2 Measure out the distance on the map's linear scale to discover the distance on the ground in kilometres.

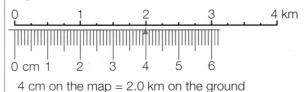

4 cm on the map = 2.0 km on the ground

Curved-line distance

1 Place the straight edge of a piece of paper along the route to be measured. Mark the start with the letter S. Look along the paper and mark off the point where the route moves away from the straight edge.

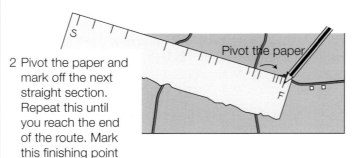

2 Pivot the paper and mark off the next straight section. Repeat this until you reach the end of the route. Mark this finishing point with the letter F.

3 Place the edge of the marked paper alongside the linear scale on the map and convert the total length to kilometres. Remember to always give the units when writing your answer!

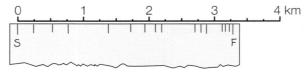

Compass directions

On most maps the direction 'north' is straight up – but not always! You should always check this in the key when using maps and diagrams, and make sure you understand the eight points of the compass (diagram **D**). Always use compass directions carefully and precisely, for example, 'Settlement X is to the north-west of Settlement Y'.

Identifying and describing landscape and relief features

Contour patterns on maps can be used to identify basic physical features, such as river valleys, ridges and plateaus (diagram **E**). Having identified these features, you need to be able to describe them by referring to size, shape, height and orientation (direction).

Example

'The ridge is about 2 km wide, is orientated roughly north-south and rises to a maximum height of 232 m at GR 376490. It has a steep eastern side and gentle western side.'

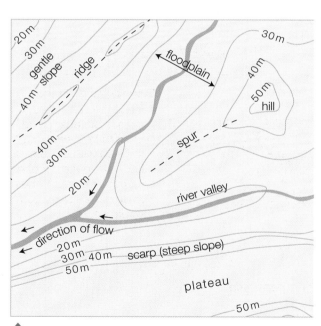
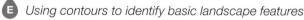

E *Using contours to identify basic landscape features*

Contours, spot heights and gradient

The height of the land is indicated by:

◆ *contours* – lines on the map (usually brown) joining points of equal height above sea level

◆ *spot heights* – usually indicated by black dots with a height above sea level written alongside.

D *The eight points of the compass*

Skills in context

You can find out more about identifying and describing relief features in Chapters 9–12. Turn to page 330 to see how cross-sections can be used to identify physical features.

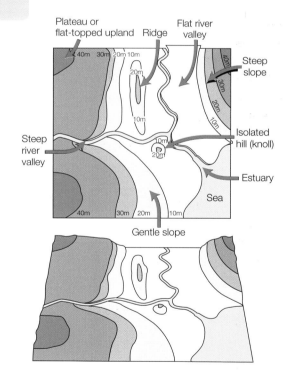

Skills in context

Interpreting cross-sections and transects of physical and human landscapes:

You can find out about how to interpret cross-sections and transects in Chapters 9–15.

The closer the contours, the steeper the gradient of the slope. Gradient can be calculated by measuring the change in height over a known distance:

1 Measure distance and height change using the same units (e.g. metres).

2 Divide the height change by the distance.

3 Express gradient as a percentage or ratio.

Example

Height change (H) of 20m over a distance (D) of 100m:

H/D = 20m/100m = 0.2

Ratio is 20% (or 1:5).

Numerical and statistical information

OS maps contain numerical and statistical information ranging from road numbers to values of height on contours and alongside spot heights. Grid references also provide numerical detail when locating a place. Be sure to use this information to add extra depth and detail to your map interpretation.

Drawing cross-sections

A cross-section is an imaginary 'slice' through a landscape. It helps to visualise what a landscape actually looks like. Make sure you can identify and label the main physical features of a landscape, for example, steep and gentle slopes, ridge, escarpment and valley. Drawing a cross-section is an important skill for a geographer. You need a piece of paper, a sharp pencil, a ruler and an eraser (diagram **F**). When you complete your section, check that you have:

◆ copied height values accurately

◆ made your vertical scale as realistic as possible (don't exaggerate it so much that you create a totally unreal landscape!)

◆ completed the section to both vertical axes by carrying on the trend of the landscape

◆ labelled any features

◆ labelled axes and given grid references for each end of your section

◆ given your section a title.

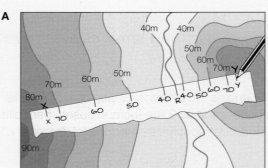

A

- Place the straight edge of a piece of paper along the chosen line of section.

- Mark the start and finish of your section.

- Mark contours and features, e.g. rivers.

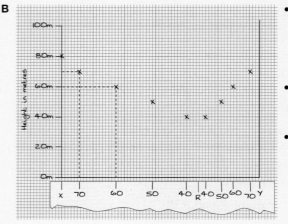

B

- Draw the axes of a graph, and choose an appropriate vertical scale.

- Lay your paper along the horizontal axis.

- Mark each contour value on the graph paper with a cross.

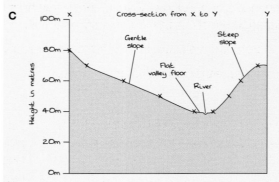

C

- Join the crosses with a freehand curve.

- Label any features.

- Give your cross-section a title.

 How to draw a cross-section

Interpreting physical and human features

Relief

Relief is the geographical term used to describe the height of the land and the different landscape features created by changes in height. When describing relief it is important to refer to simple landforms such as river valleys, hills and ridges. You should use adjectives to develop your description, for example: 'There is a *steep* river valley with *asymmetrical* (not the same on each side) valley sides'. To give a good answer you should comment on certain features:

◆ The height of the land, using actual figures taken from contours or spot heights to support your points. *Using words like 'high' and 'low' is meaningless without using actual figures.*

◆ The slope of the land. Is the land flat, or sloping? Which way do the slopes face? Are the slopes gentle or steep? Are there exposed, bare cliffs? *Remember to give precise information such as grid references and compass directions.*

◆ Features such as valleys or ridges. *Refer to names and grid references.*

Drainage

Drainage is the presence (or absence) and flow of water. When describing the drainage of an area, you should comment on:

◆ The presence or absence of rivers. Which way are they flowing? (Hint: look at the contours.) Are the rivers single or multi-channelled? *Give names of the rivers, and use distances, heights and directions to add depth to your description.*

◆ Drainage density – the total length of rivers in an area, usually expressed as 'km per km^2'. High drainage densities are typically found on impermeable rocks, whereas low densities suggest permeable rocks.

◆ The pattern of rivers (diagram **G**).

◆ The influence of people on drainage channels, for example straightening channels or building embankments. Straight channels are rare in nature and usually indicate human intervention.

◆ Evidence of underground drainage, in the form of springs or wells.

◆ The presence of lakes, either artificial or man-made.

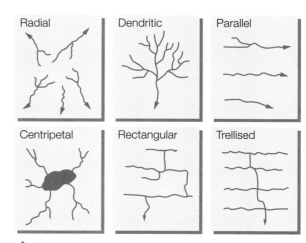

 *Drainage patterns*

Settlement

When describing patterns and types of settlement (diagram **H**) you should understand the following geographical terms:

◆ *Dispersed settlements* – low-density settlements spread out over a large area and typical of rural agricultural regions.

◆ *Nucleated settlements* – high-density settlements, tightly packed and often focused on a central point such as a major road intersection. The settlement typically spreads out in all directions.

◆ *Linear settlements* – these typically extend alongside a road, railway or canal. 'Linear' means 'line', so a linear settlement tends to be long and narrow.

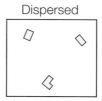
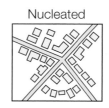
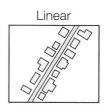

 Settlement types

Communication

Communication networks include many kinds of transport, such as:

◆ roads (of various types)
◆ railways and footpaths
◆ ferries (river and cross-Channel)
◆ airports
◆ cycleways.

You should be able to describe these networks, giving locational details such as length and orientation or compass direction and referring to patterns and density. For example, roads may radiate out from a settlement or form a series of concentric ring roads and by-passes around it.

Communication networks frequently reflect the relief of an area.

◆ Major transport arteries such as roads, canals and railways tend to follow flat, low ground, which explains why they are often located in river valleys.

◆ Footpaths often follow river valleys, as well as linking settlements and following ridge-lines or escarpments. Look out for named footpaths, such as the Pennine Way, and remember to refer to them by name when answering a question.

Land use

Land use refers to the way in which land is used or has been modified or managed by people. In writing a good answer about land use you should always refer to the map key and try to give specific examples from the map to support your statements. A typical land use map may contain information about the following land uses:

◆ different types of woodland (for example, coniferous or non-coniferous)

◆ coastal deposits (mud, sand or shingle)

◆ vegetation (for example, scrub, bracken or marsh)

◆ urban areas (be prepared to describe settlement patterns)

◆ fields (often just shown white on maps)

◆ quarries

◆ industrial areas

◆ tourist sites

◆ recreation areas.

Indeed, land use includes all aspects of the Earth's surface! When describing land use you should refer to:

◆ the specific location (don't forget to use grid references)

◆ the size and shape of the area.

Inferring human activity from map evidence

As a geographer, you will be expected to describe and interpret features on a map extract. You can use map evidence to infer human activity as well as simply identifying it. For example, you might use map evidence to infer what type of settlement or what type of urban zone you are looking at.

Remember!

You may be asked to explain why land at a specific location is used in a particular way, for example why an area or slope has been planted with coniferous trees. Remember to use the word 'because' when you are asked to explain a land use.

Skills in context

Describing physical features of coastlines, fluvial and glacial landscapes

You need to be able to identify and describe two of these three landscape types with reference to OS maps. You can find out more about how to do this in Chapters 10–12.

Inference is all about reaching informed conclusions using the evidence available to you. For example:

◆ If you are asked to identify the 'inner city', look for the appropriate evidence on the map and then use it to support your suggestion (Map **I**).

◆ At the coast, the presence of a sandy beach, sand dunes and clifftop footpaths infer that tourism may well be important in the area. Look for the blue symbols that indicate tourist facilities

◆ In a glacial landscape, the presence of a lake could be used to infer that people might take part in water sports, fishing or bird watching. The same applies to woodlands or mountains.

Skills in context

Comparing maps

Two or more maps can be compared to see how things have changed over a period of time, for example, the growth of a settlement. The similarities and differences between two maps can also be considered where there may be an association, for example earthquake epicentres and plate boundaries.

Remember!

- the term 'compare' means similarities and differences
- contrast' means differences only!

Drawing sketch maps

A sketch map is drawn to produce a simplified version of an OS map. It should focus on just a few key elements, such as patterns of roads or rivers, without lots of other information.

To draw a sketch map, follow these steps:

1 Start by drawing a frame, either to the same scale as the map or enlarging/reducing it as required.

2 Divide the frame into grid squares as they appear on the map and write the grid reference numbers around the edges of your frame. These will act as your guidelines when you draw your sketch.

3 Using a pencil, carefully draw just the features that you need onto your sketch.

4 When your sketch is complete, use colour and shading if you wish, although black and white sketches are often the most successful.

5 Label and annotate your sketch as necessary.

6 Don't forget to include a scale (which can be approximate), a north point and a title.

You will gain the most credit for your labels and annotations (detailed labels with some explanation), which show your ability to interpret the map.

I *Characteristics of a small market town – Alnwick (a nucleated settlement)*

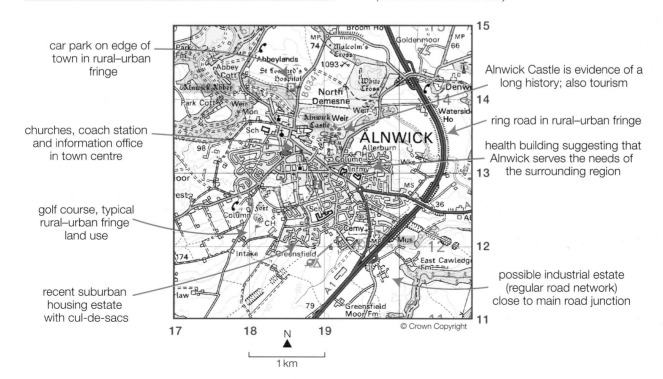

car park on edge of town in rural–urban fringe

churches, coach station and information office in town centre

golf course, typical rural–urban fringe land use

recent suburban housing estate with cul-de-sacs

Alnwick Castle is evidence of a long history; also tourism

ring road in rural–urban fringe

health building suggesting that Alnwick serves the needs of the surrounding region

possible industrial estate (regular road network) close to main road junction

© Crown Copyright

1 km

Using photos

Photos are widely used in the study of geography. They can be used on their own or in association with maps.

Geographers make use of three different types of photo.

Ground photos

Photos taken on the ground (**J**) are the most common types of photo and are usually used to focus on a particular physical feature or characteristic, such as a building or a waterfall.

Aerial photos

These are usually taken from aeroplanes, helicopters or drones, looking down on a landscape. They often show large areas that can be related directly to OS maps – for example showing settlements or stretches of coastline. There are two kinds of aerial photos:

◆ *Vertical* aerial photos (**K a**) look directly down onto the ground and therefore give no indication of relative height – so everything looks flat!

◆ *Oblique* aerial photos (**K b**) give a sideways view of the landscape. They are used more often than vertical aerial photos. They can distort size, with objects in the foreground appearing larger than those in the distant background.

Satellite photos

Like vertical aerial photos, these look directly down onto the Earth (**L**). Satellite photos may be digitally processed with enhanced colours to make certain land uses and features show up more clearly. These 'false-colour' images can be used to show environmental factors such as pollution and deforestation. They are also widely used to show weather features, such as hurricanes.

J *A ground photo*

K *a* *A vertical aerial photo*

b *An oblique aerial photo*

L *Satellite photo*

Describing human and physical landscapes

Photos are widely used by geographers to record, investigate and understand physical and human landscapes such as landforms, natural vegetation, land use and settlement.

When describing what a photo shows, you should:

◆ use directional language – for example, 'in the foreground', 'in the background' as well as 'right' and 'left'

◆ use juxtaposition – for example, 'just behind and to the right of the stack', to enable you to identify and describe features accurately.

You may be asked to use photos and maps alongside each other. You could, for example, be asked to identify the direction that a photo is looking. Take your time to orientate the photo on the map, possibly by turning it to line it up correctly – look for evidence in the foreground and the background on the photo to help you do this. Once you have orientated the photo, it should be quite easy to work with both resources.

Drawing sketches from photos

The purpose of a sketch is to identify the main geographical characteristics of the landscape (figure **M**). It is not necessary to produce a brilliant artistic drawing – clarity and accuracy are all that are needed so that labels and annotations can be added.

To draw a sketch, follow these steps:

◆ Draw a frame that is the same shape as the photograph.

◆ Draw one or two major lines that will subsequently act as guidelines for the rest of your sketch. For example, you could draw the profile of a slope, a hilltop, a road or river.

◆ Consider what it is that you are trying to show and concentrate on that feature or aspect – it may be river features or the pattern of settlements.

Labels and annotations

You should always add labels or annotations to diagrams, maps, graphs, sketches and photos, where appropriate.

◆ Labels – these are usually single words or phrases identifying features, for example the peak on a hydrograph, a river cliff on a meander or a roundabout in a town.

◆ Annotations – these give more detail and may include some detailed description and explanation, for example, 'The cliff is vertical and high, probably because it is made of hard rock and is being actively eroded and undercut at its base by the sea'.

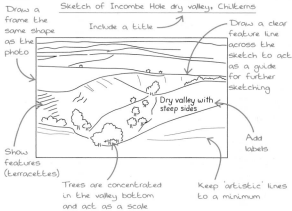

Draw a frame the same shape as the photo
Include a title
Sketch of Incombe Hole dry valley, Chilterns
Draw a clear feature line across the sketch to act as a guide for further sketching
Dry valley with steep sides
Show features (terracettes)
Trees are concentrated in the valley bottom and act as a scale
Add labels
Keep 'artistic' lines to a minimum

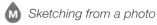 *Sketching from a photo*

Remember!

- Don't waste time drawing a lot of unnecessary detail.

- Always use a good sharp pencil and don't be afraid to rub things out as you go along.

- Always use labels or annotations (detailed labels) on your sketch, to identify the features.

- Give your sketch a title.

In this unit you will find out how to interpret and construct graphs and diagrams

As a geographer, you need to be able to read and interpret information that is presented in a variety of ways. This includes written text, photos and maps of different types and scale. You should also be able to interpret and construct a variety of basic graphs and diagrams.

How to interpret a graph or diagram

Graph **A** is a line graph showing world population growth. Notice the following points:

◆ The scales have equal intervals between each line (2 billion people on the *y* axis and every 50 years on the *x* axis). If different intervals were used, the graph would be distorted.

◆ The top line of the graph shows total world population. It increases slowly from about 1 billion in 1800 to about 3 billion in 1950. From 1950 it increases very rapidly before levelling off at about 10 billion by 2050.

◆ The graph has been subdivided into *developing* regions and *developed* regions. It can be called a 'divided' or 'compound' graph. Take care to read values from the correct section of the graph. So, population for developing regions in 1950 is about 2 billion.

Remember!

It can be easy to miss these practical activities in an exam because there are no lines for writing an answer!

Remember!

The *three* stages of a description:

1 Describe the overall trends and patterns – the 'big picture'.

2 Provide some evidence to support your description (quote a few facts and figures).

3 Consider any anomalies (exceptions) to the overall trends and patterns.

When explaining:

◆ give reasons for trends, patterns and anomalies (use the word 'because')

◆ consider links and connections between different variables that might help your explanation (for example, a war might affect the shape of a population pyramid).

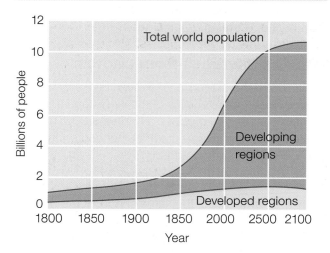

 World population growth line graph

How to construct graphs and diagrams

You may be asked to construct or complete a partly-drawn graph, diagram or map. When doing this, make sure you look closely at the scales and plot the data accurately and precisely. Use a sharp pencil and double-check that you have plotted the information correctly.

Line graphs

A line graph shows *continuous* changes over a period of time – for example, stream flow (hydrograph), traffic flow or population change. Remember that time, shown on the horizontal axis, must have an equal spacing – for example for periods of time (graph **A**).

Bar graphs and histograms

Bar graphs and histograms are one of the most common methods of displaying statistical information. You should use the one best suited to your data.

Bar graphs

A bar graph is a way of comparing quantities or frequencies in different categories, such as types of vegetation, or places. The bars are drawn in different colours with gaps between them, because they are unconnected (graph **B**).

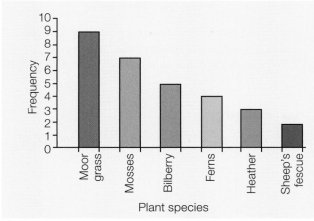

B *A bar graph*

It is possible to subdivide individual bars in order to show multiple data. In graph **C** the bars are sub-divided to show different species. This type of graph is called a *divided* bar graph.

C *A divided bar graph*

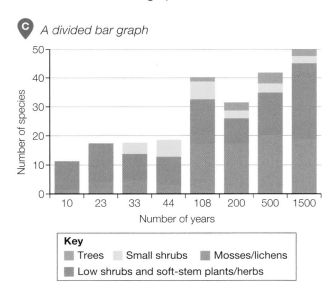

Key
- Trees
- Small shrubs
- Mosses/lichens
- Low shrubs and soft-stem plants/herbs

Histograms

A histogram also uses bars but with no gaps between them. This is because:

◆ a histogram represents *continuous* data (such as daily rainfall values over a period of a month)

◆ the values may be all part of a single sample, for example, the sizes of particles in a sediment sample (graph **D**). As the bars are effectively connected, a single colour or type of shading is used. Notice that there are equal class intervals between the bars.

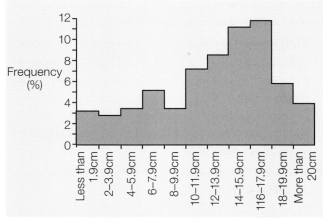

D *A histogram*

Pie charts

A pie chart is a simple circle divided into segments, rather like the slices of a pie. It shows the proportions of a total. Percentage figures are written alongside the segments to help interpret the diagram. For example, chart **E** shows that in 2012, 39 per cent of UK electricity was generated by burning coal, and 11.9 per cent from renewable sources.

Pie charts work best when there are between four and six 'segments'. Don't draw a pie chart as just one segment!

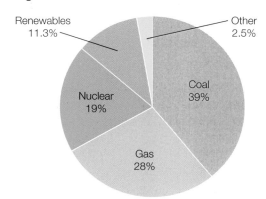

E *Pie chart showing fuel used for UK energy generation*

Remember!

To convert percentages into degrees for pie charts, multiply the value by 3.6.

Pictograms

A pictogram uses a pictorial symbol or icon instead of a bar (graph **F**). It is an effective visual technique although it may be difficult to extract precise data from the diagram. All icons must be the same size. Fractions of icons can be used to represent values in between those represented by a full icon. For example, if an icon of one person represents 100 tourists, then 50 tourists would be represented by half a person!

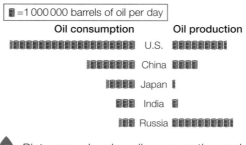

F Pictogram showing oil consumption and production in the top five oil-consuming countries

Scattergraphs

If two sets of data are thought to be related, they can be plotted on a graph called a scattergraph. To complete a scattergraph follow these steps:

1 Draw two axes in the normal way, with the variable thought to be causing the change (called the independent variable) on the horizontal (*x*) axis. On graph **G**, GNP (Gross National Product, the wealth of a country) is thought to be responsible for average car ownership.

2 Use each pair of values to plot a single point on the graph using a cross.

3 Draw a best-fit line to show the trend of the points if there is one. Your best-fit line should pass approximately through the middle of the points, with roughly the same number of points on either side of the line. Use a ruler to draw a straight line through the points. Best-fit lines may also be curved, and for these you need to draw a line by hand. Remember: the best-fit line does not need to pass through the origin.

4 Describe the resulting pattern (diagram **H**).

H ▶ *Interpreting scattergraph patterns*

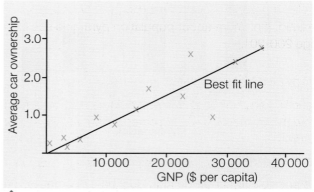

G *Scattergraph and best-fit line*

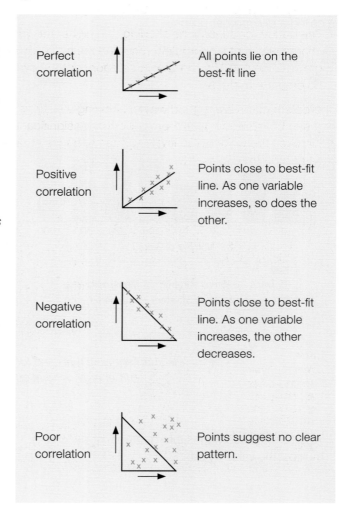

Population pyramids

A population pyramid is a type of histogram showing the proportions of a population in different age and gender categories. Graph **I** shows males on the left and females on the right. Each age group – with equal intervals – is represented by a bar.

A population pyramid represents the structure of a population and its shape provides valuable information for a government planning for the future provision of health care, schooling and housing. The labels on graph **I** show how to interpret a population pyramid. (For more about population pyramids see page 200-201.)

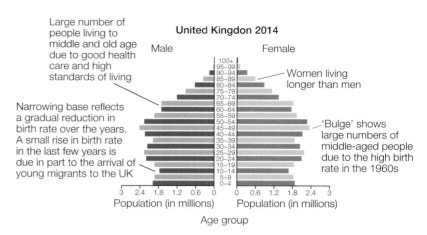

Large number of people living to middle and old age due to good health care and high standards of living

Women living longer than men

Narrowing base reflects a gradual reduction in birth rate over the years. A small rise in birth rate in the last few years is due in part to the arrival of young migrants to the UK

'Bulge' shows large numbers of middle-aged people due to the high birth rate in the 1960s

United Kingdon 2014

Male   Female

Population (in millions)   Population (in millions)

Age group

I *Population pyramid for the UK, 2014*

Choropleth maps

A choropleth map uses different colours or different densities of the same colour to show the distribution of data categories (map **J**). A choropleth map has the following features:

◆ The base map shows regions or areas – in map **J** it is countries of the world.

◆ Data are divided into groups or categories. Ideally there should be 4–6 categories. Notice that the category values do not overlap and that the intervals are equal (for example, 2–2.9, 3–3.9, etc.).

◆ The darker (or denser) the shading, the higher the values. This is a key characteristic of a choropleth map and makes it easy to interpret. White (blank) or grey can be used for the lowest category or to show areas on the map for which no data are available.

Choropleth maps are a good way of showing variations between areas. However, they can be misleading, as there will often be significant variations at a local scale within each area. They also imply sudden changes at area boundaries, which is often not the case in reality.

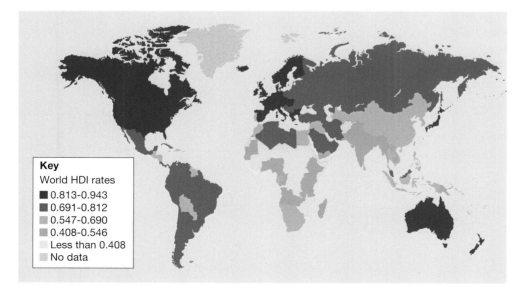

Key
World HDI rates
- 0.813-0.943
- 0.691-0.812
- 0.547-0.690
- 0.408-0.546
- Less than 0.408
- No data

J *Choropleth map showing HDI rates*

Remember!

Note the spelling of choropleth – don't make the common mistake of spelling it 'chloropleth'!

Isoline maps

An isoline map uses lines of equal value to show patterns ('iso' means 'equal'). Some of the most common types of isoline map show aspects of weather and climate. For example, isobars show pressure and isotherms show temperature.

Isoline maps can be tricky to draw but they are a good way of showing patterns when superimposed on a base map, for example, pedestrian counts at different places in a town (map **K**).

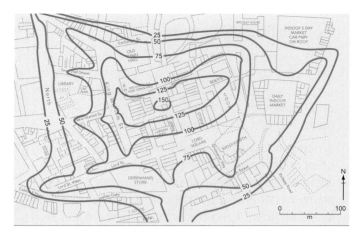

K *Isoline map showing pedestrian counts in Blackburn's CBD*

To draw an isoline map, mark your observed data onto a base map or sheet of tracing paper. Then consider how many lines to draw and at what intervals you will draw them, for example every ten units. This decision is largely 'trial and error' and you might like to draw the lines in rough first.

Map **L** shows how isolines are drawn. They pass between values that are higher and lower than the value of the line. All values to one side of a line will be higher and all those to the other side will be lower.

Each map will be individual – don't worry if yours looks different to other people's.

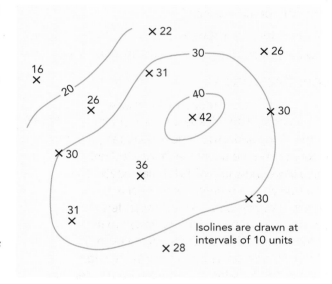

L *Drawing isolines*

Dot maps

Dots are used to represent a particular value or number (for example, a population of 1 million) and are located accurately on a map. The number and density of the dots conveys the information on the map (map **M**), but it can be difficult to extract accurate information from them.

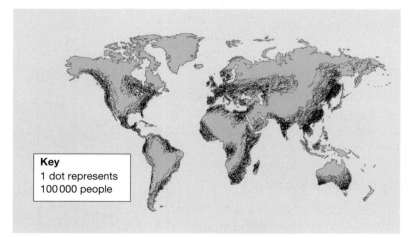

M *Dot map showing global population distribution*

Desire line maps

Desire lines show movement of people or goods between places (map **N**), for example commuters travelling to a nearby town or city from the surrounding area. They may also be proportional, and show trends in the distance travelled between places (for example, most people might commute short distances) and the spatial density of the travellers (for example, where people commute from). They are similar to flow lines, but show only direct movement from A to B, while flow lines show the exact path of movement.

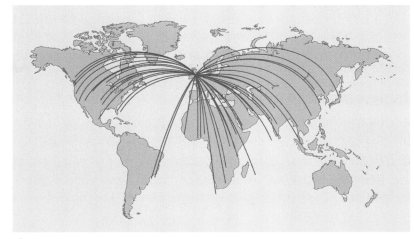

N *Desire line map of international flights from Heathrow*

When drawing a desire line map, each line should be positioned accurately to show its source and its destination.

Flow line maps

Flow lines indicate direction and volume of movement, with thickness representing volume. They show movement between places by connecting the source with the destination. Desire lines show lots of individual movements, while flow lines are drawn proportionately to show grouped data. For example, ten separate desire lines might be drawn to show the movement of commuters from Village A to Town B; or a single flow line with the width drawn proportionately to show ten commuters.

Map **O** shows the origin of tourists visiting Kenya. Each flow line connects the continent of origin and the destination (Kenya). The width of each line is in proportion to the percentage of tourists.

Flow lines can be drawn on a base map, but an appropriate scale is needed to avoid flow lines crossing over each other. Don't forget to write the scale on your map.

O *Flow line map showing the origin of tourists to Kenya*

P *Map with proportional circles*

Proportional symbols

Proportional symbols (for example, circles) are a useful way to show data on a base map where spatial variations can be seen (map **P**). They can be difficult to draw and you will need to choose your scale carefully.

When using proportional circles, select a scale for the radius of your circle. The area of the circle should be proportional, so use the square root value as your radius distance.

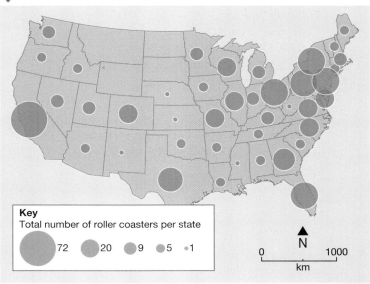

In this unit you will find out how to use a range of statistical methods to measure the average and spread of a data set, calculate percentage change over time, and describe relationships between variables

Using statistics in geography

Geographers frequently use numbers and data sets. The use of statistics is an important part of any geographical investigation. They can help you to interpret patterns and trends.

In an exam you should be prepared to spot weaknesses in the presentation of selected data. This might involve spotting incorrect labelling of axes or inaccurate interpretation of trends.

Measures of central tendency

Central tendency is a description of the 'average' within a data set. There are three ways of measuring central tendency: mean, median and mode.

Mean

Mean is calculated very simply, by summing (adding up) all the values in a data set and then dividing by the number of values. This is the most commonly used measure of central tendency and the one that is normally associated with the term 'average'. However, it can be skewed by one very high or one very low value and doesn't always fairly represent the majority of the values.

Example

10, 14, 8, 16, 14, 9, 12, 18, 10, 9

Mean is 120 divided by 10 = 12

Median

This is the central (middle) value in a ranked data set. If there is an odd number of values, identifying the middle value is easy. If there is an even number of values, the median lies mid-way between the two central values.

Example

Odd number of values:

13, 13, 13, 13, **14**, 14, 16, 18, 21

The median is the 5th value = 14

Even number of values:

8, 9, 11, 13, **14**, **16**, 16, 18, 21, 24

The median lies mid-way between the 5th value (14) and 6th value (16). This can be calculated by adding the two values together and dividing by two.

14 + 16 = 30 /2 = **15**

Mode

This is the most common value in a data set. If there are no repeated values then there is no mode in the data set.

Example

13, 13, 13, 13, 14, 14, 16, 18, 21

Here the value 13 is stated four times whereas the value 14 is only stated twice. The mode value is **13**.

Modal class

If data are grouped into categories, the category with the highest frequency is called the modal class. Look back to the histogram on page 337 (**D**). The modal class is 16–17.9cm, with the highest bar representing the greatest frequency.

Measures of spread

The measures of central tendency are useful in identifying 'average' values. However, they give no indication of how the values in a data set are spread around the average. Look at the data sets in the example. They both have the same mean but very different values!

Range

The *range* is the difference between the highest and lowest values. It gives us a good measure of the spread of the values in a data set and provides another means of description to use alongside the 'average'.

Quartiles and inter-quartile range

The median is the middle value in a ranked data set. It splits the data set into two halves – an upper half and a lower half. These two halves can be split again into quarters at a value called the quartile.

There is an upper quartile value (that splits the upper half of the values) and a lower quartile (that splits the lower half). So, if there are 20 values, the median would lie midway between the 10th and 11th values. The upper quartile value lies between the 15th and 16th values. The lower quartile value lies between the 5th and 6th values. With the median, this divides the data set (20) into four groups of five.

The *inter-quartile range* (IQR) is the difference between the upper quartile and the lower quartile. It is useful for showing degree of clustering or dispersal of values around the median.

Turn to page 344 (figure **A**) to see how the range can be applied to values displayed on a dispersion graph.

Example

Data set A: 11, 15, 17, 22, 24, 26, 34, 35

Mean = 184/8 = **23**

Data set B: 2, 3, 4, 6, 6, 7, 56, 100

Mean = 184/8 = **23**

Example

Data set A: 35 − 11 = **24**

Data set B: 100 − 2 = **98**

Example

3, 3, 3, 4, 6, 7, 7, 8, 8, 10, **14**, **14**, 16, 20, 21, 21, 22, 24, 25, 30

Median = 10 + 14 = 24/2 = **12**

Lower quartile = 6 + 7 = 13/2 = **6.5**

Upper quartile = 21 + 21 = 42/2 = **21**

Inter-quartile range = 21 − 6.5 = **14.5**

Using a formula:

Assuming the highest value is ranked as 1, the upper quartile (UQ) is calculated using the formula:

UQ = n+1/4 th position in the rank order.

The lower quartile (LQ) is calculated using the formula:

LQ = 3(n+1)/4 th position in the rank order.

LQ is then subtracted from UQ to give the IQR.

Dispersion graphs

Dispersion graphs shows the *spread* of data. They are a useful way to make comparisons between different sites, for example locations along a river or across a beach. They can be used to identify a number of statistical measures including the range, median, quartiles and inter-quartile range.

Graph **A** shows comparisons between pebble sizes across a beach. The degree of overlap of the graphs can be used to assess difference between data sets.

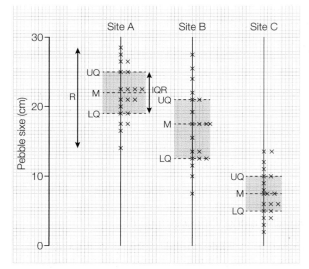

R = Range
M = Median
UQ = Upper quartile
LQ = Lower quartile
IQR = Inter-quartile range

'Box' indicating the spread of the central 50% of data set

(Notice that there is *no* overlap of the box for site C with sites A and B. This means there is a significant *difference* between the data for site C and data for sites A and B.)

A   *Dispersion graphs showing pebble sizes across a beach*

Calculating percentage change and using percentiles

Percentage change

You have probably heard comments like 'the cost of a ticket has increased by 20 per cent' or 'there has been a 50 per cent fall in attendance'. Geographers often want to make comparisons between things or describe how something has changed over a period of time. Percentage change is a good way to do this.

To calculate the percentage increase between two numbers:

1 Work out the difference (increase) between the two numbers you are comparing.

2 Divide the increase by the *original* number.

3 Multiply the answer by 100 to give you a percentage.

If your answer is a negative number then this is a *percentage decrease*.

Example

The number of bus routes in a town increased from 24 in 2010 to 31 in 2016.

$31 - 24 = 7$

$7/24 = 0.29$

$0.29 \times 100 = 29\%$

There has been a 29% increase (just less than one third) in the number of bus routes between 2010 and 2016.

Percentiles

A *percentile* is used to indicate the value below which a given percentage of observations fall. So, for example, the 80th percentile is the value in a data set below which 80 per cent of the observations occur and above which 20 per cent of the observations occur.

You have already come across the term *median*. This is a point half way along a ranked data set, so is in effect the 50th percentile (50 per cent). The upper quartile is the 75th percentile and the lower quartile the 25th percentile.

Describing relationships in bivariate data

The term *bivariate data* means data for two variables that may be considered to be related, for example GDP and energy consumption. (*Univariate* data involve a single set of data and might be displayed in a dispersion graph.) In the case of GDP and energy consumption, the amount of energy consumed might be expected to increase as the wealth of a country (GNP) increases. Energy consumption is dependent on GNP. Therefore GNP can be said to be the dependent variable and GNP the *independent* variable.

Bivariate data are usually plotted as a scattergraph (see graph **G** on page 338). The dependent variable is plotted along the side (*y* axis) and the independent variable along the bottom (*x* axis).

A *best-fit* line can be drawn to indicate a relationship (if one exists) and the relationship can be described. In most cases best-fit lines are drawn with a ruler to show a linear relationship. However, a curved line may be used to show an *exponential* relationship.

Remember!

Using a small data set may be unreliable. For example, if a scattergraph only has four or five points, the trend or relationship shown by a best-fit line could be misleading. You should try to plot at least ten points on your graph if possible.

Trend lines

A sketch trend line can also be used to suggest a trend in bivariate data. This line is drawn through the scatter plots to suggest the overall trend, such as an increase or a decrease. It is commonly used on a graph where changes take place over time (graphs **A** and **B**). A stretch trend line may be a curved line.

B ▶ *Trend line showing a decrease in Arctic sea ice 1980–2014*

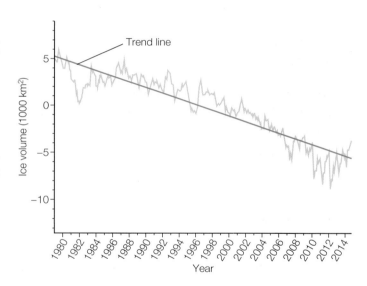

How to interpolate and extrapolate

To estimate an unknown value you first need to *interpolate* or *extrapolate* a trend. Then you can use the trend line or best-fit line to estimate the unknown value (graph **C**).

◆ Interpolation involves estimating an unknown value from *within* the data set.

◆ Extrapolation involves estimating an unknown value that is *outside* the data set.

Graph **C** shows how the use of a best-fit line enables interpolation and extrapolation to take place.

C ▶ *Using a best-fit line to interpolate and extrapolate*

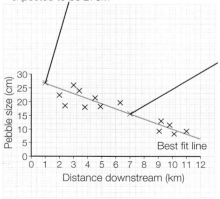

By *extrapolating* the trend it is possible to estimate that at 1km a pebble would be expected to be 27cm

By *interpolating* the trend it is possible to estimate that at 7km a pebble would be expected to be 16cm

Glossary

Abrasion (1) Rocks carried along a river wear down the river bed and banks; (2) the sandpaper effect of glacial ice scouring a valley floor and sides

Adaptation Actions taken to adjust to natural events such as climate change, to reduce damage, limit the impacts, take advantage of opportunities, or cope with the consequences

Aeroponics Growing plants in an air or mist environment without the use of soil

Appropriate (or intermediate) technology Technology suited to the needs, skills, knowledge and wealth of local people and their environment

Arch A wave-eroded passage through a small headland. This begins as a cave which is gradually widened and deepened until it cuts through

Arête A sharp, knife-like ridge formed between two corries cutting back by processes of erosion and freeze thaw

Attrition Rocks being carried by the river smash together and break into smaller, smoother and rounder particles

Bar Where a spit grows across a bay, a bay bar can eventually enclose the bay to create a lagoon

Beach A zone of deposited material that extends from the low water line to the limit of storm waves

Beach nourishment Adding new material to a beach artificially, through the dumping of large amounts of sand or shingle

Beach re-profiling Changing the profile or shape of the beach

Biodiversity The variety of life in the world or a particular ecosystem

Biomass Renewable organic materials, such as wood, agricultural crops or wastes, especially when used as a source of fuel or energy

Biotechnology The genetic engineering of living organisms to produce useful commercial products

Birth rate The number of births a year per 1000 of the total population

Brownfield site Land that has been used, abandoned and now awaits reuse; often found in urban areas

Bulldozing The pushing of deposited sediment by the snout (front) of the glacier as it advances

Business park An area of land occupied by a number of businesses

Carbon footprint Measurement of the greenhouse gases individuals produce, through burning fossil fuels

Cave A large hole in a cliff caused by waves forcing their way into cracks in the cliff face

Channel straightening Removing meanders from a river to make it straighter

Chemical weathering The decomposition (or rotting) of rock caused by a chemical change within that rock

Cliff a steep high rock face formed by weathering and erosion

Climate change A long-term change in the earth's climate, especially a change due to an increase in the average atmospheric temperature

Commercial farming Growing crops or raising livestock for profit, often involving vast areas of land

Commonwealth The Commonwealth is a voluntary association of 53 independent and equal sovereign states, most being former British colonies

Conservation Managing the environment in order to preserve, protect or restore it

Conservative plate margin Two plates sliding alongside each other, in the same or different directions

Constructive plate margin Tectonic plate margin where rising magma adds new material to plates that are diverging or moving apart

Consumer Organism that eats herbivores and/or plant matter

Corrie or cirque Armchair-shaped hollow in the mountainside formed by glacial erosion, rotational slip and freeze-thaw weathering – this is where the valley glacier begins

Cross profile The side-by-side cross section of a river channel and/or valley

Dam and reservoir A barrier built across a valley to interrupt river flow and create a man-made lake to store water and control river discharge

Death rate The number of deaths in a year per 1000 of the total population

Debt crisis When a country cannot pay its debts, often leading to calls to other countries for assistance

Debt relief Cancellation of debts to a country by a global organisation such as the World Bank

Decomposer Organisms such as bacteria or fungi that break down plant and animal material

Deforestation The cutting down and removal of forest

De-industrialisation The decline of a country's traditional manufacturing industry due to exhaustion of raw materials, loss of markets and overseas competition

Deposition Occurs when material being transported by the sea is dropped due to the sea losing energy

Dereliction Abandoned buildings and wasteland

Desertification The process by which land becomes drier and degraded, as a result of climate change or human activities, or both

Destructive plate margin Tectonic plate margin where two plates are converging and oceanic plate is subducted – there could be violent earthquakes and explosive volcanoes

Development The progress of a country in terms of economic growth, the use of technology and human welfare

Development gap Difference in standards of living and wellbeing between the world's richest and poorest countries

Discharge Quantity of water that passes a given point on a stream or river-bank within a given period of time

Drumlin Egg-shaped hill of moraine material deposited in a glacial trough

Dune regeneration Building up dunes and increasing vegetation to prevent excessive coastal retreat

Earthquake A sudden or violent movement within the Earth's crust followed by a series of shocks

Economic impact Effect of an event on the wealth of an area or community

Economic opportunities Chances for people to improve their standard of living through employment

Ecosystem A community of plants and animals that interact with each other and their physical environment

Ecotourism Nature tourism usually involving small groups with minimal impact on the environment

Embankments Artificially raised river banks often using concrete walls

Energy conservation Reducing energy consumption by using less energy and existing sources more efficiently

Energy exploitation Developing and using energy resources to the greatest possible advantage, usually for profit

Energy mix Range of energy sources of a region or country, both renewable and non-renewable

Energy security Uninterrupted availability of energy sources at an affordable price

Environmental impact Effect of an event on the landscape and ecology of the surrounding area

Erratics Rocks transported and dumped by glacial ice to a different location, often hundreds of kilometres away

Erosion Wearing away and removal of material by a moving force, such as a breaking wave

Estuary Tidal mouth of a river where it meets the sea – wide banks of deposited mud are exposed at low tide

European Union A politico-economic union of 28 European countries – the UK is a member state

Extreme weather When a weather event is significantly different from the average or usual weather pattern, and is especially severe or unseasonal

Fair trade Producers in LICs given a better price for their goods such as cocoa, coffee and cotton

Famine Widespread, serious, often fatal shortage of food

Flood Where river discharge exceeds river channel capacity and water spills onto the floodplain

Floodplain Relatively flat area forming the valley floor either side of a river channel that is sometimes flooded

Floodplain zoning Identifying how a floodplain can be developed for human uses

Flood relief channels Artificial channels that are used when a river is close to maximum discharge; they take the pressure off the main channels when floods are likely

Flood warning Providing reliable advance information about possible flooding

Fluvial processes Processes relating to deposition, erosion, and transport by a river

Food chain Connections between different organisms (plants and animals) that rely upon one another as their source of food

Food insecurity Being without reliable access to enough affordable, nutritious food

Food miles The distance covered supplying food to consumers

Food security Access to sufficient, safe, nutritious food to maintain a healthy and active life

Food web A complex hierarchy of plants and animals relying on each other for food

Formal economy the type of employment where people work to receive a regular wage, pay tax, and have certain rights, i.e. paid holidays, sickness leave

Fossil fuel A natural fuel such as coal or gas, formed in the geological past from the remains of living organisms

Fragile environment An environment that is both easily disturbed and difficult to restore

Freeze-thaw weathering (or frost shattering) A common process of weathering in a glacial environment involving repeated cycles of freezing and thawing that can make cracks in rock bigger

Gabion Steel wire mesh filled with boulders used in coastal defences

Glacial trough Wide, steep-sided valley eroded by a glacier

Geothermal energy Energy generated by heat stored deep in the Earth

Globalisation Process creating a more connected world, with increases in the global movements of goods (trade) and people (migration & tourism)

Gorge A narrow steep-sided valley – often formed as a waterfall retreats upstream

Greenfield site A plot of land, often in a rural or on the edge of an urban area that has not been built on before

Green revolution An increase in crop production, especially in poorer countries, using high-yielding varieties, artificial fertilisers and pesticides

Grey water Recycled domestic waste water

Gross national income (GNI) Measurement of economic activity calculated by dividing the gross (total) national income by the size of the population

Groundwater management Regulation and control of water levels, pollution, ownership and use of groundwater

Groyne A wooden barrier built out into the sea to stop the longshore drift of sand and shingle, and allow the beach to grow

Hanging valley A tributary glacial trough on the side of a main valley often with a waterfall

Hard engineering Using concrete or large artificial structures to defend against natural processes, either coastal, fluvial or glacial

Hazard risk Probability or chance that a natural hazard may take place

Headlands and bays A rocky coastal promontory (highpoint of land) made of rock that is resistant to erosion: headlands lie between bays of less resistant rock where the land has been eroded by the sea

High income country (HIC) A country with GNI per capita higher than $12 746 (World Bank, 2013)

Hot desert Parts of the world that have high average temperatures and very low precipitation

Human Development Index (HDI) A method of measuring development where GDP per capita, life expectancy and adult literacy are combined to give an overview

Hydraulic action Power of the water eroding the bed and banks of a river

Hydraulic power Process where breaking waves compress pockets of air in cracks in a cliff; the pressure may cause the crack to widen, breaking off rock

Hydroelectric power (HEP) Electricity generated by turbines that are driven by moving water

Hydrograph A graph which shows the discharge of a river, related to rainfall, over a period of time

Hydroponics Growing plants in water using nutrient solutions, without soil

Immediate responses Reaction of people as the disaster happens and in the immediate aftermath

Industrial structure Relative proportion of the workforce employed in different sectors of the economy

Inequalities Differences between poverty and wealth, as well as wellbeing and access to jobs, housing, education, etc.

Infant mortality Number of babies that die under one year of age, per 1000 live births

Informal economy employment outside the official knowledge of the government

Information technologies Computer, internet, mobile phone and satellite technologies

Infrastructure The basic equipment and structures (such as roads, utilities, water supply and sewage) that are needed for a country or region to function properly

Integrated transport system Different forms of transport are linked together to make it easy to transfer from one to another

Interlocking spurs Outcrops of land along the river course in a valley

Intermediate (or appropriate) technology Simple, easily learned and maintained technology used in LICs for a range of economic activities

International aid Money, goods and services given by single governments or an organisation like the World Bank or IMF to help the quality of life and economy of another country

Irrigation Artificial application of water to the land or soil

Landscape An extensive area of land regarded as being visually and physically distinct

Land use conflicts Disagreements between interest groups who do not agree on how land should be used

Lateral erosion Erosion of river banks rather than the bed – helps to form the floodplain

Levee Raised bank found on either side of a river, formed naturally by regular flooding or built up by people to protect the area against flooding

Life expectancy The average number of years a person is expected to live

Literacy rate Percentage of people in a country who have basic reading and writing skills

Local food sourcing Food production and distribution that is local, rather than national and/or international

Logging The business of cutting down trees and transporting the logs to sawmills

Long profile The gradient of a river, from its source to its mouth

Longshore drift Transport of sediment along a stretch of coastline caused by waves approaching the beach at an angle

Long-term responses Later reactions that occur in the weeks, months and years after the event

Low income country (LIC) A country with GNI per capita lower than $1045 (World Bank, 2013)

Managed retreat Controlled retreat of the coastline, often allowing flooding to occur over low-lying land

Management strategies Techniques of controlling, responding to, or dealing with an event

Mass movement Downhill movement of weathered material under the force of gravity

Meander A wide bend in a river

Mechanical weathering Physical disintegration or break up of exposed rock without any change in its chemical composition, i.e. freeze–thaw

Megacity An urban area with a total population of more than ten million people

Microfinance loans Very small loans which are given to people in the LICs to help them start a small business

Migration When people move from one area to another; in many LICS people move from rural to urban areas (rural–urban migration)

Mineral extraction Removal of solid mineral resources from the earth

Mitigation Action taken to reduce the long-term risk from natural hazards, such as earthquake-proof buildings or international agreements to reduce greenhouse gas emissions.

Monitoring (1) Recording physical changes, i.e. tracking a tropical storm by satellite, to help forecast when and where a natural hazard might strike; (2) using scientific methods to study coastal processes to help inform management options.

Moraine Frost-shattered rock debris and material eroded from the valley floor and sides, transported and deposited by glaciers

Natural increase Birth rate minus the death rate of a population

Newly-Emerging Economies Countries that have begun to experience high rates of economic development, usually along with rapid industrialisation

North-south divide (UK) Economic and cultural differences between southern England and northern England

Nuclear power Energy released by a nuclear reaction, especially by fission or fusion

Nutrient cycling On-going recycling of nutrients between living organisms and their environment

Orbital change Changes in the pathway of the Earth around the Sun

Organic produce Food produced without the use of chemicals such as fertilisers and pesticides

Outwash Sediment deposited by meltwater that is well sorted and rounded in front of a glacier

Over abstraction When water is used more quickly than it is being replaced

Over-cultivation Where the intensive growing of crops exhausts the soil leaving it barren

Overgrazing Feeding too many livestock for too long on the land, so it is unable to recover its vegetation

Oxbow lake An arc-shaped lake on a floodplain formed by a cut-off meander

Permafrost Permanently frozen ground, found in polar and tundra regions

Planning Actions taken to enable communities to respond to, and recover from, natural disasters

Plate margin The border between two tectonic plates

Plucking A process of erosion – rocks are pulled from the valley floor as water freezes them to a glacier

Polar The most extreme cold environment with permanent ice, i.e. Greenland and Antarctica

Pollution Chemicals, noise, dirt or other substances which have harmful or poisonous effects on an environment

Post-industrial economy The shift of some HIC economies from producing goods to providing services

Precipitation Moisture falling from the atmosphere – rain, sleet or snow

Primary effects Initial impact of a natural event on people and property, caused directly by it, i.e. the buildings collapsing following an earthquake

Producer An organism or plant that is able to absorb energy from the sun through photosynthesis

Protection Actions taken before a hazard strikes to reduce its impact, such as educating people or improving building design

Pyramidal peak Where several corries cut back to meet at a central point, the mountain takes the form of a steep pyramid

Renewable energy sources A resource that cannot be exhausted, i.e. wind, solar and tidal energy

Resource management Control and monitoring of resources so that they do not become exhausted

Ribbon lake A long narrow lake in the bottom of a glacial trough

Rock armour Large boulders deliberately dumped on a beach as part of coastal defences

Rotational slip Slippage of ice along a curved surface

Rural-urban fringe A zone of transition between a built-up area and the countryside, where there is often competition for land use

Saltation Hopping movement of pebbles along a river or sea bed

Sand dune Coastal sand hill above the high tide mark, shaped by wind action

Sanitation Measures designed to protect public health, such as providing clean water and disposing of sewage and waste

Science park A collection of scientific and technical knowledge-based businesses located on a single site

Sea wall Concrete wall aiming to prevent erosion of the coast by reflecting wave energy

Secondary effects After-effects that occur as indirect impacts of a natural event, sometimes on a longer timescale, i.e. fires due to ruptured gas mains, resulting from the ground shaking

Selective logging sustainable forestry management where only carefully selected trees are cut down

Service (tertiary) industries The economic activities that provide various services – commercial, professional, social, entertainment and personal

Shale gas Natural gas that is found trapped within shale formations of fine-grained sedimentary rock

Sliding Loose surface material becomes saturated and the extra weight causes the material to become unstable and move rapidly downhill

Slumping Rapid mass movement where a whole segment of a cliff moves down-slope along a saturated shear-plane or line of weakness

Social deprivation The extent an individual or an area lacks services, decent housing, adequate income and employment

Social impact The effect of an event on the lives of people or community

Social opportunities The chances available to improve quality of life, i.e. access to education, health care, etc.

Soft engineering Managing erosion by working with natural processes to help restore beaches and coastal ecosystems or to reduce the risk of river flooding

Soil erosion Removal of topsoil faster than it can be replaced, due to natural (water and wind action), animal, and human activity

Solar energy Sun's energy exploited by solar panels, collectors or cells to heat water or air or to generate electricity

Solution (or corrosion) Chemical erosion caused by the dissolving of rocks and minerals by river or sea water

Spit Depositional landform formed when a finger of sediment extends from the shore out to sea, often at a river mouth

Squatter settlement An area of (often illegal) poor-quality housing, lacking in services like water supply, sewerage and electricity

Stack Isolated pillar of rock left when the top of an arch has collapsed

Subsistence farming A type of agriculture producing only enough food and materials for the benefit of a farmer and their family

Suspension Small particles carried in river flow or sea water, i.e. sands, silts and clays

Sustainability Actions that meet the needs of the present without reducing the ability of future generations to meet their needs

Sustainable energy supply Energy that can potentially be used well into the future without harming future generations

Sustainable food supply Food production that avoids damaging natural resources, providing good quality produce and social and economic benefits to local communities

Sustainable water supply Meeting the present-day need for safe, reliable and affordable water without reducing supply for future generations

Tectonic hazard Natural hazard caused by the movement of tectonic plates (i.e. volcanoes and earthquakes)

Till Sediment deposited by a glacier that is unsorted and angular

Traction Where material is rolled along a river bed or by waves

Trade Buying and selling of goods and services between countries

Traffic congestion When there is too great a volume of traffic for roads to cope with, and traffic slows to a crawl

Transnational corporation (TNC) A company that has operations (factories, offices, research and development, shops) in more than one country

Transportation The movement of eroded material

Tropical storm (hurricane, cyclone, typhoon) An area of low pressure with winds moving in a spiral around a calm central point called the eye of the storm – winds are powerful and rainfall is heavy

Truncated spur A former river valley spur which has been sliced off by a valley glacier, forming steep edges

Tundra A vast, flat, treeless Arctic region of Europe, Asia, and North America where the subsoil is permanently frozen

Undernutrition When people do not eat enough nutrients to cover their needs for energy and growth, or to maintain a healthy immune system

Urban farming Growing food and raising animals in towns and cities; processing and distributing food; collecting and re-using food waste

Urban greening Process of increasing and preserving open space in urban areas, i.e. public parks and gardens

Urbanisation When an increasing percentage of a country's population comes to live in towns and cities

Urban regeneration Reversing the urban decline by modernising or redeveloping, aiming to improve the local economy

Urban sprawl Unplanned growth of urban areas into the surrounding rural areas

Urban sustainability A city organised without over reliance on the surrounding rural areas and using renewable energy

Vertical erosion Downward erosion of the river bed

Volcano An opening in the Earth's crust from which lava, ash and gases erupt

Waste recycling Process of extracting and reusing useful substances found in waste

Waterfall A step in the long profile of a river usually formed when a river crosses over a hard (resistant) band of rock

Waterborne diseases Diseases like cholera and typhoid caused by micro-organisms in contaminated water

Water conflict Disputes between different regions or countries about the distribution and use of freshwater

Water deficit When demand for water is greater than supply

Water insecurity When water availability is insufficient to ensure the good health and livelihood of a population, due to short supply or poor quality

Water security Availability of a reliable source of acceptable quantity and quality of water

Water quality Measured in terms of the chemical, physical and biological content of the water

Water stress When the demand for water exceeds supply in a certain period or when poor quality restricts its use

Water surplus When water supply is greater than demand

Water transfer Matching supply with demand by moving water from an area with water surplus to another with water deficit

Wave cut platform Rocky, level shelf at or around sea level representing the base of old, retreated cliffs

Waves Ripples in the sea caused by the transfer of energy from the wind blowing over the surface of the sea

Wilderness area A natural environment that has not been significantly modified by human activity

Wind energy Electrical energy produced from the power of the wind, using windmills or wind turbines

Index

Symbols on Ordnance Survey maps (1:50 000 and 1:25 000)

ROADS AND PATHS

M I or A 6(M)	Motorway
A 35	Dual carriageway
A 31(T) or A 35	Trunk or main road
B 3074	Secondary road
	Narrow road with passing places
	Road under construction
	Road generally more than 4 m wide
	Road generally less than 4 m wide
	Other road, drive or track, fenced and unfenced
	Gradient: steeper than 1 in 5; 1 in 7 to 1 in 5
Ferry	Ferry; Ferry P – passenger only
	Path

PUBLIC RIGHTS OF WAY

(Not applicable to Scotland)

1:25 000	1:50 000	
		Footpath
		Road used as a public footpath
+++++++		Bridleway
		Byway open to all traffic

RAILWAYS

	Multiple track
	Single track
	Narrow gauge/Light rapid transit system
	Road over; road under; level crossing
	Cutting; tunnel; embankment
	Station, open to passengers; siding

BOUNDARIES

	National
	District
	County, Unitary Authority, Metropolitan District or London Borough
	National Park

HEIGHTS/ROCK FEATURES

50	Contour lines
·144	Spot height to the nearest metre above sea level

outcrop   cliff   scree

ABBREVIATIONS

P	Post office	PC	Public convenience (rural areas)
PH	Public house	TH	Town Hall, Guildhall or equivalent
MS	Milestone	Sch	School
MP	Milepost	Coll	College
CH	Clubhouse	Mus	Museum
CG	Coastguard	Cemy	Cemetery
Fm	Farm		

ANTIQUITIES

VILLA	Roman	✗	Battlefield (with date)
Castle	Non-Roman	✡	Tumulus/Tumuli (mound over burial place)

© Crown copyright.

LAND FEATURES

ruin	Buildings
	Public building
	Bus or coach station
Place of Worship	with tower
	with spire, minaret or dome
	without such additions
°	Chimney or tower
	Glass structure
H	Heliport
△	Triangulation pillar
	Mast
	Wind pump / wind generator
	Windmill
	Graticule intersection
	Cutting, embankment
	Quarry
	Spoil heap, refuse tip or dump
	Coniferous wood
	Non-coniferous wood
	Mixed wood
	Orchard
	Park or ornamental ground
	Forestry Commission access land
	National Trust – always open
	National Trust, limited access, observe local signs
	National Trust for Scotland

WATER FEATURES

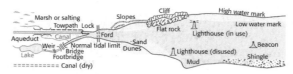

Marsh or salting   Towpath Lock   Slopes   Cliff   High water mark
Aqueduct   Canal   Ford   Flat rock   Low water mark   Lighthouse (in use)
Weir   Normal tidal limit   Sand   Dunes   Lighthouse (disused)   Beacon
Lake   Bridge   Footbridge   Mud   Shingle
Canal (dry)

TOURIST INFORMATION

P	Parking
P&R	Park & Ride
V	Visitor centre
i	Information centre
	Telephone
	Camp site/ Caravan site
	Golf course or links
	Viewpoint
PC	Public convenience
	Picnic site
	Pub/s
	Museum
	Castle/fort
	Building of historic interest
	Steam railway
	English Heritage
	Garden
	Nature reserve
	Water activities
	Fishing
☆	Other tourist feature
	Moorings (free)
	Electric boat charging point
	Recreation/leisure/ sports centre